Methods in Molecular Biology

For further volumes:
http://www.springer.com/series/7651

For over 35 years, biological scientists have come to rely on the research protocols and methodologies in the critically acclaimed Methods in Molecular Biology series. The series was the first to introduce the step-by-step protocols approach that has become the standard in all biomedical protocol publishing. Each protocol is provided in readily-reproducible step-by step fashion, opening with an introductory overview, a list of the materials and reagents needed to complete the experiment, and followed by a detailed procedure that is supported with a helpful notes section offering tips and tricks of the trade as well as troubleshooting advice. These hallmark features were introduced by series editor Dr. John Walker and constitute the key ingredient in each and every volume of the Methods in Molecular Biology series. Tested and trusted, comprehensive and reliable, all protocols from the series are indexed in PubMed.

Manuscript Instructions https://drive.google.com/file/d/1K4Az5vpAFVzYadJ3jKx2-QSo7Dqa_SNA/view?usp=sharing

Methods for The Conservation of Microorganisms

Edited by

Marco A. Ramírez-Mosqueda

Centro Nacional de Recursos Genéticos, INIFAP, Tepatitlán de Morelos, Jalisco, Mexico

Jorge David Cadena-Zamudio

Centro Nacional de Recursos Genéticos, INIFAP, Tepatitlán de Morelos, Jalisco, Mexico

José Luis Aguirre-Noyola

Centro Nacional de Recursos Genéticos, INIFAP, Tepatitlán de Morelos, Jalisco, Mexico

Ramón Ignacio Arteaga-Garibay

Centro Nacional de Recursos Genéticos, INIFAP, Tepatitlán de Morelos, Jalisco, Mexico

Humana Press

Editors
Marco A. Ramírez-Mosqueda
Centro Nacional de Recursos Genéticos
INIFAP
Tepatitlán de Morelos
Jalisco, Mexico

Jorge David Cadena-Zamudio
Centro Nacional de Recursos Genéticos
INIFAP
Tepatitlán de Morelos
Jalisco, Mexico

José Luis Aguirre-Noyola
Centro Nacional de Recursos Genéticos
INIFAP
Tepatitlán de Morelos
Jalisco, Mexico

Ramón Ignacio Arteaga-Garibay
Centro Nacional dc Rccursos Genéticos
INIFAP
Tepatitlán de Morelos
Jalisco, Mexico

ISSN 1064-3745 ISSN 1940-6029 (electronic)
Methods in Molecular Biology
ISBN 978-1-0716-5622-8 ISBN 978-1-0716-5623-5 (eBook)
https://doi.org/10.1007/978-1-0716-5623-5

This Humana imprint is published by the registered company Springer Science+Business Media, LLC, part of Springer Nature.
The registered company address is: 1 New York Plaza, New York, NY 10004, U.S.A.

Preface

Conservation of Microbial Genetic Resources

Microbial preservation is vital for maintaining microorganisms' biotechnological potential, food sovereignty, and environmental sustainability. The main goal of effective microbial conservation is to preserve axenic (pure) cultures, ensuring their viability and genetic stability. The growing use of microorganisms in diverse fields, particularly biotechnology, underscores the need to create and maintain key strains that retain their unique characteristics.

In recent decades, the accelerated expansion of biotechnology, genomics, synthetic biology, metabolomics, and microbial engineering has significantly increased the demand for reliable microbial repositories capable of preserving strains with unique metabolic, enzymatic, ecological, and industrial properties. Consequently, microbial culture collections now play a strategic role in the conservation of global biodiversity and in the protection of valuable biological heritage. These collections serve as indispensable platforms for scientific research, academic training, innovation, industrial development, and quality assurance, while also functioning as reference centers for taxonomy, authentication, traceability, and reproducibility in microbiological studies.

The preservation of microbial diversity is particularly relevant in the context of emerging global challenges, including climate change, environmental degradation, food insecurity, antimicrobial resistance, and the growing need for sustainable biotechnological alternatives. Conserved microorganisms provide an invaluable reservoir of genetic and biochemical diversity, enabling the discovery of novel metabolites, antibiotics, enzymes, biofertilizers, biocontrol agents, and other bioactive compounds with broad scientific and commercial applications. Moreover, long-term conservation strategies contribute to preventing the irreversible loss of unique microbial taxa and their associated functional traits.

This book outlines fundamental and applied methodologies for the long-term ex situ conservation of microorganisms, including bacteria, protozoa, fungi, protists, and viruses. It provides comprehensive practical guidance on implementing these methods in various microbiology laboratories and for potential users. Additionally, the book covers standard methodologies related to microbial identification, which complement the preservation process and are vital for characterizing microorganisms with diverse biotechnological applications, with special emphasis on methodologies designed to maximize post-preservation viability, minimize genetic drift, and maintain phenotypic stability during storage.

Overall, the primary objective of this book is to disseminate foundational and advanced knowledge related to the conservation, management, and characterization of Microbial Genetic Resources, thereby contributing to the strengthening of global efforts aimed at preserving microbial biodiversity and promoting its sustainable utilization for future generations.

Jalisco, México *Marco A. Ramírez-Mosqueda*
Jalisco, México *Jorge David Cadena-Zamudio*
Jalisco, México *José Luis Aguirre-Noyola*
Jalisco, México *Ramón Ignacio Arteaga-Garibay*

Contents

Contributors

JULIO AGUILAR-CAAMAL • *Tecnológico Nacional de México, Campus Instituto Tecnológico Superior de Valladolid, Valladolid, Yucatán, Mexico*

JOSÉ LUIS AGUIRRE-NOYOLA • *Centro Nacional de Recursos Genéticos, INIFAP, Tepatitlán de Morelos, Jalisco, Mexico*

RAMÓN IGNACIO ARTEAGA-GARIBAY • *Centro Nacional de Recursos Genéticos, INIFAP, Tepatitlán de Morelos, Jalisco, Mexico*

RUBÉN BARRERA-RAMÍREZ • *Campo Experimental Uruapan, INIFAP, Uruapan, Michoacán, Mexico*

DERYA BERIKTEN • *Department of Molecular Biology and Genetics, Faculty of Engineering and Natural Sciences, Kütahya Health Sciences University, Kütahya, Türkiye*

ALEJANDRO BOLAÑOS-DIRCIO • *Laboratorio de Microbiología Molecular y Biotecnología Ambiental, Facultad de Ciencias Químico Biológicas, Universidad Autónoma de Guerrero, Chilpancingo, Guerrero, Mexico*

JORGE CADENA-IÑIGUEZ • *Postgrado en Innovación en Manejo de Recursos Naturales, Campus San Luis Potosí, Colegio de Postgraduados, San Luis Potosí, CP, Mexico*

DANIEL ALEJANDRO CADENA-ZAMUDIO • *Instituto de Ecología, A. C., Red de Ambiente y Sustentabilidad, Veracruz, Mexico*

JORGE DAVID CADENA-ZAMUDIO • *Centro Nacional de Recursos Genéticos, INIFAP, Tepatitlán de Morelos, Jalisco, Mexico*

ENRIQUE I. CANALES-ISLAS • *Centro Nacional de Recursos Genéticos, INIFAP, Tepatitlán de Morelos, Jalisco, Mexico*

ISMAEL FERNANDO CHÁVEZ-DÍAZ • *Centro Nacional de Recursos Genéticos, INIFAP, Tepatitlán de Morelos, Jalisco, Mexico*

JOSÉ DANIEL CHÁVEZ-GONZÁLEZ • *Escuela superior de Ciencias de la Tierra, Doctorado en Recursos Naturales y Ecología. Universidad Autónoma de Guerrero, Taxco el Viejo, Guerrero, Mexico*

IO HONG CHEONG • *State Key Laboratory of Oncogenes and Related Genes, Centre for Single-Cell Omics, School of Public Health, Shanghai Jiao Tong University School of Medicine, Shanghai, China*

CARLOS ALBERTO CRUZ-CRUZ • *Facultad de Ciencias Químicas, Universidad Veracruzana, Orizaba, Veracruz, Mexico*

GUSTAVO CUAXINQUE-FLORES • *Doctorado en Recursos Naturales y Ecología. Universidad Autónoma de Guerrero, Acapulco de Juárez, Guerrero, México*

AYUSHI DALMIA • *302, Tandice 69, F residency, Mumbai, India*

ANA M. ESCALANTE-MAÑÉ • *Facultad de Ingeniería, Universidad Autónoma de Yucatán, Mérida, Yucatán, Mexico*

IDOLINA FLORES-CORTEZ • *Instituto de Investigaciones Químico Biológicas, Universidad Michoacana de San Nicolás de Hidalgo, Edificio B1, Ciudad Universitaria, Morelia, Michoacán, Mexico*

HUGO ERNESTO FLORES-LÓPEZ • *Campo Experimental Centro Altos de Jalisco, INIFAP, Tepatitlán de Morelos, Jalisco, Mexico*

Emma O. Fuentes-Ramírez • *BIOTECHVERA, AC. Dirección de Investigación Científica, Orizaba, Veracruz, Mexico*
Tania García-Herrera • *Facultad de Ciencias Químicas, Universidad Veracruzana, Orizaba, Veracruz, Mexico*
Verónica García-Iglesias • *Maestría en Recursos Naturales y Ecología. Universidad Autónoma de Guerrero, Acapulco de Juárez, Guerrero, Mexico*
Lorena Jacqueline Gómez-Godínez • *Centro Nacional de Recursos Genéticos, INIFAP, Tepatitlán de Morelos, Jalisco, Mexico*
Francisco Erik González-Jiménez • *Facultad de Ciencias Químicas, Universidad Veracruzana, Orizaba, Veracruz, Mexico*
Ning Guo • *Institute of Biothermal Science and Technology, University of Shanghai for Science and Technology, Shanghai, China*
Gabriel R. Hernández-Martínez • *Facultad de Ciencias Químicas, Universidad Veracruzana, Orizaba, Veracruz, Mexico*
María Isabel Iñiguez-Luna • *Universidad de Guadalajara, Centro Universitario de los Altos, Tepatitlán de Morelos, Jalisco, Mexico*
Huiling Jin • *Department of Public Health, Medical College of Yanbian University, Yanji, Jilin, China*
Zisis Kozlakidis • *Laboratory Services and Biobank Group, International Agency for Research on Cancer (IARC), World Health Organization, Lyon, France*
Carlos Eduardo Lara-Hidalgo • *Facultad de Ciencias de la Salud, Universidad Anáhuac, Ciudad de México, Mexico*
Ge Liu • *Department of Public Health, Medical College of Yanbian University, Yanji, Jilin, China*
Yang Liu • *National Pathogen Resource Center, Chinese Center for Disease Control and Prevention (Chinese Academy of Preventive Medicine), Beijing, China*
Geny I. López-Arellano • *Tecnológico Nacional de México, Campus Instituto Tecnológico Superior de Valladolid, Valladolid, Yucatán, Mexico*
Sergio De los Santos-Villalobos • *Centro Nacional de Recursos Genéticos, INIFAP, Tepatitlán de Morelos, Jalisco, Mexico; Instituto Tecnológico de Sonora, Sonora, Mexico*
Mauricio Luna-Rodríguez • *Facultad de Ciencias Agrícolas, Universidad Veracruzana, Xalapa, Veracruz, Mexico*
Guadalupe Luna-Solano • *Instituto Tecnológico de Orizaba, Tecnológico Nacional de México, Orizaba, Veracruz, Mexico*
Esperanza Martínez-Romero • *Centro de Ciencias Genómicas, Universidad Nacional Autónoma de México Universidad Autónoma del Estado de Morelos, Cuernavaca, Morelos, Mexico*
Sergio Montoya-Montoya • *Instituto de Investigaciones Químico Biológicas, Universidad Michoacana de San Nicolás de Hidalgo, Edificio B1, Ciudad Universitaria, Morelia, Michoacán, Mexico*
Hipólito Jesús Muñoz-Flores • *Campo Experimental Uruapan, INIFAP, Uruapan, Michoacán, Mexico*
Dulce María Murrieta-Hernández • *Facultad de Ciencias Agrícolas, Universidad Veracruzana, Xalapa, Veracruz, Mexico*
Iván Oros-Ortega • *Tecnológico Nacional de México, Campus Instituto Tecnológico de la Zona Maya, Quintana Roo, Mexico*

ALEJANDRA ORTIZ-GASCA • *Doctorado en Micología Aplicada, Centro de Investigación en Micología Aplicada, Universidad Veracruzana, Xalapa, Veracruz, Mexico*
AUDRY PEREDO-LOVILLO • *Facultad de Ciencias Químicas, Universidad Veracruzana, Orizaba, Veracruz, Mexico*
CARLOS A. PUCH-HAU • *Tecnológico Nacional de México, Campus Instituto Tecnológico Superior de Valladolid, Valladolid, Yucatán, Mexico; Instituto de Investigaciones Oceanológicas, Universidad Autónoma de Baja California, Ensenada, Baja California, México*
MARCO A. RAMÍREZ-MOSQUEDA • *Centro Nacional de Recursos Genéticos, INIFAP, Tepatitlán de Morelos, Jalisco, Mexico*
LIZANDRO RAMÍREZ-TREJO • *Unidad de Biotecnología Vegetal, Centro de Investigación y Asistencia en Tecnología y Diseño del Estado de Jalisco, Guadalajara, Jalisco, Mexico*
JUAN RAMOS-GARZA • *Especialidad en Regulación Sanitaria de Medicamentos y Vacunas, Universidad de la Salud, Ciudad de Mexico, Mexico*
RUBÉN RÍOS-MEDINA • *Tecnológico Nacional de México, Campus Instituto Tecnológico Superior de Valladolid, Valladolid, Yucatán, Mexico*
RAFAEL RIVERA-BUSTAMANTE • *Centro de Investigación y de Estudios Avanzados del Instituto Politécnico Nacional, unidad Mérida. Depto de Recursos del Mar, Mérida, Yucatán, Mexico*
AUGUSTO ROJAS-APARICIO • *Laboratorio de Microbiología Molecular y Biotecnología Ambiental, Facultad de Ciencias Químico Biológicas, Universidad Autónoma de Guerrero, Chilpancingo, Guerrero, Mexico*
HAYDEE ELIZA ROMERO-LUNA • *Instituto Tecnológico de Orizaba, Tecnológico Nacional de México, Orizaba, Veracruz, Mexico*
YANET ROMERO-RAMÍREZ • *Laboratorio de Microbiología Molecular y Biotecnología Ambiental, Facultad de Ciencias Químico Biológicas, Universidad Autónoma de Guerrero, Chilpancingo, Guerrero, Mexico*
MAGALI RUÍZ-RIVAS • *Campo Experimental Uruapan, INIFAP, Uruapan, Michoacán, Mexico*
JOSÉ MARTÍN RUVALCABA-GÓMEZ • *Centro Nacional de Recursos Genéticos, INIFAP, Tepatitlán de Morelos, Jalisco, Mexico*
LUIS A. SÁENZ-CARBONELL • *Centro de Investigación Científica de Yucatán, Mérida, Yucatán, Mexico*
JASMÍN SALAZAR-MENDOZA • *Department of Engineering, Centro Universitario de los Altos, University of Guadalajara, Tepatitlán de Morelos, Jalisco, Mexico*
WENDY SANGABRIEL-CONDE • *Facultad de Ciencias Agrícolas Campus Xalapa, Universidad Veracruzana, Veracruz, Mexico*
OSCAR TALAVERA-MENDOZA • *Escuela Superior de Ciencias de la Tierra, Universidad Autónoma de Guerrero, Taxco el Viejo, Guerrero, Mexico*
JEIRY TORIBIO-JIMÉNEZ • *Laboratorio de Microbiología Molecular y Biotecnología Ambiental, Facultad de Ciencias Químico Biológicas, Universidad Autónoma de Guerrero, Chilpancingo, Guerrero, Mexico*
DORA TREJO-AGUILAR • *Facultad de Ciencias Agrícolas Campus Xalapa, Universidad Veracruzana, Veracruz, Mexico*
AJAY W. TUMANEY • *Department of Biochemistry, School of Life Sciences, University of Hyderabad, Hyderabad, India*

RAFAEL UZÁRRAGA-SALAZAR • *Facultad de Ciencias Químicas, Universidad Veracruzana, Orizaba, Veracruz, Mexico*
EDUARDO VALENCIA-CANTERO • *Instituto de Investigaciones Químico Biológicas, Universidad Michoacana de San Nicolás de Hidalgo, Edificio B1, Ciudad Universitaria, Morelia, Michoacán, Mexico*
IGNACIO VALENZUELA-CHAVIRA • *Departamento de Ciencias de la Salud, Centro Universitario de los Altos, Universidad de Guadalajara, Tepatitlán de Morelos, Jalisco, Mexico*
JULIO VEGA-ARREGUIN • *Escuela Nacional de Estudios Superiores-UNAM, León, Mexico*
JANETTE BETZABETH VILLARRUEL-MUÑOZ • *Department of Engineering, Centro Universitario de los Altos, University of Guadalajara, Tepatitlán de Morelos, Jalisco, Mexico*
DUOCHUN WANG • *National Institute for Communicable Disease Control and Prevention, Chinese Center for Disease Control and Prevention (Chinese Academy of Preventive Medicine), Beijing, China*
QIANG WEI • *National Pathogen Resource Center, Chinese Center for Disease Control and Prevention (Chinese Academy of Preventive Medicine), Beijing, China*
YI XU • *Institute of Biothermal Science and Technology, University of Shanghai for Science and Technology, Shanghai, China*
XIAOMEI YAN • *National Key Laboratory of Intelligent Tracking and Forecasting for Infectious Diseases, National Institute for Communicable Disease Control and Prevention, Chinese Center for Disease Control and Prevention (Chinese Academy of Preventive Medicine), Beijing, China*
LILY X. ZELAYA-MOLINA • *Centro Nacional de Recursos Genéticos, INIFAP, Tepatitlán de Morelos, Jalisco, Mexico*

Part I

Introduction

Check for updates

Chapter 1

Overview of Conservation of Microbial Genetic Resources

Marco A. Ramírez-Mosqueda, Lizandro Ramírez-Trejo, Jorge David Cadena-Zamudio, José Luis Aguirre-Noyola and Enrique I. Canales-Islas

Abstract

Microorganisms constitute a fundamental component of ecosystems and sustain essential services in agriculture, industry, and health. Despite their relevance, it is estimated that less than 1% of global microbial biodiversity has been isolated, cultivated, characterized, and preserved, highlighting a profound knowledge gap. This limitation represents a critical challenge for the sustainable use of these resources, particularly in a context of increasing biodiversity loss driven by anthropogenic pressures. The conservation of microbial genetic resources relies on complementary in situ and ex situ strategies. In situ conservation focuses on protecting natural habitats, maintaining microbial populations in their native environments, and enabling the continuity of evolutionary and ecosystem processes. Moreover, it represents the only approach capable of preserving the majority of still-undescribed microbial biodiversity. In contrast, ex situ conservation involves the preservation of microbial strains and microbiomes in specialized microbial resource centers. Rigorous strategies, such as the adoption of FAIR principles (findable, accessible, interoperable, and reusable), the implementation of standardized quality frameworks (ISO 21710, 9001, 17025, 20387), and compliance with international legal frameworks (such as the Nagoya Protocol and the Budapest Treaty), are globally prioritized to ensure the reliability, traceability, and interoperability of microbial resources. Furthermore, the integration of omics approaches and global databases is transforming how microbial diversity is documented, preserved, and shared. This chapter provides an overview of the conservation of microbial genetic resources, addressing their ecological and biotechnological relevance, as well as current strategies, global challenges, and actions undertaken by international networks.

Keywords Microbial genetic resources, In situ conservation, Ex situ conservation, Microbiome, Biobanks

1 Introduction

Microorganisms have constituted the life-support system of the planet for more than 3.5 billion years, enabling the evolution and persistence of multicellular organisms [1]. In ecosystems and food webs, they play fundamental and structural roles, with a broad spectrum of bioactivities and bioprocesses, including nutrient cycling (carbon, nitrogen,

Marco A. Ramírez-Mosqueda et al. (eds.), *Methods for The Conservation of Microorganisms*, Methods in Molecular Biology, vol. 3090, https://doi.org/10.1007/978-1-0716-5623-5_1,

phosphorus, sulfur) and the decomposition and recycling of organic compounds [2]. Given their immense biodiversity and abundance (~10^{30} prokaryotic cells on Earth), only a minute fraction (~1%) has been formally described, highlighting a significant knowledge gap regarding this group of living organisms [3, 4]. Despite their historical resilience, microbial communities are facing increasing erosion due to anthropogenic pressures, such as climate change, habitat destruction, pollution, and agricultural intensification [3, 5]. The loss of microbial diversity represents a systemic risk that destabilizes complex interaction networks within ecosystems and disrupts soil biogeochemical processes, directly affecting ecosystem functionality and resilience [6, 7]. Despite their multifunctional importance, microbial conservation has been underrepresented, prioritizing microorganisms with direct utility for humans while largely neglecting vast microbial biodiversity and its ecosystem functions, particularly those that are unculturable or yet to be characterized [8].

From a biotechnological perspective, the importance of microbial biodiversity lies in its immense potential for bioprospecting and the development of solutions that benefit society [9]. In practice, these are referred to as microbial genetic resources (MGRs), defined as any material of microbial origin containing functional units of heredity and possessing actual or potential value for humanity [10]. These resources include isolated strains of bacteria, archaea, fungi, viruses, and protists, as well as their associated genetic components and, more recently, complex microbial communities or microbiomes, including their ecological interactions and emergent functions [11]. To ensure that global microbial heritage remains a renewable resource for present and future generations, conservation strategies have been developed. These range from the identification and risk assessment of threatened microbiomes to the functional characterization of MGRs, integrating multi-omics approaches and systems biology tools [2, 12, 13].

Several national and international networks have undertaken efforts to harmonize processes, share technical knowledge, and develop standardized frameworks to improve the management, traceability, and accessibility of information associated with MGRs. Therefore, the objective of this chapter is to provide an overview of the current state of microbial genetic resource conservation, addressing its theoretical foundations, current preservation methodologies, relevant regulatory frameworks, and future challenges to ensure that this microbial heritage remains a renewable, strategic, and sustainable resource in the long term.

2 Global Microbial Biodiversity

For approximately 3.5–3.8 billion years, microorganisms have shaped the biosphere and enabled the evolution and persistence of all other forms of life, acting as the planet's primary ecological engineers [1]. It is currently estimated that there are approximately ~10^{30} prokaryotic cells on Earth [3]. In terms of biomass, bacteria, archaea, protists, and fungi together account for approximately 46 gigatons of carbon, making them the second most abundant biological component globally, after plants [4]. This vast abundance is reflected in their diversity, representing the majority of the planet's biological richness [4]. Although the magnitude of this diversity remains under discussion, statistical modeling estimates suggest that the total number could reach up to one trillion (10^{12}) microbial species, underscoring the scale of uncharacterized diversity [14]. However, a critical gap persists in the taxonomic knowledge of microorganisms, as only a minimal fraction (~1%) has been formally described, largely due to limitations in cultivation and characterization techniques. For example, approximately 22,000–30,000 species of bacteria and archaea have been validated [15, 16], while about 120,000 fungal species have been described [17], although estimates suggest that their actual diversity is significantly higher.

Microorganisms exhibit a cosmopolitan distribution, occupying virtually all ecological niches, including those hostile to macroscopic life, demonstrating their extraordinary physiological and metabolic adaptability. These are known as extremophiles and are classified as thermophiles, psychrophiles, acidophiles, alkaliphiles, halophiles, barophiles, xerophiles, and radiophiles [18]. From a phylogenetic perspective, animals, plants, and multicellular fungi represent only minor branches of the tree of life compared to the vast diversity of microbial lineages [4], highlighting their deep evolutionary divergence and antiquity. This diversity is also reflected in their global functional relevance, as microbial populations participate in a wide range of biogeochemical processes, particularly regulating the flux of essential elements such as carbon, nitrogen, phosphorus, and sulfur within ecosystems, thereby maintaining their stability and productivity [2]. Furthermore, many microorganisms establish symbiotic relationships with plants and animals that are essential for both host physiological function and ecosystem maintenance, forming highly specialized and dynamic associations [19, 20].

3 Microbial Genetic Resources

Microbial genetic resources (MGRs) are defined as any material of microbial origin containing functional units of heredity and possessing actual or potential value for humanity. Specifically, an MGR is a microbial strain (bacteria, archaea, fungi, viruses, or protists) that has been authenticated, taxonomically defined, physiologically characterized, and documented under rigorous quality management protocols, ensuring traceability, reproducibility, and genetic stability [10]. These resources include living cells and their replicable components, such as genomic DNA, plasmids, cDNA libraries, and associated digital information, including genomic maps and metabolic profiles, as well as data derived from high-resolution omics approaches [21].

These resources underpin the global bioeconomy across diverse biotechnological industries, supporting key production processes in strategic sectors [10]. In modern medicine, MGRs are used as biofactories for pharmaceutical production. Approximately 60%–75% of drugs approved in recent decades are directly or indirectly derived from natural products, many of microbial origin or inspired by microbial metabolites [22]. Most of these compounds are produced using recombinant microbial systems, including antibiotics, vaccines, hormones such as recombinant insulin, and monoclonal antibodies, which have significantly transformed the treatment of infectious and chronic diseases and expanded therapeutic possibilities in precision medicine [23].

In industry, MGRs are used as biofactories for the synthesis of enzymes and biopolymers, optimizing industrial processes under sustainable biotechnology approaches [24]. For example, *Aspergillus niger* and *Bacillus subtilis* are used for the production of amylases, enzymes that degrade starch in the food and textile industries [25, 26]. Species of *Trichoderma* are employed for the production of cellulases and hemicellulases, which are essential for processing agricultural residues and recalcitrant biomass, facilitating the valorization of lignocellulosic byproducts [27, 28]. Similarly, the bacterium *Cupriavidus necator*, considered a classical model organism for the production of polyhydroxyalkanoates (PHA), has been used to produce these biodegradable biopolymers with potential as sustainable alternatives to conventional plastics [29].

In sustainable agriculture and food security, MGRs play a strategic role as plant growth–promoting microorganisms, acting as biostimulants that enhance nutrient acquisition efficiency and tolerance to abiotic stress [30, 31]. At the same time, the use of microorganisms as biocontrol agents and biopesticides offers an ecological alternative to reduce reliance on synthetic agrochemicals, promoting more resilient and sustainable agricultural systems [32].

Despite their enormous biotechnological potential, a considerable gap remains between available microbial diversity and the extent to which it has been studied and utilized. It is estimated that less than 1% of global microbial biodiversity has been cultivated, characterized, and preserved in public collections, limiting its accessibility and biotechnological exploitation [33]. Access to these resources is being transformed by the genomic era, where the integration of biological material with omics data under FAIR principles is a priority for the international scientific community, promoting standardization, interoperability, and data reuse [12]. The utilization of MGRs is governed by international legal frameworks, such as the Convention on Biological Diversity (CBD) and the Nagoya Protocol, which ensure national sovereignty over these resources and establish mechanisms for fair and equitable benefit-sharing under Access and Benefit-Sharing (ABS) agreements. Additionally, the Budapest Treaty facilitates intellectual property protection processes through the deposit of microorganisms for patent purposes in recognized international authorities, ensuring their availability for legal and commercial use.

4 Conservation of Microbial Genetic Resources

The conservation of microbial genetic resources (MGRs) is based on ensuring their long-term availability for scientific, agricultural, and industrial innovation, as well as their sustainable use by present and future generations. Conservation strategies are categorized into two complementary approaches: in situ and ex situ. In situ conservation focuses on maintaining viable microbial populations within their natural habitats, such as soil, water, or symbiotic associations with plant and animal hosts. This approach enables the continuation of natural evolutionary processes and the maintenance of ecological interactions, thereby preserving the evolutionary and functional dynamics of microbiomes [34]. Moreover, in situ conservation represents the only viable strategy to preserve approximately 99% of microbial biodiversity that has not yet been isolated and cultivated in laboratory conditions, including the so-called "microbial dark matter." Effective in situ management requires the protection of natural refuges and the implementation of sustainable land-use practices, such as conservation tillage [35], as well as biodiversity-based soil management approaches. It also involves reducing the use of chemical inputs to prevent the disruption of native microbiomes and minimize disturbances to their ecological networks [36].

In contrast, ex situ conservation involves the controlled preservation of microorganisms outside their natural habitats within culture collections and germplasm banks. Microbial biological resource center's function both as service providers and as repositories of living cells, their components (e.g., nucleic acids, proteins, and metabolites), and associated digital information [20, 21], acting as key infrastructures for research and innovation. These centers operate under rigorous quality management systems and international standards, including Best Practice Guidelines for Biological Resource Centers [37], World Federation for Culture Collections (WFCC) guidelines [38], European Culture Collections' Organisation (ECCO) frameworks [39], and various ISO certifications.

These include ISO 9001 and ISO 21710 for management systems, ISO 17025 for experimental validity, and ISO 20387 specifically for biobanking and technical competence [20]. These standards ensure that preserved materials are authentic, pure, and genetically stable, providing the scientific community with resources that guarantee reproducibility and comparability across studies and laboratories.

Ex situ conservation is based on methods that induce a state of metabolic stasis, that is, a controlled reduction or cessation of metabolic activity, allowing the preservation of the physiological and genetic integrity of microorganisms. Among the most widely used techniques for preserving microbial resources are cryopreservation (Fig. 1) and lyophilization (Fig. 2). Cryopreservation involves storing cells in liquid nitrogen at ultralow temperatures (−196°C) and is considered the preferred method for long-term preservation due to its high efficiency in maintaining cell viability and genomic stability. In contrast, lyophilization is commonly used for bacteria, yeasts, and spore-forming fungi [24, 40], and also facilitates storage and transportation. To maximize safety and minimize the risk of strain loss, the WFCC [38] recommends preserving each strain using at least two different methods. Additionally, it suggests storing duplicates of the most important genetic resources in separate facilities to protect them from disasters, such as fires, floods, or armed conflicts, following principles of redundancy and risk mitigation.

Despite technical advances and high-quality guidelines, significant challenges remain in the conservation of MGRs. Microbial resource centers are inherently vulnerable to financial constraints, which threaten their long-term operational sustainability [41]. From a technical perspective, high staff turnover and job instability hinder

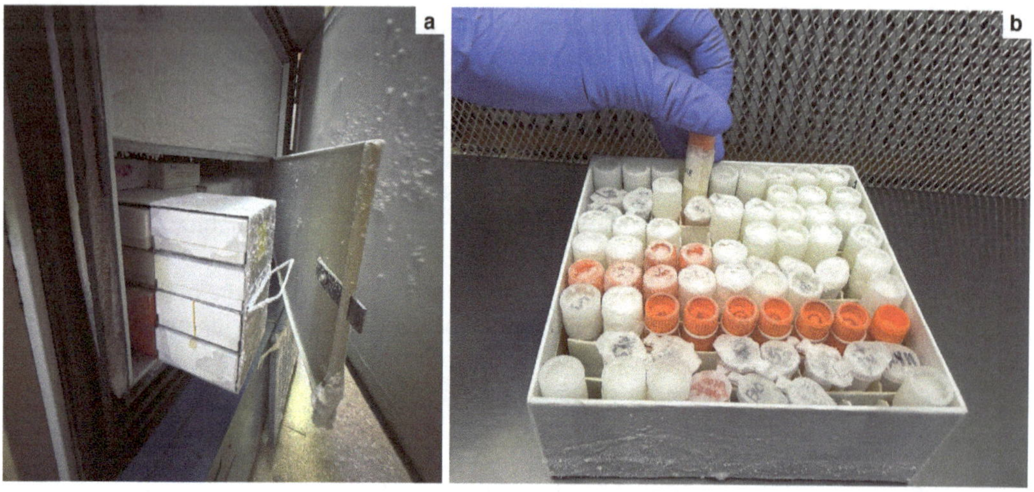

Fig. 1 Cryopreservation of bacterial cultures at −80°C. Representative images of bacterial stock storage under ultra-low temperature conditions. (**a**) Cryogenic freezer with labeled cardboard boxes containing bacterial culture collections. (**b**) Cryobox showing cryovials with bacterial cultures stored at −80°C

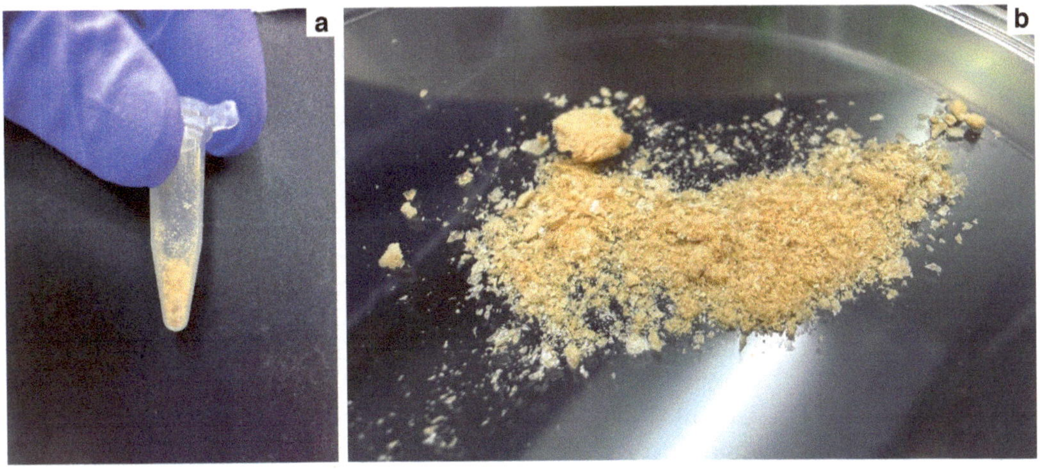

Fig. 2 Lyophilized bacterial cells as a long-term preservation method. Representative images of bacterial biomass obtained after the freeze-drying process. (**a**) Tube containing lyophilized bacterial sample. (**b**) Macro view of the lyophilized powder, exhibiting its characteristic dry, friable, and granular texture

the consolidation of specialized skills required for MGR conservation, affecting the transfer of expertise [42]. These limitations are compounded by the challenge of maintaining the genetic stability of preserved resources, particularly in collections where strains are maintained through periodic subculturing. This practice promotes the accumulation of genetic changes associated with microevolution and genetic drift, potentially altering their original phenotypic and functional properties. Therefore, it is essential to implement strategies that induce metabolic stasis, such as cryopreservation and lyophilization, to preserve genomic integrity over the long term [12, 43]. Finally, due to ongoing environmental degradation, the conservation of microbiome samples has become increasingly important, creating the need to optimize protocols that preserve their composition, functionality, and interactions, recognizing microbiomes as complex ecological units [11].

To address these challenges, a coordinated global framework is required, integrating technical expertise, sustainable funding models, and standardized data management systems, ensuring that global microbial heritage remains a renewable resource for future generations, aligned with international initiatives in open science and biodiversity conservation [21, 44].

5 International Actions for the Conservation of Microbial Genetic Resources

The current landscape of MGRs conservation is characterized by an increasingly coordinated global effort, driven by growing concern over microbial biodiversity loss and the need to achieve the highest quality standards in biological resource centers. Among the most relevant international legal frameworks are the Convention on Biological Diversity (CBD) and its Nagoya Protocol on Access and Benefit-Sharing (ABS) of MGRs. These instruments provide legal certainty for the exchange of genetic resources, support national sovereignty over MGRs, and regulate mutually agreed terms (MAT) for their access and use.

At the global level, institutional coordination is primarily led by the World Federation for Culture Collections (WFCC), which currently oversees more than 892 registered collections across 80 countries, forming a key global network for ex situ microbial conservation. The WFCC data center, the World Data Centre for Microorganisms (WDCM), has developed advanced digital infrastructures to improve accessibility and transparency of MGRs. Notable initiatives include the Global Catalogue of Microorganisms (GCM) and the Global Catalogue of Type Strains (gcType), which apply FAIR data principles by linking strain information with genomic annotations, scientific publications, and patent data, thereby facilitating global data integration and reuse [13]. To standardize these operations worldwide, the Organisation for Economic Co-operation and Development (OECD) has published guidelines on best practices for biological resource centers, establishing rigorous criteria for quality management, biosafety, and technical competence, strengthening international standardization.

In Europe, the Microbial Resource Research Infrastructure (MIRRI-ERIC) serves as a leading model by connecting more than 50 microbial resource centers to provide high-quality authenticated materials and legal guidance, integrating scientific, technical, and regulatory capacities within a single platform [20]. Similar initiatives include the United States Culture Collection Network (USCCN), the Asian Consortium for the Conservation and Sustainable Use of Microbial Resources, and the Latin American Federation for Culture Collections (FELACC), all of which focus on sharing best practices and harmonizing material transfer agreements, promoting regional and international cooperation.

More recently, the IUCN (International Union for Conservation of Nature) Microbial Conservation Specialist Group (MCSG) was established as the first official body within the International Union for Conservation of Nature dedicated to the management of all microbial life [2]. The MCSG aims to integrate microbial diversity into global conservation cycles (Assessment, Planning, Action, Networking, and Communication and Policy), promoting the systematic evaluation of microbial biodiversity through the identification of conservation hotspots and the development of conceptual frameworks for its inclusion in global conservation policies.

In Mexico, the Microorganism Collection of the National Center for Genetic Resources (CM-CNRG) of INIFAP, located in

Tepatitlán, Jalisco, safeguards more than 2650 accessions and 5000 germplasm units, representing one of the country's main infrastructures for the conservation of MGRs. Its mission is to preserve microorganisms of agro-food importance (bacteria, fungi, and viruses) to ensure biodiversity conservation and support national research, industry, and agriculture.

Finally, there is a need to develop criteria compatible with the IUCN Red List and to incorporate functional metrics that allow the assessment of the integrity and resilience of microbial communities across ecosystems, moving beyond traditional taxonomy-based approaches. Collectively, these global actions consolidate regulatory frameworks and quality management schemes aimed at safeguarding microbial resources of the biosphere, advancing toward an integrative approach that combines conservation, sustainable use, and scientific innovation.

Conflict of Interest

The authors have no conflicts of interest to declare that are relevant to the content of this chapter.

References

1. Knoll AH (2015) Paleobiological perspectives on early microbial evolution. Cold Spring Harb Perspect Biol 7(7):a018093
2. Gilbert JA, Scholz AH, Dominguez Bello MG, Korsten L, Berg G, Singh BK, Boetius A, Wang F, Greening C, Wrighton K, Bordenstein SR, Jansson J, Lennon JT, Souza V, Allard SM, Thomas T, Cowan D, Crowther TW, Nguyen N, Harper L, Haraoui LP, Ishaq SL, McFall-Ngai M, Redford KH, Peixoto R (2025) Safeguarding microbial biodiversity: microbial conservation specialist group within the species survival commission of the International Union for Conservation of Nature. FEMS Microbiol Ecol 101(12):fiaf107. https://doi.org/10.1093/femsec/fiaf107
3. Cavicchioli R, Ripple WJ, Timmis KN, Azam F, Bakken LR, Baylis M, Webster NS (2019) Scientists' warning to humanity: microorganisms and climate change. Nat Rev Microbiol 17(9):569–586
4. Crowther TW, Rappuoli R, Corinaldesi C, Danovaro R, Donohue TJ, Huisman J, van Galen LG (2024) Scientists' call to action: microbes, planetary health, and the sustainable development goals. Cell 187(19):5195–5216
5. Lanz B, Dietz S, Swanson T (2018) The expansion of modern agriculture and global biodiversity decline: an integrated assessment. Ecol Econ 144:260–277
6. Delgado-Baquerizo M, Maestre FT, Reich PB, Jeffries TC, Gaitan JJ, Encinar D, Singh BK (2016) Microbial diversity drives multifunctionality in terrestrial ecosystems. Nat Commun 7(1):10541
7. Yang X, Huang X, Cheng Z, Li S, Mahjoob HA, Xu J, He Y (2024) Response of soil general and specific functions following loss of microbial diversity: a review. Soil Secur 15:100151
8. Redford KH (2023) Extending conservation to include Earth's microbiome. Conserv Biol 37:3
9. Vuong P, Chong S, Kaur P (2022) The little things that matter: how bioprospecting microbial biodiversity can build towards the realization of United Nations Sustainable Development Goals. NPJ Biodivers 1:4
10. Overmann J (2015) Significance and future role of microbial resource centers. Syst Appl Microbiol 38(4):258–265
11. Kim Y, Cheong HM, Choi G, Choi KM, Chung EJ, Kim A, Hong SB (2024) Strengthening the Korean network of microbial culture collections in the microbiome era. Mycobiology 52(4):207–213
12. Becker P, Bosschaerts M, Chaerle P, Daniel HM, Hellemans A, Olbrechts A, Hendrickx M (2019) Public microbial resource centers:

key hubs for findable, accessible, interoperable, and reusable (FAIR) microorganisms and genetic materials. Appl Environ Microbiol 85(21):e01444–19
13. Fan G, Sun Q, Sun Y, Liu D, Li S, Li M, Wu L (2025) GCM and gcType in 2024: comprehensive resources for microbial strains and genomic data. Nucleic Acids Res 53(D1):D763–D771
14. Locey KJ, Lennon JT (2016) Scaling laws predict global microbial diversity. Proc Natl Acad Sci 113(21):5970–5975
15. Dykhuizen D (2005) Species numbers in bacteria. Proc Calif Acad Sci 56(6):62
16. Reimer LC, Sardà Carbasse J, Koblitz J, Ebeling C, Podstawka A, Overmann J (2022) Bac Dive in 2022: the knowledge base for standardized bacterial and archaeal data. Nucleic Acids Res 50(D1):D741–D746
17. Hawksworth DL, Lücking R (2017) Fungal diversity revisited: 2.2 to 3.8 million species. Microbiol Spectr 5(4):10–1128
18. Rekadwad BN, Li WJ, Gonzalez JM, Punchappady Devasya R, Ananthapadmanabha Bhagwath A, Urana R, Parwez K (2023) Extremophiles: the species that evolve and survive under hostile conditions. 3 Biotech 13(9):316
19. Kebede E (2021) Contribution, utilization, and improvement of legumes-driven biological nitrogen fixation in agricultural systems. Front Sustain Food Syst 5:767998
20. Lima N (2025) Preserving and utilizing microbial diversity for innovation and sustainability. Microbiology 171(3):001544
21. Stackebrandt E (2010) Diversification and focusing: strategies of microbial culture collections. Trends Microbiol 18(7):283–287
22. Newman DJ, Cragg GM (2012) Natural products as sources of new drugs over the 30 years from 1981 to 2010. J Nat Prod 75(3):311–335
23. Vitorino LC, Bessa LA (2017) Technological microbiology: development and applications. Front Microbiol 8:827
24. Sette LD, Pagnocca FC, Rodrigues A (2013) Microbial culture collections as pillars for promoting fungal diversity, conservation and exploitation. Fungal Genet Biol 60:2–8
25. Djekrif-Dakhmouche S, Gheribi-Aoulmi Z, Meraihi Z, Bennamoun L (2006) Application of a statistical design to the optimization of culture medium for α-amylase production by Aspergillus niger ATCC 16404 grown on orange waste powder. J Food Eng 73(2):190–197
26. Asgher M, Asad MJ, Rahman SU, Legge RL (2007) A thermostable α-amylase from a moderately thermophilic Bacillus subtilis strain for starch processing. J Food Eng 79(3):950–955
27. Druzhinina IS, Kubicek CP (2017) Genetic engineering of Trichoderma reesei cellulases and their production. Microb Biotechnol 10:1485–1499
28. Lima PC, Karimian P, Johnston E, Hartley CJ (2024) The use of Trichoderma spp. for the bioconversion of agro-industrial waste biomass via fermentation: a review. Fermentation 10(9):442
29. Zhang L, Jiang Z, Tsui TH, Loh KC, Dai Y, Tong YW (2022) A review on enhancing Cupriavidus necator fermentation for poly (3-hydroxybutyrate) (PHB) production from low-cost carbon sources. Front Bioeng Biotechnol 10:946085
30. Ruzzi M, Aroca R (2015) Plant growth-promoting rhizobacteria act as biostimulants in horticulture. Sci Hortic 196:124–134
31. Hajji-Hedfi L, Dali S, Rhouma A, Al-Ani LKT, Chihani-Hammas N, Khlif A, Bargougui O (2025) Microbial inoculants for plant resilience performance: roles, prospects and challenges. J Plant Dis Prot 132(2):88
32. Chaudhary R, Nawaz A, Khattak Z, Butt MA, Fouillaud M, Dufossé L, Mukhtar H (2024) Microbial bio-control agents: a comprehensive analysis on sustainable pest management in agriculture. J Agric Food Res 18:101421
33. Bodelier PL (2011) Toward understanding, managing, and protecting microbial ecosystems. Front Microbiol 2:80
34. Rodríguez Guzmán MP, Chávez Díaz IF, Zelaya Molina LX (2025) Reflexions on the role, diversity, conservation and management of the genetic microbial resources in agriculture. Curr Res Microb Sci 8:100365
35. Romano I, Bodenhausen N, Basch G, Soares M, Faist H, Trognitz F, Symanczik S (2023) Impact of conservation tillage on wheat performance and its microbiome. Front Plant Sci 14:1211758
36. Meena RS, Kumar S, Datta R, Lal R, Vijayakumar V, Brtnicky M, Marfo TD (2020) Impact of agrochemicals on soil microbiota and management: a review. Land 9(2):34
37. OECD (2007) Best practice guidelines for biological resource centres. Available from: https://www.oecd.org/en/publications/oecd-best-practice-guidelines-for-biological-resource-centres_9789264128767-en.html. Accessed 13 Mar 2026
38. WFCC (2010) World Federation for Culture Collection Guide-lines. Available from: https://wfcc.info/. Accessed 13 Mar 2026
39. ECCO (2022) European Culture Collections' Organisation. Available from: https://www.eccosite.org/. Accessed 13 Mar 2026
40. Sharma SK, Saini S, Verma A, Sharma PK, Lal R, Roy M, Sharma AK (2019) National agriculturally important microbial culture collection in the global context of microbial culture

collection centres. Proc Natl Acad Sci India Sect B Biol Sci 89(2):405–418
41. McCluskey K, Alvarez A, Bennett R, Bokati D, Boundy-Mills K, Brown D, Zehr S (2016) The US culture collection network lays the foundation for progress in preservation of valuable microbial resources. Phytopathology 106(6):532–540
42. Moretti M, Tartaglia J, Accotto GP, Beato MS, Bernini V, Bevivino A, Varese GC (2024) Treasures of Italian microbial culture collections: an overview of preserved biological resources, offered services and know-how, and management. Sustainability 16(9):3777
43. Wassenaar TM, Zimmermann K (2020) How industrial bacterial cultures can be kept stable over time. Lett Appl Microbiol 71(3):220–228
44. Smith D, Kermode A, Cafà G, Buddie AG, Caine TS, Ryan MJ (2020) Strengthening mycology research through coordinated access to microbial culture collection strains. CABI Agric Biosci 1(1):2

Chapter 2

Microbial Genetic Resource Conservation Techniques

Jasmín Salazar-Mendoza, Janette Betzabeth Villarruel-Muñoz, Lorena Jacqueline Gómez-Godínez, Ramón Ignacio Arteaga-Garibay, Hugo Ernesto Flores-López and Ignacio Valenzuela-Chavira

Abstract

Microbial genetic resources are a key part of global biodiversity. They are critical for biomedical innovation, environmental sustainability, and industrial biotechnology. In the past 10 years, microbiome research, synthetic biology, and systems biology have expanded how we view microorganisms. Microorganisms are now seen as more than just pathogens or environmental agents. As a result, preserving microbial diversity has become a strategic priority. The microbial resource ensures scientific reproducibility, supports biotechnology, and protects ecosystem services. This chapter gives a broad overview of microorganisms important to biomedicine and the environment. It examines the main principles, methods, and limits of their conservation. The chapter highlights the challenges of maintaining microbes' viability, genetic stability, and functional integrity. It discusses common preservation methods, such as subculturing, refrigeration, ultra-freezing, cryopreservation, and lyophilization. It also reviews which methods fit best based on biology and intended use. Finally, this chapter highlights the need for advanced preservation strategies that combine cryobiology, materials science, and synthetic biology. Microbial conservation is presented as a crucial part of building a sustainable bioeconomy and strong global health systems.

Keywords Microbial conservation, Cryopreservation, Microbial biodiversity, Biobanking, Functional stability

1 Introduction

Microbial genetic resources constitute the fundamental bedrock of modern biotechnology and are essential for maintaining biogeochemical cycles, human health, and global ecosystem services [1, 2]. Despite their critical role, microbial diversity faces increasing threats from human activity and climate change, intensifying the need for robust conservation strategies that ensure long-term viability and genetic stability [1, 3]. The imperative to safeguard these "invisible architects" is no longer merely a scientific interest but a priority for global sustainability and planetary health [1, 4].

Marco A. Ramírez-Mosqueda et al. (eds.), *Methods for The Conservation of Microorganisms*, Methods in Molecular Biology, vol. 3090, https://doi.org/10.1007/978-1-0716-5623-5_2,

This chapter explores the intersection of cutting-edge microbiology and preservation methodologies, focusing on microorganisms of high biomedical and environmental value. In the biomedical domain, the conservation of priority pathogens, such as the SKAPE group, and the surveillance of antimicrobial resistance are key pillars of global biosecurity and vaccine development [5, 6]. Furthermore, the rise of next-generation biotherapeutics and postbiotics derived from commensal bacteria, such as *Akkermansia muciniphila*, requires specialized anaerobic preservation protocols to maintain their functional integrity and therapeutic potency [6, 7]. Additionally, extremophiles from hydrothermal springs and polar environments represent a unique frontier of genetic resources; their "extreme enzymes" provide invaluable tools for molecular biology (e.g., Taq and Pfu polymerases) and industrial biocatalysis [7].

The preservation of these resources is primarily articulated through ex situ strategies, such as Microbial Biological Resource Centers (mBRCs), which provide the infrastructure necessary to standardize, store, and distribute strains under rigorous quality control [1]. However, the vast complexity of microbial diversity suggests that culture collections must be complemented by in situ conservation and advanced bioprospecting to capture genetic variation in natural habitats [3]. By adhering to international regulatory frameworks, such as the Nagoya Protocol, MGR management ensures equitable access and fosters innovation in synthetic biology and bioengineering [2].

This chapter provides a detailed analysis of the classification of microorganisms of interest and examines the fundamental principles of conservation, including viability, genetic stability, and functional integrity. By bridging traditional methods such as cryopreservation with emerging stabilization technologies, this work aims to establish a comprehensive framework for the responsible management of global microbial heritage.

2 Microorganisms of Biomedical and Environmental Interest

Microbial diversity includes many biological functions across all taxa. Over the past decade, microorganisms have played significant roles as public health pathogens and as contributors to environmental processes [1]. This section categorizes microbial resources by their impact on the bioeconomy and global health and examines challenges in conservation and applied research [2].

2.1 Microorganisms of Biomedical Interest

Medical microbiology has shifted from focusing only on pathogenicity. Now, it takes a comprehensive approach that includes the microbiome and genetic engineering as key parts of modern medicine [3]. Under the One Health paradigm, important microorganisms include both those that require strict containment and those that form beneficial groups [2]. Additionally, synthetic biology tools are being designed for precision therapies [4].

Priority Pathogens, Antimicrobial Resistance, and Biosecurity

Bacterial and viral pathogens are the focus of surveillance due to their global impact. The World Health Organization lists antimicrobial resistance, in which microorganisms no longer respond to drugs, as one of the ten greatest health threats [2]. Inappropriate and excessive use of antimicrobials (antibiotics, antivirals, and similar drugs) in humans, animals, and agriculture accelerates the emergence of multidrug-resistant strains [4]. Inappropriate and excessive use of antimicrobials in humans can turn treatable infections into lethal risks.

Preserving pathogenic strains in Microbial Biological Resource Centers (mBRC) facilities that collect, store, and distribute microorganisms is strategic for drug and vaccine development [5]. Priority goes to clinical isolates from the ESKAPE group (*Enterococcus faecium, Staphylococcus aureus, Klebsiella pneumoniae, Acinetobacter baumannii, Pseudomonas aeruginosa, and Enterobacter* spp.). These bacteria are known for their involvement in hospital infections and for their resistance to multiple drugs [6]. Both multidrug resistance and extended drug resistance are considered [4].

However, preserving these biological resources requires specialized infrastructure. According to biosafety guidelines, handling and storing agents with a high risk of aerosol transmission or lethal effects requires biosafety level 3 or 4 facilities (BSL-3/BSL-4) [4]. Examples include Mycobacterium tuberculosis and hemorrhagic fever viruses. These facilities require negative directional airflow, HEPA filtration, and strict decontamination protocols. Consequently, such requirements greatly increase operating costs and technical complexity for biobanks [4].

Beyond bacterial strains, contemporary epidemiological surveillance also requires preserving mobile genetic elements, such as plasmids and transposons [4]. Because these elements can transfer between genomes, recent studies on antimicrobial use in pig farming show that selective pressure accelerates the evolution of resistance genes on farms. These genes have implications for human clinical practice and highlight the need for biobanks that integrate both environmental and clinical surveillance [3].

Next-Generation Biotherapeutics and Postbiotics

Concurrently, advances in microbiome research have facilitated the development of next-generation probiotics. In contrast to conventional probiotics, such as *Lactobacillus*, NGPs are generally strict anaerobic commensal bacteria that constitute the core of the healthy gut microbiota [3]. Notably, recent studies have demonstrated the clinical potential of genera, including *Faecalibacterium, Roseburia, Eubacterium, and Akkermansia* [3, 6]. For instance, *Faecalibacterium prausnitzii* exhibits anti-inflammatory properties by producing butyrate [6]. Similarly, *Roseburia intestinalis* contributes to metabolic homeostasis and the degradation of complex polysaccharides. The primary technological challenge in preserving NGPs is their high sensitivity to oxygen [7] and complex nutritional requirements, including mucin. However, an important consideration is that evidence indicates cell viability is not always required to achieve therapeutic benefits. As a result, this finding has led to the development of postbiotics, inert components or derivatives of microorganisms that confer health benefits [6].

Next-Generation Biotherapeutics and Postbiotics

Concurrently, advances in microbiome research have led to the development of next-generation probiotics. These differ from conventional probiotics, such as Lactobacillus. NGPs are usually strict anaerobic commensal bacteria and form the core of healthy gut microbiota [3]. Recent studies have shown clinical potential for genera such as *Faecalibacterium*, *Roseburia*, *Eubacterium*, and *Akkermansia* [6]. For example, *Faecalibacterium prausnitzii* has anti-inflammatory properties because it produces butyrate. *Roseburia intestinalis* also supports metabolic homeostasis and degrades complex polysaccharides.

The main technological challenge in maintaining NGPs' viability is their high sensitivity to oxygen [6, 7]. NGPs also exhibit complex nutritional requirements, including mucin dependence. However, accumulating evidence indicates that microbial viability is not always necessary to achieve therapeutic benefits. The recognition that therapeutic effects can occur in the absence of viable cells has led to the development of postbiotics, defined as inactivated microbial cells or their bioactive derivatives that confer health benefits [6].

For example, recent studies suggest that pasteurized *Akkermansia muciniphila* may be more effective than live bacteria in reducing obesity and insulin resistance [6]. This benefit stems from the thermal stability of the outer membrane protein Amuc-1100, which retains its function after pasteurization. Amuc-1100 interacts with the host's TLR2 receptor, strengthens the intestinal barrier, and reduces metabolic endotoxemia [6]. These findings support the formulation of products that remain stable at room temperature, thus removing cold-chain challenges for live bacteria.

2.2 Microorganisms of Environmental Interest

Environmental microorganisms are no longer considered passive elements of biodiversity; they are now recognized as engineers of biogeochemical cycles and as key tools for the restoration of anthropized ecosystems [8, 9]. Their study and conservation are essential both for the ecosystem services they provide and for the potential risks they pose [10, 11].

Soil Microbiomes: Functional Diversity and Associated Risks

Soil environments, meaning habitats made up of soil and its components, host the most complex microbiomes in the biosphere and serve as the foundation for terrestrial agricultural productivity [8]. Strategic conservation efforts in this context prioritize the preservation of key functional microbial groups [10].

Plant growth-promoting Rhizobacteria stimulate plant growth. These include free-living and symbiotic diazotrophic species [12]. They can fix atmospheric nitrogen independently or with plants. Notable examples are *Azospirillum brasilense*, *Rhizobium leguminosarum*, and *Burkholderia vietnamiensis* [13]. These bacteria are essential for biological nitrogen fixation, converting atmospheric nitrogen into plant-available forms [12]. Phosphate-solubilizing bacteria, such as *Pseudomonas fluorescens* and *Bacillus megaterium*, also. They reduce reliance on synthetic fertilizers by releasing soil-bound phosphate for plant uptake [14]. Biocontrol agents include bacteria from the genus Streptomyces, such as *Streptomyces griseus*, a soil bacterium [15]. Fungi like *Trichoderma harzianum*, a beneficial fungus, also act as biocontrol agents [16]. These organisms are recognized for producing secondary metabolites [17]. These metabolites suppress phytopathogens and enhance general crop resistance [15].

It is crucial to identify and manage adverse soil microbiomes. These are microorganisms in agricultural soils that negatively impact crops. Recent studies highlight the risks associated with the uncontrolled application of organic amendments, such as manure and sludge [18]. These amendments can turn agricultural soils into reservoirs for persistent human pathogens [19]. Such pathogens include *Salmonella enterica* serovar Typhimurium, *Listeria monocytogenes*, and *Escherichia coli* O157:H7 [20]. The accumulation of antibiotic resistance genes and mobile genetic elements, such as integrons, in soil bacteria like *Acinetobacter* spp. further increases risks [21]. This facilitates horizontal transfer of resistance within the food chain and threatens global food safety [19].

Environmental Biotechnology and Data-Driven Synthetic Microbes (DDSM)

Environmental conservation now includes synthetic biology (SynBio), which applies engineering to living things [4]. In SynBio, data-driven synthetic microbes are artificial organisms made using advanced data analysis. DDSM uses SynBio tools to redesign strong bacteria that can break down recalcitrant pollutants, substances that are hard to degrade and not removed efficiently by nature [22]. In addition, synthetic consortia—engineered assemblies of interacting microorganisms—are being designed to emulate natural ecosystems and enhance pollutant degradation through functional specialization and metabolic cooperation. When integrated with biosensing mechanisms, these systems are capable of detecting contaminants and responding in a dynamic manner, which broadens their potential for environmental monitoring and in situ bioremediation. Collectively, these advances highlight SynBio and DDSM as valuable approaches for developing sustainable solutions in environmental conservation and ecosystem restoration (Table 1).

Extremophiles: Genetic Resources from Extreme Environments

Extremophile microorganisms are a key resource for bioprospecting [23] (Table 2). These organisms develop unique molecular adaptations, known as "extreme enzymes," that enable them to thrive in extreme conditions and confer significant industrial value [3].

From an application standpoint, extremozymes have attracted significant interest because of their stability and high performance under demanding industrial conditions. For instance, enzymes derived from thermophiles are extensively employed in polymerase chain reactions (PCR), whereas those from halophilic and alkaliphilic organisms are useful in sectors such as detergents, textiles, and leather processing. In the same way, psychrophilic enzymes, which retain activity at low temperatures, are especially valuable for energy-efficient applications in food processing and for bioremediation in cold ecosystems (Table 2).

3 Fundamental Principles and Limitations of Microbial Conservation

Ex situ conservation of microbial resources is a critical infrastructure for scientific research and the bioeconomy [3]. Beyond simply keeping cells alive, this process represents a biophysical challenge to ensure the "biological authenticity" of the specimen over time [10]. The fundamental objective is to decouple biological time from chronological time, thereby inducing an anabiotic state, or reversible metabolic dormancy [3]. This state must be achieved without compromising structural integrity or genomic information, ensuring that the recovered material is phenotypically identical to the original for experimental reproducibility, industrial application, and epidemiological surveillance [7].

On the other hand, the gap between the microbial diversity discovered and the technical capacity to conserve it faithfully requires a transition to new-generation preservation strategies that integrate principles of advanced cryobiology, materials science, and molecular biology [3, 7].

3.1 Viability, Genetic Stability, and Functional Integrity

The quality of preserved microbial resources is evaluated using three key criteria: viability, genetic stability, and functionality [6]. Viability was once defined solely by the ability to form colonies. Now, it also includes membrane integrity and the ability to resume metabolic activity after storage [6, 7]. Recognizing intermediate physiological states, including the "viable but non-culturable" condition, is crucial [7]. This state is often triggered by stress when preservation conditions are not ideal. Overlooking the VBNC state can lead to mistaking such cells for dead, resulting in the unintentional loss of valuable strains.

Genetic stability means keeping the original genome unchanged. Genetic stability excludes point mutations, recombination events, and the loss of plasmids and other mobile genetic elements [4]. Recent research shows that, even during cryopreservation at −80°C, failing to reach the glass transition temperature (Tg) can allow DNA repair

Table 1
Microorganisms in environmental biotechnology and data-driven synthetic microbes (DDSM)

Biotechnological strategy	Species	Type of modification/ function	Target contaminant	Environmental application	Regulatory advantages	Biosafety challenges
Bacterial chassis engineering	*Pseudomonas putida* KT2440	CRISPR–Cas editing; heterologous catabolic pathways; solvent tolerance	PET and chlorinated aromatics	Synthetic bioremediation of plastics and xenobiotics	Well-characterized and non-pathogenic; established laboratory safety profile	Gene transfer risk; potential environmental spread; need for containment controls
Consortia for persistent pollutants (PFAS)	*Acidimicrobium* sp. + *Geobacter sulfurreducens*	Defluorination + electron transfer consortium	PFAS	Bioelectrochemical degradation of persistent pollutants	Mimics natural interactions; suitable for controlled bioreactors	Consortium stability; ecological complexity; monitoring of byproducts
Living biosensors	*Escherichia coli* + *Shewanella oneidensis*	Engineered genetic circuits for signal detection	Heavy metals and emerging pollutants	Real-time environmental monitoring	High sensitivity; low cost; applicable in confined systems	Containment requirements; risk of release; variable regulatory frameworks

Table 2
Examples of extremophiles for bioprospecting

Microbial group	Species	Extreme environment type	Key adaptation	Main biotechnological application
Thermophiles	*Thermus aquaticus*	Hydrothermal springs (≈ 70°C–75°C)	Thermostable DNA polymerase (Taq) resistant to high temperatures	Polymerase chain reaction (PCR) and molecular biology
Hyperthermophiles	*Pyrococcus furiosus*	Marine volcanic environments (>100°C)	Pfu polymerase with high fidelity and thermal stability	High-precision DNA amplification; genetic engineering applications
Thermophiles	*Thermotoga maritima*	Geothermal marine environments	Thermostable hydrolytic enzymes	Lignocellulosic biomass hydrolysis in high-temperature biorefineries
Extreme halophiles	*Haloferax volcanii*	Hypersaline environments	Salt-adapted metabolism; PHA production	Production of biodegradable bioplastics (polyhydroxyalkanoates) with reduced contamination risk
Extreme halophiles	*Halobacterium salinarum*	Salt flats and hypersaline lakes	Halotolerant proteins and enzymes	Production of biomolecules and bioplastics under saline industrial conditions
Radiotolerant organisms	*Deinococcus radiodurans*	High-ionizing radiation environments	Highly efficient DNA repair systems	Bioremediation of mixed nuclear (radioactive and chemical) waste
Psychrophiles	*Pseudoalteromonas* spp.	Polar environments and cold oceans	Cold-active enzymes (psychroenzymes)	Low-energy industrial processes (detergents, food processing, cold biocatalysis)

processes that introduce errors [7]. These errors threaten genomic fidelity and can affect comparative or evolutionary analyses.

Functional stability means maintaining the phenotype of interest. This stability includes the production of secondary metabolites, pathogenic potential, or bioremediation capacity [6]. "Laboratory domestication" can occur when repeated subculturing results in the loss of virulence or of complex metabolic clusters. Laboratory domestication usually happens because natural environmental pressures are missing [24]. These changes challenge culture collections and biobanks. Rigorous preservation strategies are needed [7].

3.2 Factors Affecting Preservation

The effectiveness of preservation protocols depends on both intrinsic and extrinsic factors [7] (Fig. 1). Intrinsic factors are the microorganism's genotypic and physiological traits, such as membrane robustness, metabolic state, and growth phase [11]. Extrinsic factors include physicochemical conditions of storage, such as temperature, osmolarity, and the presence of cryoprotective agents. Understanding and controlling these variables is critical for minimizing damage and maintaining the stability of biological resources [10].

3.3 Traditional Methods and Their Limitations

The cells' initial physiological state influences the survival and stability of microorganisms during preservation, the physical processes involved in freezing, and the occurrence of oxidative stress [25]. Optimization of these factors is critical for maintaining viability, genetic stability, and functional integrity in microbial collections [3]. Conventional preservation techniques still exhibit substantial limitations in addressing contemporary scientific and technological requirements [7] (Fig. 2).

Subculturing (Serial Transfers)

Subculturing involves maintaining the microbial population in an active vegetative state by periodically transferring a fraction of the culture into fresh medium [7]. From a physiological perspective, this

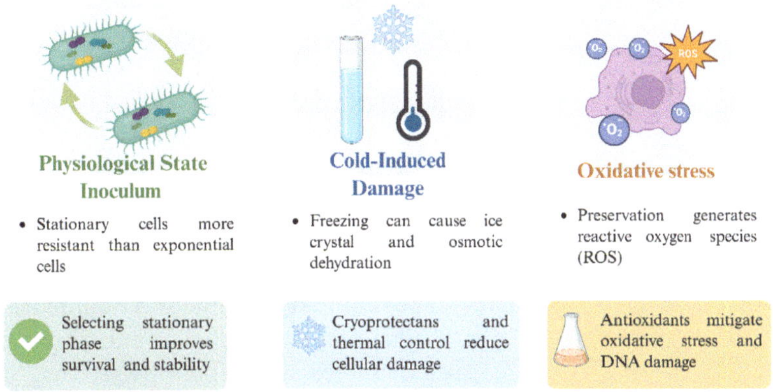

Fig. 1 Key factors influencing microbial preservation

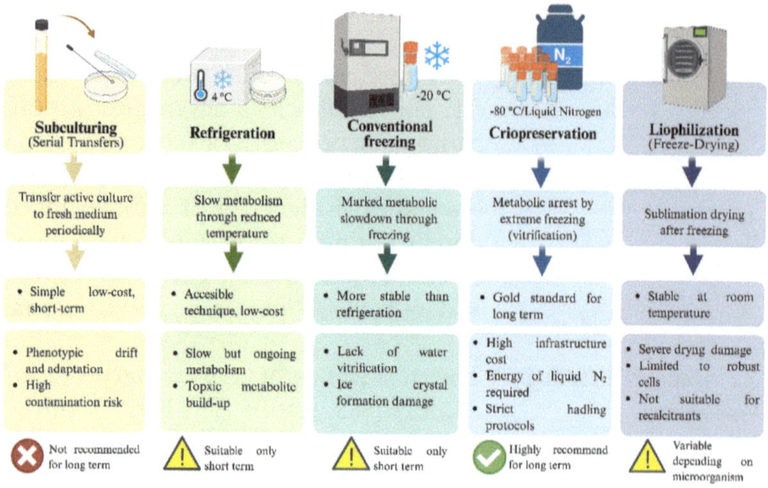

Fig. 2 Comparative summary of traditional microbial preservation methods

approach does not arrest metabolism; instead, it replenishes nutrients and prevents system exhaustion [26]. Consequently, it is useful as a short-term maintenance technique; however, biologically, it represents the least conservative preservation strategy, as the population remains subject to continuous cell division, selective pressure, and adaptive drift [7, 27].

In bacteria, such drift may manifest as changes in substrate utilization, plasmid loss, variation in colonial phenotypes, reduced virulence, or the selection of resistant subpopulations when the medium contains antibiotics or other selective agents [7]. In hemoflagellates, such as *Leishmania*, continuous in vitro passaging is also associated with attenuation and reduced infectivity; therefore, clinical isolates should be cryopreserved early and not subjected to excessive passages.

Methodologically, the operational scheme is straightforward. A colony or pure inoculum is aseptically transferred to an appropriate solid or liquid medium. It is incubated until growth is achieved, then subcultured again before deterioration (loss of viability or contamination) occurs. Commonly used media include nutrient agar, tryptic soy agar2, or broth (TSA/TSB), brain heart infusion (BHI), media supporting specific clinical bacteria, and Sabouraud dextrose agar (SDA), or potato dextrose agar (PDA) for fungi. In medical mycology, specialized media may be needed depending on the taxon. Examples include Dixon or Leeming–Notman media for *Malassezia* (lipid-dependent yeast), Sabouraud for *Candida*, *Cryptococcus*, and dermatophytes, and specific formulations for dimorphic fungi.

The required equipment is minimal: a biosafety cabinet or aseptic workspace, an incubator, inoculating loops or micropipettes, tubes or plates, and labeling systems. Storage duration is short. Bacteria, such as *Escherichia coli*, and yeasts, such as *Saccharomyces cerevisiae*, typically need daily or frequent subculturing. *Aspergillus niger* requires weekly transfers. Many filamentous fungi need passaging every 2–4 weeks, depending on growth rate and whether vegetative (actively growing) or sporulated (spore-forming) structures are available [7].

While appropriate for working collections and teaching laboratories, this method is unsuitable for master culture collections or pathogens whose virulence must remain stable [26, 27]. The main advantages of subculturing are its low cost and simplicity; however, its major disadvantages include being labor-intensive, posing a high risk of contamination, and causing cumulative biological changes with each passage [7]. In modern practice, subculturing should be regarded as a temporary bridge toward more robust preservation strategies [27]

Refrigeration

Refrigeration is based on a simple physical principle: lowering the temperature to 2–8°C slows biochemical reactions, reduces growth rates, and delays medium depletion [7]. Unlike cryopreservation, it does not induce complete metabolic arrest. Therefore, during storage, cellular maintenance processes, environmental acidification, and the accumulation of potentially toxic metabolites continue [7, 28]. This behavior explains why refrigeration is useful as a temporary strategy but not equivalent to frozen storage [3, 7].

Methodologically, cultures are prepared on agar slants or in liquid media and stored in temperature-controlled refrigerators under continuous monitoring [7]. Commonly used media include TSA, BHI, or nutrient agar for bacteria and PDA or Sabouraud for fungi [7]. The required equipment consists of a stable laboratory refrigerator and a calibrated thermometer to minimize temperature fluctuations. The recommended storage duration for bacteria ranges from days to weeks. In some filamentous fungi, it may last months to up to 1–2 years at 4°C. Recent evaluations in fungal models indicate that while refrigeration is a viable working alternative for 6–24 months, its stability remains inferior to cryopreservation [6].

Refrigeration is recommended for short-term use and internal transport of non-fastidious bacteria and yeasts [7]. However, it is suboptimal for highly fastidious pathogens; for instance, respiratory bacteria, such as *Streptococcus pneumoniae*, *Haemophilus influenzae*, and *Neisseria meningitidis*, are particularly sensitive to cold and are not suitable for prolonged refrigerated storage (Senaratne & Jayaweera [7]). The primary advantage is accessibility, but limitations include residual metabolism and progressive viability loss [28].

Conventional Ultra-Freezing

Conventional ultra-freezing at −80°C relies on water solidification and the reduction in molecular mobility to halt growth and reduce biological damage [29]. It differs from liquid nitrogen cryopreservation in that it is often performed without controlled-rate freezing and is primarily used for working stocks rather than long-term master collections. A critical factor in this method is balancing sufficient cellular dehydration by minimizing intracellular ice formation. Rapid freezing may induce lethal ice formation, whereas slow cooling promotes solution effects and osmotic stress [29]. The standard protocol involves mixing an exponentially growing culture with cryoprotectants, typically 15%–20% glycerol for bacteria and yeasts, sometimes supplemented with stabilizers like skim milk to reduce osmotic damage [7].

Ultra-freezing is widely utilized for university collections and working stocks of pathogens, such as Listeria monocytogenes and *Staphylococcus* spp. [7]. However, recent evaluations using ex vivo microbiome assays and metaproteomics have shown that functionality can be compromised if storage conditions are not strictly stabilized [6]. Furthermore, the freezing process can induce a Viable but Non-Culturable state, necessitating the use of resuscitation-promoting factors to ensure full population recovery [30]. While operationally simpler than liquid nitrogen storage, ultra-freezing remains vulnerable to power failures and may exhibit lower long-term genetic stability, especially in complex synthetic microbial consortia [31, 32].

Cryopreservation

Cryopreservation represents the standard methodology for long term storage, as it aims to achieve near-complete metabolic arrest and maximal structural protection [7, 28]. It is routinely performed at −80°C or, preferably, at −196°C in liquid nitrogen to reach a stable vitrified state [3, 28]. Its underlying principles are based on three key mechanisms: the drastic reduction of molecular kinetics, the control of water transfer during cooling to prevent intracellular ice formation, and the use of cryoprotective agents that mitigate damage from solution effects and recrystallization [3, 7, 28]. The most used penetrating cryoprotectants are dimethyl sulfoxide and glycerol, whereas non-penetrating agents include trehalose, sucrose, and complex matrices like skim milk that stabilize membranes and proteins [25]

In bacteria, yeasts, and many simple eukaryotic cells, typical concentrations include 5%–15% DMSO or 10%–20% glycerol [7]. The importance of the physiological phase and equilibration is well documented; for instance, in axenic Entamoeba histolytica, recent protocols utilize TYI-S-33 medium with 10% DMSO and bovine serum, emphasizing controlled-rate cooling (1°C/min) and rapid dilution after thawing to prevent cell lysis [33]. The flexibility of cryopreservation is also evident in complex parasites. In multicellular helminths like *Echinococcus multilocularis*, metacestodes and protoscoleces have demonstrated viability after 35 years of storage in liquid nitrogen using DMSO as a stabilizer. Similarly, the cryopreservation of *Dirofilaria immitis* microfilariae has been experimentally validated to maintain the full life cycle, reducing the reliance on continuous donor animals [34].

Cryopreservation is particularly valuable for taxa recalcitrant to lyophilization, such as Basidiomycetes, which are better preserved as mycelium on specific supports (beads, agar discs, or grains) rather than as dried material [35]. Although no universal protocol exists for these fungi, liquid nitrogen is preferred to ensure long-term genetic and physiological stability. Thus, cryopreservation is the method of choice for master culture collections and reference strains susceptible to subculture-induced drift [3, 7].

Lyophilization

Lyophilization, or freeze-drying, is based on the sublimation of water from a previously frozen material, reducing water activity and limiting degradation reactions [7, 28]. The procedure comprises three stages: freezing, primary drying (sublimation), and secondary drying (desorption) [7]. Because dehydration imposes severe stress on cellular structures, the protective matrix (lyoprotectant) plays a critical role in determining survival [25, 36].

Methodologically, the process requires a high-density suspension ($>10^7$ cells/mL) mixed with agents such as skim milk (10%–20%) or trehalose (5%–10%), the latter being widely recognized for its ability to stabilize membranes during the drying–rehydration cycle [7]. This method is particularly effective for bacteria, yeasts, and spore-forming fungi, facilitating strain distribution and transport due to its independence from continuous ultra-low temperature storage [3]. However, it is less effective for microorganisms with fragile vegetative mycelia (like many Basidiomycetes) or those with high desiccation sensitivity, such as *Helicobacter pylori* and *Clostridium* spp. [36].

The logistical advantages of lyophilization are substantial, but its success depends on optimizing the culture's physiological state and rehydration protocols [6, 7]. Recent biobanking evaluations suggest that post-rehydration functionality must be experimentally verified using advanced assays to ensure the preservation of essential metabolic traits [3]. Therefore, lyophilization is recommended when the primary objective is logistical stability, provided the taxon tolerates the process [7].

3.4 Integrated Comparison and Selection Criteria

When the five preservation methods are compared on a common scale, the primary distinction lies not only in storage duration but also in the degree of control over the physical state of water and residual metabolic activity. Subculturing and refrigeration maintain microorganisms in biologically active or semi-active states and are therefore suitable for immediate use but offer limited long-term stability. In contrast, ultra-freezing and cryopreservation provide much better control over cumulative damage and are preferred for working and master culture collections, respectively [7, 29]. Lyophilization, however, sacrifices biological universality in favor of exceptional logistical advantages [3]. In practical terms, labs handling routine bacteria and yeasts may use cryopreserved master stocks and secondary stocks stored at −80°C. Mycological collections containing Basidiomycetes should prioritize species-specific cryopreservation protocols, often utilizing specialized supports like beads or agar discs to ensure long-term viability [35, 36]. Parasitology labs must validate preservation strategies for each species and life stage, without assuming general protocol transferability, as evidenced by specific successes in models like *Echinococcus* and *Dirofilaria* [34, 37]. Post-preservation recovery must not be judged solely by visible growth. For high-value cultures, it is essential to assess purity, morphology, growth kinetics, enzymatic or metabolic activity, and molecular identity [6, 7]. This is especially crucial for filamentous fungi and parasites whose life cycles or virulence factors may subtly change after the stress of freezing or drying; hence, advanced assays like metaproteomics and ex vivo microbiome analysis are increasingly used to verify functional stability [3].

4 Conclusion

The conservation of microbial genetic resources presents a critical challenge at the intersection of microbiology, biotechnology, and global sustainability. This review demonstrates that traditional preservation methods differ fundamentally. They vary in their ability to control metabolic activity, maintain structural integrity, and preserve genetic and functional stability. These differences highlight that microbial preservation is not just a technical procedure. It is a complex biophysical process aimed at decoupling biological time from chronological time while safeguarding the authenticity of microbial resources.

Short-term strategies, such as subculturing and refrigeration, remain essential for routine laboratory use and immediate experimental needs. However, they have inherent limitations. The persistence of metabolic activity and the risk of adaptive drift make these methods unsuitable for long-term conservation. In contrast, ultra-freezing and cryopreservation provide improved control over cellular damage. These methods are therefore more appropriate for working with and mastering culture collections. Cryopreservation is the most robust method for maintaining viability, genetic fidelity, and functional integrity. It works across diverse microbial taxa, including bacteria, fungi, protozoa, and certain parasitic organisms.

Lyophilization offers a complementary approach characterized by exceptional logistical advantages, including long-term stability in ambient conditions and ease of transport. Nevertheless, its applicability is taxon-dependent and requires careful optimization of protective matrices and process parameters. Consequently, no single preservation method can be considered universally optimal; rather, selection must be tailored to the organism's biological characteristics, its physiological state, and the intended application.

Importantly, this review underscores that successful preservation should not be evaluated solely by post-storage viability. A comprehensive assessment must include genetic stability, phenotypic consistency, and functional performance. These factors are especially relevant in clinically significant pathogens, industrial strains, and environmentally important microorganisms. Failing to consider these dimensions may compromise reproducibility, biosafety, and translational potential of preserved resources.

Finally, the growing gap between the vast microbial diversity being discovered and the current capacity to preserve it effectively calls for integrating advanced approaches. Emerging strategies combining cryobiology, materials science, and synthetic biology are expected to redefine microbial conservation by improving preservation efficiency, reducing cellular damage, and enabling the stabilization of complex microbiomes. In this context, microbial biobanking must evolve from a passive storage system into an active, standardized, and globally coordinated infrastructure capable of safeguarding microbial biodiversity for future scientific, medical, and environmental applications.

Conflict of Interest

The authors have no conflicts of interest to declare that are relevant to the content of this chapter.

References

1. Ervens B, Amato P, Aregahegn K et al (2025) Ideas and perspectives: microorganisms in the air through the lenses of atmospheric chemistry and microphysics. Biogeosciences 22(1):243–256. https://doi.org/10.5194/bg-22-243-2025
2. Rappuoli R, Young P, Ron E et al (2023) Save the microbes to save the planet. A call to action of the International Union of the Microbiological Societies (IUMS). One Health Outlook 5(1):5. https://doi.org/10.1186/s42522-023-00077-2
3. Dominguez-Bello MG, Steiger D, Fankhauser M et al (2025) The microbiota vault initiative: safeguarding Earth's microbial heritage for future generations. Nat Commun 16(1):5373
4. Chemla Y, Sweeney CJ, Wozniak CA et al (2025) Design and regulation of engineered bacteria for environmental release. Nat Microbiol 10(2):281–300. https://doi.org/10.1038/s41564-024-01918-0
5. Schultz, M. G. (2024). SIVB's 2024 In Vitro Biology Meeting Proceedings. In Vitro Cellular and Developmental Biology-Animal, 60 (1 Supplement).
6. Zhang X, Walker K, Mayne J et al (2022) Evaluating live microbiota biobanking using an ex vivo microbiome assay and metaproteomics. Gut Microbes 14(1):2035658. https://doi.org/10.1080/19490976.2022.2035658
7. Senaratne W, Jayaweera J (2024) Comparison of microbial preservation methods: a narrative review. Germs 14(4):375–386. https://doi.org/10.18683/germs.2024.1447
8. Greening C, Grinter RA, Chiri E (2019) Uncovering the metabolic strategies of the dormant microbial majority: towards integrative approaches. mSystems 4(3):10–1128. https://doi.org/10.1128/mSystems.00107-19
9. Zaman W, Ayaz A, Puppe D (2025) Biogeochemical cycles in plant-soil systems: significance for agriculture, interconnections,

and anthropogenic disruptions. Biology (Basel) 14(4):433. https://doi.org/10.3390/biology14040433
10. Junker RR, Farwig N (2025) Microbial conservation is essential for sustaining ecosystem functions and services. Proc Natl Acad Sci U S A 122(38):e2503013122. https://doi.org/10.1073/pnas.2503013122
11. Pooja D, Prasad S, Dinesh GK, Avinash C (2024) Role of Environmental Factors Influencing Microbes and Microbiomes for Ecosystem Restoration. In: Prasad S, Dinesh GK, Sinduja M, Sathya V, Poornima R, Karthika S (Eds.) The Role of Microbes and Microbiomes in Ecosystem Restoration. Bentham Science Publishers, brpp, 44–69
12. Singh A, Yadav VK, Chundawat RS et al (2023) Enhancing plant growth promoting rhizobacterial activities through consortium exposure: a review. Front Bioeng Biotechnol 11:1099999. https://doi.org/10.3389/fbioe.2023.1099999
13. Aloo BN, Tripathi V, Makumba BA et al (2022) Plant growth-promoting rhizobacterial biofertilizers for crop production: the past, present, and future. Front Plant Sci 13:1002448. https://doi.org/10.3389/fpls.2022.1002448
14. Fouda A, Eid AM, Elsaied A et al (2021) Plant growth-promoting endophytic bacterial community inhabiting the leaves of Pulicaria incisa (Lam.) DC inherent to arid regions. Plants 10(1):76. https://doi.org/10.3390/plants10010076
15. El-Saadony MT, Saad AM, Soliman SM et al (2022) Plant growth-promoting microorganisms as biocontrol agents of plant diseases: mechanisms, challenges and future perspectives. Front Plant Sci 13:923880. https://doi.org/10.3389/fpls.2022.923880
16. Jin S, Alberti F (2025) Advances in the discovery and study of Trichoderma natural products for biological control applications. Nat Prod Rep 42(8):1367–1386. https://doi.org/10.1039/d5np00017c
17. Lakhdari W, Benyahia I, Bouhenna MM et al (2023) Exploration and evaluation of secondary metabolites from Trichoderma harzianum: GC-MS analysis, phytochemical profiling, antifungal and antioxidant activity assessment. Molecules 28(13):5025. https://doi.org/10.3390/molecules28135025
18. Black Z, Balta I, Black L et al (2021) The fate of foodborne pathogens in manure treated soil. Front Microbiol 12:781357. https://doi.org/10.3389/fmicb.2021.781357
19. Zalewska M, Błażejewska A, Czapko A et al (2021) Antibiotics and antibiotic resistance genes in animal manure: consequences of its application in agriculture. Front Microbiol 12:610656. https://doi.org/10.3389/fmicb.2021.610656
20. Abraham A, Mtewa AG, Chiutula C et al (2025) Prevalence of antibiotic resistance bacteria in manure, soil, and vegetables in urban Blantyre, Malawi, from a farm-to-fork perspective. Int J Environ Res Public Health 22(8):1273. https://doi.org/10.3390/ijerph22081273
21. Liu B, Shen Z, Zhou Q et al (2025) Antibiotic resistance genes in the animal manure-amended soil-plant system: occurrence, transmission, bacterial hosts, and human health risks. Ecotoxicol Environ Saf 303:118890. https://doi.org/10.1016/j.ecoenv.2025.118890
22. Mariam I, Rova U, Christakopoulos P et al (2025) Data-driven synthetic microbes for sustainable future. NPJ Syst Biol Appl 11(1):74. https://doi.org/10.1038/s41540-025-00556-4
23. Morales-Mancera LT, Diaz Ortiz LL, Gutiérrez-Fonseca JE et al (2025) Restoring nature with microbes: bioremediation in the world's biodiversity hotspots. Appl Environ Microbiol 91(10):e01442–25. https://doi.org/10.1128/aem.01442-25
24. Giangeri G, Campanaro S, Kyrpides NC, Angelidaki I (2025) Unlocking the potential of designed microbial consortia: a breakthrough for sustainable waste management and climate resilience. Environ Sci Ecotechnology 25, 100558. https://doi.org/10.1016/j.ese.2025.100558
25. Senaratne W, Jayaweera J (2024) Comparison of microbial preservation methods: a narrative review. Germs 14(4):375–386. https://doi.org/10.18683/germs.2024.1447
26. Sharma SK, Singh SK, Ramesh A, Sharma PK, Varma A, Ahmad E, Khande R, Singh UB, Saxena AK (2018) Microbial genetic resources: status, conservation, and access and benefit-sharing regulations. In: Sharma SK, Varma A (Eds.) Microbial resource conservation: conventional to modern approaches. Springer International Publishing, Cham, pp 1–33
27. Overmann J (2015) Significance and future role of microbial resource centers. Syst appl microbiol 38(4):258–265
28. Hallsworth JE (2022) Water is a preservative of microbes. Microb Biotechnol 15(1):191–214.
29. Lemke M, DeSalle R (2023) The next generation of microbial ecology and its importance in environmental sustainability. Microb Ecol 85(3):781–795. https://doi.org/10.1007/s00248-023-02185-y

30. Lennon JT, Lehmkuhl BK, Chen L et al (2025) Resuscitation-promoting factor (Rpf) terminates dormancy among diverse soil bacteria. mSystems 10(5):e01517–e01524
31. He Y, Wang S, Mi Y et al (2025) Adaptive laboratory evolution of a microbial consortium enhancing non-protein nitrogen assimilation for feed protein production. Microorganisms 13:1416. https://doi.org/10.3390/microorganisms13061416
32. Zhou Y, Zeeshan Ul Haq M (2025) Engineering of synthetic microbial consortia for sustainable management of wastewater and polyethylene terephthalate: a comprehensive review. Int J Mol Sci 26(23):11623. https://doi.org/10.3390/ijms262311623
33. Suleiman RL, Ralston KS (2022) Growth and genetic manipulation of Entamoeba histolytica. Curr protoc 2(1), e327
34. Zinser EW, McTier TL, Kernell NS, Woods DJ (2021) Cryogenic preservation of Dirofilaria immitis microfilariae, reactivation and completion of the life-cycle in the mosquito and vertebrate hosts. Parasites vectors 14(1):367
35. Corrales C, Astrin JJ (2023) Biodiversity biobanking – a handbook on protocols and practices. Pensoft Publishers
36. Castro-Rios K, Bermeo-Escobar LP (2021) Methods for the culture conservation of edible and medicinal fungi: review. J Microbiol Biotechnol Food Sci 10(4):620–625. https://doi.org/10.15414/jmbfs.2021.10.4.620-625
37. Hemphill A, Lundström-Stadelmann B (2021) Echinococcus: the model cestode parasite. Parasitology 148(12):1401–1405. https://doi.org/10.1017/S003118202100113X

Part II

Conservation Methods of Bacteria

Check for updates

Chapter 3

Technical Guidelines for Cryopreservation and Lyophilization of Pathogenic Microorganism Strains

Duochun Wang, Xiaomei Yan, Yi Xu, Zisis Kozlakidis, Io Hong Cheong, Ning Guo, Ge Liu, Yang Liu, Huiling Jin and Qiang Wei

Abstract

Pathogenic microorganisms, defined as microbial entities capable of invading the host and causing infection or communicable disease, encompass a taxonomically diverse assemblage that includes viruses, bacteria, fungi, and rickettsiae.

Cryopreservation is a technique that stores microbial cultures in a low-temperature environment, reducing their metabolism to the minimum or even to a complete standstill so that the organisms enter a semi-dormant or fully dormant state. Lyophilization (freeze-drying) is a preservation technique in which microorganisms are first frozen in a cryoprotective medium. Water is then removed by sublimation under vacuum and reduced pressure. Since the process avoids abrupt changes, cellular damage is minimized, and microbes are kept in a dry, oxygen-free state where their physiological activities are markedly inhibited.

Standardized, scientifically sound low-temperature and freeze-drying preservation techniques for pathogenic microorganism strains are essential for ensuring their long-term, stable, and high-quality storage. This chapter outlines the procedures for low-temperature and freeze-drying preservation of microorganism strains, providing technical guidance for institutions engaged in the storage, preservation, utilization, and production of pathogenic microorganisms.

Keywords Pathogenic microorganisms, Strains, Cryopreservation, Lyophilization, Technical Guidelines

1 Introduction

Pathogenic microorganisms, defined as microbial entities capable of invading the host and inducing infection or communicable disease, encompass a taxonomically diverse assemblage that includes viruses, bacteria, fungi, and rickettsiae.

Cryopreservation refers to the technique of storing microorganism strains at low temperatures, reducing their metabolic activity to the point where growth and reproduction cease entirely, thereby inducing a semi-dormant or fully dormant state. The underlying principle is to arrest cellular metabolism by exposing the strains to

Marco A. Ramírez-Mosqueda et al. (eds.), *Methods for The Conservation of Microorganisms*, Methods in Molecular Biology, vol. 3090, https://doi.org/10.1007/978-1-0716-5623-5_3,

sub-zero conditions, thus enabling their long-term conservation. This low-temperature approach plays a pivotal role in maintaining strain quality. Research has shown that the efficiency of microbial cryopreservation is influenced by cell size and morphology, growth phase, culture conditions, cooling rate, storage temperature, and thawing rate [1, 2].

Lyophilization (freeze-drying) is a preservation technique in which microorganisms are first frozen in a cryoprotective medium and then dehydrated by sublimation under vacuum and reduced pressure [3, 4]. By avoiding harsh conditions, the process minimizes direct cellular damage, leaving the microbes in a dry, oxygen-free state where their physiological activities are markedly inhibited. By combining the effects of desiccation, low temperature, and oxygen exclusion, lyophilization slows metabolic rates to a semi-permanent dormant level, enabling long-term storage while preventing contamination and genetic drift [3, 4].

High-quality microorganism strains are of great significance to scientific research and biological production [5]. Practical low-temperature preservation and freeze-drying not only help maintain the viability, purity, and genetic stability of these strains but also directly affect their post-revival activity. Standardized, scientific low-temperature preservation freeze-drying techniques for pathogenic microorganism strains are fundamental to achieving long-term, stable, and high-quality preservation [5]. This chapter mainly introduces the reagents, equipment, and consumables involved in low-temperature and freeze-drying preservation processes, along with the specific operational procedures. These include preparing cryoprotective and lyoprotective agents, preparing strains, mixing, cooling or freeze-drying, storage, and revival. It aims to provide technical guidance to institutions engaged in the conservation, storage, utilization, and production of pathogenic microorganism strains.

2 Materials

2.1 Reagents

1. Cryoprotectants [2]:
   - (a) Permeable small-molecule cryoprotectants:
   - (b) Alcohols: glycerol, methanol, ethylene glycol, propanediol. This chapter uses glycerol as an example to describe the corresponding operating procedures.
   - (c) Amides/Sulfoxides: dimethyl sulfoxide, formamide.
   - (d) Non-permeable macromolecular cryoprotectants:
   - (e) Polyols: sorbitol, mannitol.
   - (f) Sugars: glucose, sucrose, lactose, trehalose.
   - (g) Proteins: skim milk, gelatin, serum albumin.

   (h) Polymers: polyvinylpyrrolidone, Polyethylene glycol, polyethylene oxide, dextran, and Hydroxyethyl starch.
   (i) Complex natural macromolecules: gum arabic, serum.
   (j) Antioxidants:
   (k) Vitamin C, L-glutamic acid, L-Proline.

2. Lyoprotectants [6]:
   (a) Sugars/Polyols: glucose, galactose, mannose, D-fructose, D-ribose, xylose, sucrose, lactose, maltose, trehalose, raffinose, D-mannitol, glycerol, sorbitol, xylitol, myo-inositol.
   (b) Polymers: polyethylene glycol, dextran, hydroxyethyl starch, Ficoll, gum arabic, soluble starch, gelatin, pectin, polyvinylpyrrolidone, cellulose, β-cyclodextrin, methylcellulose, maltodextrin DE 18–20, Sephadex G-200.
   (c) Amino acids: L-proline, glycine, L-glutamic acid, histidine, arginine, L-4-hydroxyproline, L-serine, β-alanine, lysine hydrochloride, L-lysine, sarcosine, γ-aminobutyric acid (GABA), L-cysteine.
   (d) Antioxidants: vitamin E, vitamin C, lecithin, D-(−)-erythorbic acid, sodium L-ascorbate, sodium thiosulfate, butylated hydroxytoluene, propyl gallate, disodium ethylenediaminetetraacetate dihydrate.
   (e) Buffering agents: citric acid monohydrate, phosphoric acid, ethylenediaminetetraacetic acid, 4-(2-hydroxyethyl)-1-piperazineethanesulfonic acid, tartaric acid, potassium acetate, potassium citrate, dibasic potassium hydrogen phosphate, sodium acetate, sodium carbonate, sodium citrate, sodium dihydrogen phosphate.
   (f) Surfactants: Tween 80, Triton X-100, sucrose fatty-acid esters, 3-[(3-cholamidopropyl) dimethylammonio]-1 propanesulfonate, Hydroxypropyl-β-cyclodextrin, sodium dodecyl sulfate, fatty-alcohol ethoxylates (Brij 35, Brij 30), Lubrol-px, Pluronic F127.
   (g) Natural Complexes: skimmed milk, serum, gelatin, bovine serum albumin.

3. Media
   (a) Solid media: Blood Agar Plate (BAP), Tryptic Soy Agar (TSA).
   (b) Liquid medium: Tryptic Soy Broth (TSB).
4. General Use: deionized water (dH_2O), phosphate-buffered saline (PBS).
5. Identification: Gram stain reagent kit;
   (a) Lysostaphin, agarose, Tris-acetate-EDTA buffer.

2.2 Culture

A pathogenic microorganism strain that has been clearly isolated, cultured, and identified; this chapter uses *Staphylococcus aureus* as an example.

2.3 Equipment and Consumables

1. Bacterial culture: Disposable sterile inoculating loops (1 μL, 10 μL), sterile glass or magnetic beads, constant-temperature incubator, constant-temperature shaking incubator.
2. Mixing/centrifugation: 1 mL pipettor and tips, magnetic stirrer, magnetic stirrer bar, beaker; sterile centrifuge tubes, centrifuge.
3. Sterilization/filtration: Autoclave, 0.22 μm filter membrane.
4. Cooling/pre-cooling: controlled-rate freezer (or freezing container).
5. Cryo-storage: Sterile cryovials, storage boxes; sterilized ampoules or vials.
6. Low-temperature/freeze-drying equipment: Refrigerators (4°C, –20°C, –80°C), liquid-nitrogen tank (vapor or liquid phase); freeze-dryer (lyophilizer).
7. Revival/thawing: Constant-temperature water bath.
8. Identification: Optical microscope; steel target plate for grinding, micro-pipettor, MALDI-TOF mass spectrometer (with reference database); 0.2 mL PCR tubes, thermal cycler.

3 Methods

3.1 Preparation of Cryoprotectants and Lyoprotectants

1. Select appropriate, target-specific cryoprotectants and lyoprotectants according to the type of microorganism strains [7].
2. This chapter uses 20% (v/v) glycerol as the exemplar for low-temperature storage [8] and 10% (w/v) skim milk for freeze-drying [3].
   (a) 40% (v/v) glycerol stock solution:
   (b) 40 mL glycerol + 60 mL PBS; stir magnetically for 5 min. Autoclave at 121°C for 15 min [9]; store at 4°C.
   (c) 20% (w/v) skim milk:
   (d) Make a paste with 20 g skim-milk powder and about 60 mL de-ionized water; bring to 100 mL with additional water while stirring until completely homogeneous. Autoclave at 121°C for 15 min; store at 4°C.
3. If the protectant is heat-labile, sterilize by passing through a 0.22 μm membrane filter under aseptic conditions and store at 4°C (*see* **Note 1**).

3.2 Preparation of the Strains

Revival

Inside a biosafety cabinet, dip a sterile inoculating loop into the *Staphylococcus aureus* stock suspension and streak four quadrants on a TSA plate (*see* **Note 2**). Incubate at 37°C for 18~24 h.

Identification

Pick a single Staphylococcus aureus colony from the TSA plate and streak it onto a BAP (*see* **Note 3**). Incubate at 37°C for 24~48 h (*see* **Note 4**).

(a) Presumptive identification: After 24 h at 37°C, *Staphylococcus aureus* forms circular, smooth, large, convex, moist, opaque colonies 2~3 mm in diameter; the pigment is usually golden-yellow but may be white. Most colonies exhibit a clear zone of β-hemolysis.

(b) Confirmatory identification:

1. Gram-stain microscopy:

   Staphylococcus aureus appears as Gram-positive cocci arranged in grape-like clusters; no spores or capsules are observed.

2. MALDI-TOF colony identification [10]:

   With a sterile loop, smear a thin film of the single colony onto the steel MALDI target. Overlay the dried film with the manufacturer-recommended MALDI matrix. Allow the sample–matrix mixture to dry at room temperature, acquire the mass spectrum, and compare the fingerprint with the reference library for definitive identification (*see* **Note 5**).

3. 16S rRNA identification [11]:

   a) Total DNA extraction:
   With a sterile disposable loop, suspend three to five colonies in 1 mL PBS and vortex to homogeneity. Add 5 μL lyostaphin (10 mg mL^{-1}) and incubate at 37°C for 30 min (*see* **Note 6**). Extract genomic DNA using a commercial kit according to the manufacturer's instructions.

   b) PCR amplification:
   Amplify the 16S rRNA gene with primers 27 F (5'-AGAGTTTGATYMTGGCTCAG-3') and 1492R (5'-GGTTACCTTGTTACGACTT-3') using the following thermal profile: initial denaturation 95°C 30 s; 35 cycles of 95°C 20 s, 58°C 1 min, 72°C 2 min; final extension 72°C 10 min; hold at 4°C.

   c) Agarose-gel electrophoresis:
   Check the PCR product on a 1.5% agarose gel; the 16S rRNA amplicon should appear as a single band of 1500 bp (*see* **Note 7**).

d) PCR product purification
Purify the amplicon using a commercial PCR cleanup kit to remove residual primers, dNTPs, and other contaminants (*see* **Note 8**).

e) Sanger sequencing:
Perform cycle-sequencing reactions with the Applied Biosystems Cycle Sequencing Kit using the same primers (*see* **Note 9**).

f) Sequence analysis:
Separate the extension products on a genetic analyzer. After base-calling and assembly with appropriate software, construct a consensus contig. Submit the final sequence to GenBank for taxonomic assignment and comparative analysis.

Cultivation

(a) Inoculate *Staphylococcus aureus* into 10 mL TSB and incubate at 37°C, 200 rpm for 20~24 h. Transfer 10 μL of this culture into 10 mL fresh sterile TSB and incubate again at 37°C, 200 rpm for 12~16 h.

(b) Cell harvesting: Transfer the TSB culture to a sterile centrifuge tube and pellet the cells by centrifugation at 3000 × *g* for 15 min. Discard the supernatant, resuspend the pellet in sterile PBS, wash, and centrifuge; repeat the wash step twice more (*see* **Note 10**). Finally, prepare a bacterial suspension at 10^8 CFU mL^{-1} [12].

3.3 Mixing

1. Mix the bacterial suspension with cryoprotectant at a 1:1 (v/v) ratio, aliquot into sterile cryovials, and store at low temperature [9] (*see* **Note 11**).
2. Mix the bacterial suspension with lyoprotectant at a 1:1 (v/v) ratio, dispense into sterilized ampoules or vials within 1~2 h while avoiding splashing onto the upper walls, seal aseptically, and proceed to freeze-drying (*see* **Note 12**).

3.4 Cryopreservation Procedure

Controlled-Rate Freezing

Place the cryovials containing the microbial/viral suspension in a controlled-rate freezer and cool at −1°C min^{-1} until the temperature is 5°C~10°C below the suspension's freezing point, then hold for 1~2 h.

Storage

1. Remove the cryovials from the controlled-rate freezer and immediately transfer them to the appropriate low-temperature storage device—e.g., an ultra-low-temperature freezer at −80°C, or a deep-cryogenic storage unit in either the vapor phase (−150°C) or liquid phase (−196°C) (*see* **Note 13**).
2. Viability of the stored microorganism strains must be verified periodically using methods such as morphological examination and molecular assays.

a) For ultra-low-temperature stocks, perform spot checks at least once every 3~5 years on ≥1‰ of the total inventory.
b) For deep-cryogenic stocks, perform spot checks at least once every 5~10 years on ≥1‰ of the total inventory.

Revival

Retrieve the cryovial(s) from the low-temperature storage unit, open them inside a biosafety cabinet, and inoculate the microbial/viral content onto the appropriate culture medium for recovery (*see* **Note 14**).

3.5 Lyophilization Procedure

Freeze-drying

1. Pre-freezing: Pre-cool the shelf to 25°C, load the ampoules or vials, then lower the temperature to –30°C at 1°C min^{-1}; hold at –30°C for 80 min.
2. Primary drying: keep the product at –30°C under 0.3 mbar for 1000 min [13].
3. Secondary drying: ramp the temperature from –30°C to 25°C over 550 min, then maintain 25°C for a further 360 min; keep the chamber pressure at ≤0.05 mbar throughout this step.

Vacuum/Seal Integrity Determination

Before flame-sealing or crimp-sealing, verify the vacuum level or integrity of each ampoule/vial by an appropriate method [7]. For example, when using a high-frequency spark vacuum gauge, a blue glow inside the tube indicates adequate vacuum, while a red glow shows that vacuum has not yet been achieved (*see* **Note 15**).

Flame-Sealing or Crimp-Sealing

After drying, constrict and flame-seal the neck of each ampoule under a vacuum of ≥1 mbar (100 Pa), or crimp-seal the vials (*see* **Note 16**).

Storage

1. Flame-sealed ampoules or crimp-sealed vials must be stored in the dark at 2°C~8°C or at –20°C (or lower).
2. Viability of the lyophilized strains must be verified periodically using methods such as morphological examination and molecular assays. Spot checks must be carried out at least once every 10~15 years on ≥1‰ of the total inventory.

Rehydration

Remove the ampoule or vial from low-temperature stor age, open it aseptically, and add 1~3 mL of TSB medium. Gently rotate the ampoule/vial to dissolve the pellet (*see* **Note 17**), then transfer the rehydrated microorganism strains suspension to fresh medium and incubate.

Technology Outlook

Despite the maturity of current cryopreservation and lyophilization practices, these technical requirements remain constrained by variations in strain-specific responses, equipment performance, and operator expertise. Emerging technologies, such as automated freezing systems, advanced cryoprotective formulations, and real-time monitoring tools, may necessitate periodic revisions to ensure alignment with best practices. Accordingly, regular evaluation and updating of preservation protocols are essential to maintain accuracy, safety, and long-term biological integrity as scientific knowledge continues to advance.

4 Notes

1. If the compound protective agent is not heat-resistant and highly viscous, it is recommended to filter first and then mix it into the compound protective agent to avoid clogging and rendering filtration impossible.
2. Using a 1 μL inoculation loop, let only the inner surface of the loop touch the bacterial suspension for 1~2 mm to avoid forming a hanging droplet. Lightly touch the loop to the starting point on the agar for 5 mm; the capillary action of the agar's micro-pores will wick away excess liquid, preventing confluent growth. After finishing each zone, rotate the plate 30° clockwise so the new streak zone intersects the end of the previous zone at an angle ≤15°, preventing overlap between zones and ensuring that the final fourth zone yields well-isolated single colonies.
3. Place the BAP upright first; wait until surface moisture has evaporated, then close the lid and invert for continued incubation. *Staphylococcus aureus* needs high humidity to produce hemolysins, but an overly wet surface can cause "water hemolysis" artifacts. Starting upright lets any droplets sink in, preventing false-positive hemolysis.
4. Place the BAP on the middle shelf of the incubator, at least 5 cm away from the heating wall, to prevent localized overheating that could suppress hemolysin expression and lead to a misinterpretation of "weak hemolysis."
5. Pick a single colony 1 mm in diameter; touch it lightly with the tip of a 1 μL loop and make two to three circular strokes to form a thin, even film. A too-thick smear can crystallize unevenly and suppress ionization. Add 1 μL of matrix to completely cover the film; avoid repeated pipetting that creates ring-like deposits. After the target is dry, store the plate in a desiccator or at room temperature in the dark and run it within 24 h.
6. Select three to five single colonies with intact margins and 0.5~1 mm in diameter; discard over-aged, confluent colonies whose cell walls are highly cross-linked, as the compacted peptidoglycan sharply reduces lyostaphin efficiency.

Immediately after adding lyostaphin, mix by gently flicking the tube; avoid high-speed vortexing, which generates heat and shear that can inactivate the enzyme and damage protoplasts. Halfway through the 30-min water-bath incubation, flick the tube once more to resuspend the cells and ensure uniform enzyme exposure. When digestion is complete, centrifuge at 4°C, 10,000 × g for 1 min and discard the supernatant to terminate the reaction. Proceed immediately to DNA extraction or electroporation to prevent protoplast autolysis at room temperature and consequent loss of yield.

7. Prior to loading, denature the samples at 65°C for 5 min and immediately transfer to ice for 2 min to disrupt 16S rRNA secondary structure and eliminate "double-band" artifacts. Insert the pipette tip vertically above the well and deliver the sample slowly without touching the bottom, preventing well puncture and sample leakage. Maintain a field strength of 5~10 V·cm^{-1}; higher values overheat the gel and can partially denature DNA, producing smeared bands, whereas lower values excessively prolong run time and should be adjusted according to fragment size. Do not reposition the electrophoresis tank during the run, as this causes gel displacement and distorted bands. If progress must be checked, turn off the power supply before opening the lid.
8. For PCR products >10 μL or with high salt, dilute with an equal volume of nuclease-free water before loading the column to improve binding efficiency. After discarding the final wash, spin the column at 12,000 × g for 1 min to completely remove residual ethanol from the membrane and prevent inhibition of downstream enzymatic or sequencing reactions.
9. Following PCR, residual primers must be completely eliminated. In the sequencing reaction, use a final primer concentration of 0.3~0.5 μM and a template amount of 1~5 ng per 100 bp. Excess of either component promotes nonspecific extension and signal saturation, whereas insufficient quantities yield weak fluorescence. This concentration range ensures well-resolved sequencing peaks with optimal signal-to-noise ratio.
10. Before harvesting the cells by centrifugation, place the culture on ice for 5 min and then centrifuge at 4°C. This pretreatment keeps the cell pellet loose, making it easier to resuspend, while also reducing the amount of PBS buffer and vortexing time required and minimizing cell damage.
11. Slowly pipette 0.5 mL of cell suspension and 0.5 mL of CPA down the inner wall of the cryovial, leaving ≥1/3 headspace to prevent tube rupture from liquid expansion at low temperature. After filling, equilibrate the vial at 4°C for 30 min

to allow the cryoprotectant to fully penetrate the cells before starting stepwise cooling.

12. Slowly load the sample along the wall at a 20°~30° angle, leave ≥1/3 headspace, and keep the vial neck within 5 mm free of liquid to ensure a tight seal; incubate at 4°C for 30 min to allow the cryoprotectant to fully penetrate the cells, then proceed with controlled cooling and lyophilization.
13. Operators must complete specialized training covering sample handling, biosafety, and low-temperature preservation protocols; they are authorized to work only after passing both the theoretical and practical examinations and receiving a competence certificate. Select and don PPE according to the risk level of the samples being handled, and strictly follow established procedures. When working with liquid nitrogen or other cryogenic agents, always wear approved low-temperature gloves to prevent cold burns.
14. Based on the laboratory's activities and specific circumstances, the host institution must conduct a risk assessment, develop a biological security plan, provide security training, investigate any breaches, and implement corrective actions [14].
15. Switch on the high-frequency spark generator and maintain the discharge at the tip for 5~10 s; allow the discharge pattern to stabilize before testing to eliminate any color bias from the initial transient white flash. Keep the electrode tip 5~10 mm from the ampoule wall and traverse the neck at a constant speed with a 30°~45° tilt; under these conditions the observed glow originates predominantly from ionization of the residual gas inside the ampoule rather than from glass-surface fluorescence. For each batch, introduce a negative control consisting of one ampoule of identical dimensions that has not been evacuated; only if the control shows a red glow and the sample a pale-blue glow may the batch be approved for sealing. Ensure the work area is well ventilated to prevent ozone accumulation; keep the electrode tip sharp and periodically restore its profile with Silicon Carbide abrasive paper to avoid misinterpretation from a broadened discharge.
16. Focus the flame on the upper third of the neck, rotate and pull to form a 1 mm filament within 3 s, then fuse; prolonged heating raises the cake temperature above its glass-transition temperature, inducing a glass-to-rubber transition and subsequent moisture uptake, increasing residual moisture after sealing.
17. Open ampoules or vials containing liquid or lyophilized cultures only inside a BSC to contain any aerosol generated. Staff must be trained in the safe opening of glass ampoules. Wear gloves and safety goggles when opening sealed ampoules or cryovials. Score the neck with a file, wrap the ampoule in a

disinfectant-soaked disposable towel, hold it vertically, and snap at the score line. Rehydrate slowly to prevent aerosolization of the dry content; avoid bubbling while mixing and transfer the suspension to a new vessel.

Acknowledgments

This work was supported by the National Key Research and Development Project (2022YFC2602200) and the National Science.

Conflict of Interest The authors have no conflicts of interest to declare that are relevant to the content of this chapter.

References

1. Guo N, Wei Q, Xu Y (2020) Optimization of cryopreservation of pathogenic microbial strains. J Biosaf Biosecur 2:66–70
2. Hubálek Z (2003) Protectants used in the cryopreservation of microorganisms. Cryobiology 46:205–229
3. Yuste A, Arosemena EL, Calvo M (2021) Study of the probiotic potential and evaluation of the survival rate of Lactiplantibacillus plantarum lyophilized as a function of cryoprotectant. Sci Rep 11: 19078
4. Rocha IV, Bezerra MF, Sobreira M, Leal NC, de Almeida AMP (2024) Lyophilization for bacteria preservation: a promising approach for Yersinia pestis strains from an unique collection in Brazil (Fiocruz-CYP). Antonie Van Leeuwenhoek 117:61
5. Zhao Y, Guo N, Jiang M, Xu Y, Wei Q (2023) Interpretation for the standard of technical guidelines for low temperature preservation of pathogenic microorganisms for prophylactic biological products. Dis Surveill 38:1432–1435
6. Butreddy A, Dudhipala N, Janga KY, Gaddam RP (2020) Lyophilization of small-molecule injectables: an industry perspective on formulation development, process optimization, scale-up challenges, and drug product quality attributes. AAPS PharmSciTech 21:252
7. Cheong IH, Zhao Y, Jiang M, Lei S, Wang H, Kozlakidis Z et al (2025) Technical guideline for low-temperature preservation of pathogenic microorganisms in China and its relevance to developing countries. Biopreserv Biobank 23:152–156
8. Senaratne W, Jayaweera J (2024) Comparison of microbial preservation methods: a narrative review. Germs 14:375–386
9. Guo N, Song Y, Yan J, Jiang M, Xu Y, Li Z et al (2023) The effect of cryopreservation on the survival of Nocardia farcinica and Yersinia pestis vaccine strains. Biopreserv Biobank 21: 397–406
10. Croxatto A, Prod'hom G, Greub G (2012) Applications of MALDI-TOF mass spectrometry in clinical diagnostic microbiology. FEMS Microbiol Rev 36:380–407
11. YalÇIn S, TÜRk AH, Mehmetzade HF, Aydin SK, SaĞTaŞ E (2025) Comparison of 16S rRNA sequencing methods for bacterial identification in clinical microbiology laboratories: Sanger sequencing vs. third-generation sequencing. Turk Bull Hyg Exp Biol 82:311–322
12. Xiao Y, Huang P, Huang Z, Yu K, Song Y, Guo N et al (2022) Influencing factors on the preservation of lytic bacteriophage VP3. Biosaf Health 4:314–320
13. Merabishvili M, Vervaet C, Pirnay JP, De Vos D, Verbeken G, Mast J et al (2013) Stability of Staphylococcus aureus phage ISP after freeze-drying (lyophilization). PLoS One 8:e68797
14. Ficociello B, Giordano D, Incoronato F, Farinella A, Pietrangeli B (2023) WHO Laboratory Biosafety Manual: A New Approach to Security. Ann Work Expo Health 67:425–429

Chapter 4

Cryopreservation of *Rhizobium* Strains Associated with Common Bean (*Phaseolus vulgaris* L.): Evaluation of Functional Stability and Nodulation Capacity

José Luis Aguirre-Noyola, Lorena Jacqueline Gómez-Godínez, Gustavo Cuaxinque-Flores and Esperanza Martínez-Romero

Abstract

The conservation of rhizobial diversity is essential to sustain the functionality of symbiotic associations that underpin biological nitrogen fixation in ecologically and agriculturally important legumes. In particular, *Rhizobium* strains associated with the common bean (*Phaseolus vulgaris* L.) represent a strategic genetic resource in both productive systems and natural ecosystems of Mexico. However, the loss of viability or symbiotic plasmids during laboratory maintenance may compromise their nodulation capacity and nitrogen-fixing ability. In this study, we describe a standardized glycerol-based cryopreservation protocol at −80°C for the long-term preservation of *Rhizobium* strains symbiotic with *P. vulgaris*, along with the evaluation of their functional stability after thawing. The method relies on the use of glycerol as a penetrating cryoprotectant, which minimizes cellular damage associated with ice crystal formation and osmotic stress, thereby preserving cell viability and genetic integrity. Furthermore, the capacity of preserved strains to form nitrogen-fixing nodules is assessed as an indicator of symbiotic stability. This protocol provides a reliable tool for the preservation of microbial genetic resources of high ecological and agronomic value, contributing to the safeguarding of rhizobial biodiversity and to its potential application in sustainable agriculture.

Keywords Rhizobia, Biological nitrogen fixation, Symbiotic plasmid, Microbial preservation, Genetic stability

1 Introduction

Rhizobia constitute a functional group of soil bacteria capable of establishing a highly specialized symbiosis with legumes (family Fabaceae) [1]. Although the term "rhizobia" does not represent a formal taxonomic category, it encompasses phylogenetically diverse bacteria that share the ability to induce nodulation and convert atmospheric nitrogen into ammonium in association with plant roots [2]. This process represents one of the major natural inputs of nitrogen

Marco A. Ramírez-Mosqueda et al. (eds.), *Methods for The Conservation of Microorganisms*, Methods in Molecular Biology, vol. 3090, https://doi.org/10.1007/978-1-0716-5623-5_4,

into terrestrial ecosystems and constitutes a central component of sustainable agricultural productivity. Most rhizobia belong to the phylum Pseudomonadota, primarily within the classes *Alphaproteobacteria* and *Betaproteobacteria*. The most extensively studied genera include *Rhizobium*, *Bradyrhizobium*, *Ensifer* (formerly Sinorhizobium), *Mesorhizobium*, and *Azorhizobium* [3]. This phylogenetic diversity reflects a dynamic evolutionary history characterized by horizontal transfer of symbiotic genetic determinants—such as nodulation (nod) and nitrogen fixation (*nif* and *fix*) genes—which has facilitated the expansion of the symbiotic trait, the emergence of distinct symbiovars, and adaptation to multiple hosts and environments [4].

In economically important legumes—such as faba bean, soybean, lentil, chickpea, and alfalfa—rhizobia play a decisive role in plant development and nutrition [2]. Particularly, common bean (*Phaseolus vulgaris* L.) is among the most relevant crops in which this symbiosis plays a pivotal role, substantially reducing the need for synthetic nitrogen fertilizers through biological nitrogen fixation (Fig. 1a, b) [5]. In Mexican soils, the symbionts nodulating *P. vulgaris* roots primarily belong to the genus *Rhizobium*, particularly *R. etli*, *R. phaseoli*, and *R. gallicum*, which constitute a reservoir of genetic diversity adapted to local conditions [6]. Nevertheless, the importance of rhizobia extends beyond agricultural systems. In natural ecosystems, rhizobia–legume symbiosis contributes to soil fertility, facilitates plant establishment in nitrogen-limited environments, and promotes the structural stability of plant communities [7].

In this context, the conservation of rhizobial diversity acquires ecological and strategic relevance. The loss of native strains not only implies a reduction in microbial biodiversity but may also decrease symbiotic efficiency and resilience under changing environmental conditions. Cryopreservation with glycerol represents a fundamental tool for the long-term preservation of these genetic resources [8]. This method maintains cell viability and genetic stability by effectively arresting metabolic activity and minimizing cellular damage caused by ice crystal formation and osmotic stress, due to the action

Fig. 1 Nitrogen-fixing nodules of common bean (*Phaseolus vulgaris*) growing in soils from Mexico. (**a**) Whole plants and (**b**) symbiotic root nodules of *P. vulgaris*

of glycerol as a penetrating cryoprotectant that stabilizes membranes and proteins [9]. Its main advantages include high post-thaw recovery rates, reduced probability of mutations or plasmid loss compared with continuous subculturing, and broad applicability in environmental and agricultural microbiology [10]. In rhizobia, the loss of plasmids—particularly the symbiotic plasmid—directly compromises their ability to establish an effective symbiosis with their legume host [11].

The present work describes a glycerol-based cryopreservation protocol for *Rhizobium* strains symbiotic with *P. vulgaris*, together with the evaluation of their functional stability in the formation of nitrogen-fixing nodules, in order to ensure the effective preservation of this microbial genetic resource of high ecological and agronomic value.

2 Materials

2.1 Bacterial Strains

1. The rhizobial strains used in this protocol correspond to *Rhizobium etli* and *Rhizobium phaseoli*, isolated from root nodules of *P. vulgaris* in Mexico. Molecular identification was performed through amplification and sequencing of the 16S rRNA gene.

2.2 Culture Medium

1. Tryptone–Yeast Extract (TY) medium. This medium is suitable for rhizobia because it promotes rapid growth and high cellular viability due to its rich nutritional composition. Additionally, $CaCl_2$ contributes to membrane stability, enhancing cellular resistance and post-storage recovery [12].
2. Bacteriological agar as a gelling agent.

2.3 Solutions

1. Sterile glycerol solution at 40%–50% (v/v) prepared in sterile distilled water (*see* **Note 1**).
2. pH adjustment solutions (1 N): hydrochloric acid (HCl) and sodium hydroxide (NaOH).
3. Calcium chloride stock solution (100×): 87 g/L $CaCl_2{\cdot}2H_2O$ prepared in distilled water, sterilized by filtration (0.22 μm).
4. Gram staining solutions: crystal violet (1%, w/v), Gram's iodine solution (Lugol's iodine; 1% iodine and 2% potassium iodide, w/v), decolorizer (95% ethanol or ethanol–acetone, 1:1 v/v), and safranin (0.5%–1%, w/v)

2.4 Equipment

1. Microbial culture room: horizontal laminar flow cabinet, Bunsen burner or glass bead sterilizer, spectrophotometer, and incubator shaker.
2. Media preparation room: analytical balance, spatula, autoclave, pH meter, hot plate stirrer, filters (0.22 μm), micropipettes

(100–1000 μL), plastic Petri dishes (100 mm × 15 mm) and cryogenic tubes (2 mL).

3. Glassware: beakers (500 mL and 1 L), screw-cap bottles (1 L), graduated cylinders (50 mL, 100 mL, and 1 L), and Erlenmeyer flasks (250 mL and 500 mL).
4. Others: plastic sealing film, magnetic stir bar, and indelible marker.

3 Methods

3.1 Preparation of Culture Medium

1. Prepare TY medium by dissolving the following components in approximately 900 mL of distilled water: tryptone (5 g), yeast extract (3 g), and agar (15 g, for solid medium).
2. Adjust the pH to 7.0 using 1 N NaOH or 1 N HCl.
3. Adjust the final volume to 990 mL with distilled water.
4. Sterilize by autoclaving at 121°C for 15–20 min.
5. After cooling the medium to approximately 50°C, aseptically add 10 mL of sterile calcium chloride stock solution (100×) to obtain a final concentration of 0.87 g/L (*see* **Note 2**), and homogenize gently.
6. Dispense approximately 25 mL of TY medium into sterile Petri dishes under a laminar flow cabinet and allow the medium to solidify.

3.2 Bacterial Culture

1. Under aseptic conditions in a laminar flow cabinet, streak the rhizobial strains (*R. etli* and *R. phaseoli*) onto TY agar plates using a sterile inoculation loop and the cross-streak method.
2. Incubate the plates at 30°C for 48 h.
3. After incubation, observe isolated colonies displaying typical rhizobial morphology: mucoid, convex, and pearly white in appearance (Fig. 2).
4. Select a well-isolated colony and verify purity by Gram staining (*see* **Note 3**).

3.3 Cryopreservation of Rhizobial Strains

1. Inoculate a fresh single colony into 150 mL Erlenmeyer flasks containing 50 mL of TY broth.
2. Incubate at 30°C with shaking at 150–180 rpm for 24–36 h.
3. Monitor bacterial growth by measuring the optical density at 600 nm (OD_{600}) using a spectrophotometer. Proceed when OD_{600} reaches approximately 0.8–1.2, corresponding to late exponential or early stationary phase.
4. Mix equal volumes of bacterial culture and sterile glycerol solution to obtain a final glycerol concentration of 20%–25% (v/v).

5. Aliquot the suspension into sterile cryogenic tubes (2 mL per tube) and rapidly freeze at −80°C for long-term storage (Fig. 3).

3.4 Reactivation of Rhizobial Strains by "Scraping from Frozen Stock"

1. Remove the cryovial from −80°C storage and transport it to the laminar flow cabinet (*see* **Note 4**).
2. Open the vial and, working rapidly (≤15 s), collect a small portion of the frozen material using a sterile inoculation loop previously cooled (*see* **Note 5**).

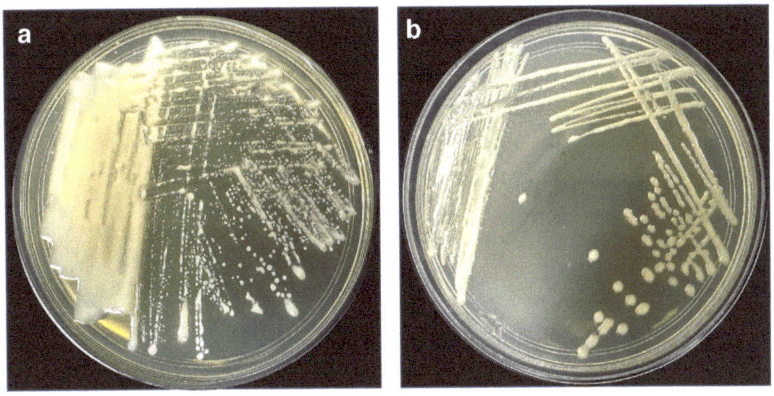

Fig. 2 Colonial morphology of symbiotic rhizobial strains isolated from *P. vulgaris*. (**a**) *Rhizobium etli* and (**b**) *Rhizobium phaseoli* grown on TY medium

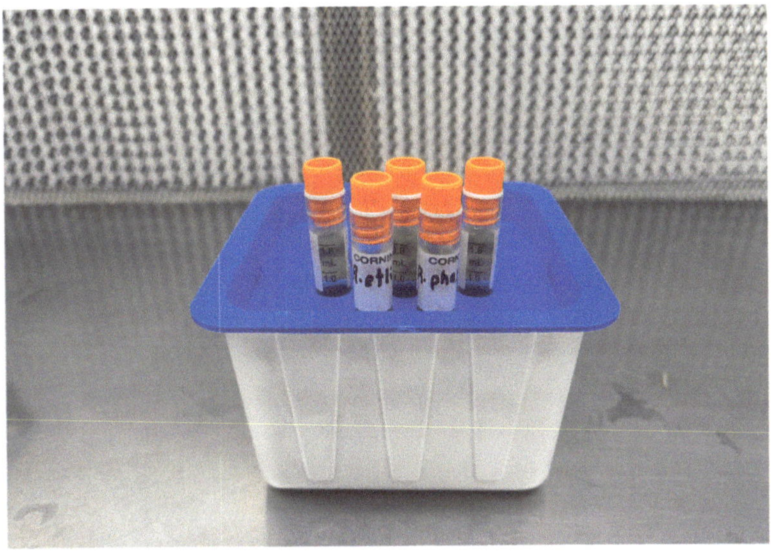

Fig. 3 Glycerol-based cryopreservation of symbiotic rhizobial strains (*R. etli* and *R. phaseoli*) in cryogenic tubes

3. Immediately streak the collected material onto a TY agar plate using the cross-streak method.
4. Close the cryovial and promptly return it to −80°C storage, avoiding partial or complete thawing of the contents.
5. Incubate the plates at 30°C for 48 h and evaluate bacterial growth, colony morphology, and purity by Gram staining. Furthermore, the symbiotic competence of the strain should be confirmed through a host interaction assay *P. vulgaris*, assessing its ability to induce effective nodule formation on the roots.

3.5 Evaluation of Symbiotic Effectiveness Stability

1. Surface-sterilize *Phaseolus vulgaris* seeds and germinate them under sterile conditions on water–agar plates, incubating in darkness at 28°C for 48 h.
2. From reactivated cultures, prepare a rhizobial suspension in sterile 0.89% saline solution, adjusted to a final concentration of 1×10^6 cells mL^{-1}.
3. Inoculate 50 μL of the bacterial suspension directly onto the radicle of each germinated bean seedling.
4. Transfer the inoculated seedlings into sterile 250 mL flasks containing N-free semisolid Fahraeus medium, as previously described [13].
5. Maintain plants under controlled growth conditions with a 16 h light/8 h dark photoperiod at 26°C.
6. After 15–21 days of cultivation, evaluate root nodulation (*see* **Note 6**). Mature and functional nodules exhibit a reddish internal coloration, indicative of leghemoglobin presence and nitrogenase activity, responsible for biological nitrogen fixation (Fig. 4).

Fig. 4 Root nodules of *P. vulgaris* formed by rhizobial strains reactivated after cryopreservation. (**a**) In vitro nodulation assay. (**b**) Reddish determinate nodules indicating leghemoglobin production and active biological nitrogen fixation

4 Notes

1. Prepare a sterile 40%–50% (v/v) glycerol solution by diluting molecular-grade glycerol in distilled water. Sterilize by autoclaving (121 °C, 15 min) or by sterile filtration (0.22 μm). Store at room temperature until use.
2. Allow TY medium to cool to room temperature before adding the 100× $CaCl_2$ solution; otherwise, a white precipitate may form due to salt destabilization.
3. Microscopic observation should be performed using a 100× objective with immersion oil. Strains of the genus *Rhizobium* should appear as short Gram-negative rods (pink-stained), straight or slightly curved, typically observed singly or in pairs. The presence of atypical morphologies or Gram-positive bacteria suggests culture contamination.
4. Cryovials should be handled in a precooled rack or on dry ice to minimize thawing of the contents.
5. The "scraping from frozen stock" procedure allows recovery of cultures without completely thawing the cryovial, thereby reducing stress associated with repeated freeze–thaw cycles. However, this method should be performed rapidly and the number of uses limited to minimize potential effects on cell viability and genetic stability, particularly in strains harboring symbiotic plasmids.
6. If the reactivated strain fails to induce nodulation or forms white, ineffective, or degenerated nodules, it may have lost the symbiotic plasmid. In such cases, the strain should be recovered from a properly preserved backup cryovial.

Conflict of Interest

The authors have no conflicts of interest to declare that are relevant to the content of this chapter.

References

1. Poole P, Ramachandran V, Terpolilli J (2018) Rhizobia: from saprophytes to endosymbionts. Nat Rev Microbiol 16(5):291–303
2. Lindström K, Mousavi SA (2020) Effectiveness of nitrogen fixation in rhizobia. Microb Biotechnol 13(5):1314–1335
3. Willems A (2006) The taxonomy of rhizobia: an overview. Plant Soil 287(1):3–14
4. Martinez-Romero E, Peix A, Hungria M, Mousavi SA, Martinez-Romero J, Young P (2024) Guidelines for the description of rhizobial symbiovars. Int J Syst Evol Microbiol 74(5):006373
5. Shamseldin A, Velázquez E (2020) The promiscuity of *Phaseolus vulgaris* L. (common bean) for nodulation with rhizobia: a review. World J Microbiol Biotechnol 36(5):63

6. Silva C, Vinuesa P, Eguiarte LE, Martínez-Romero E, Souza V (2003) *Rhizobium etli* and *Rhizobium gallicum* nodulate common bean (*Phaseolus vulgaris*) in a traditionally managed milpa plot in Mexico: population genetics and biogeographic implications. Appl Environ Microbiol 69(2):884–893
7. Moura EG, Carvalho CS, Bucher CP, Souza JL, Aguiar AC, Ferraz Junior AS, Coelho KP (2020) Diversity of rhizobia and importance of their interactions with legume trees for feasibility and sustainability of the tropical agrosystems. Diversity 12(5):206
8. Gómez-Godínez LJ, Hernández-Cruz E, Guzmán-González RD, Ruiz-Sandoval G, Ruiz Rivas M, Aguirre-Noyola JL, De los Santos Villalobos S. Preservation of microorganisms and microbiomes: methods, impacts, and future prospects. Biopreserv Biobank, 19475535251401795, doi: https://doi.org/10.1177/19475535251401795
9. Whaley D, Damyar K, Witek RP, Mendoza A, Alexander M, Lakey JR (2021) Cryopreservation: an overview of principles and cell-specific considerations. Cell Transplant 30:0963689721999617
10. Singh SK, Baghela A (2017) Cryopreservation of microorganisms. Modern tools and techniques to understand microbes. Springer International Publishing, Cham, 321–333
11. López-Guerrero MG, Ormeño-Orrillo E, Acosta JL, Mendoza-Vargas A, Rogel MA, Ramírez MA,..., Martínez-Romero E (2012) Rhizobial extrachromosomal replicon variability, stability and expression in natural niches. Plasmid 68(3):149–158
12. Bromfield ESP, Wheatcroft R, Barran LR (1994) Medium for direct isolation of Rhizobium meliloti from soils. Soil Biol Biochem 26(4):423–428
13. Aguirre-Noyola JL, Rosenblueth M, Santiago-Martínez MG, Martínez-Romero E (2021) Transcriptomic responses of Rhizobium phaseoli to root exudates reflect its capacity to colonize maize and common bean in an intercropping system. Front Microbiol 12:740818

Chapter 5

Glycerol-Based Cryopreservation of *Sinorhizobium/Ensifer* Strains from Sweet Acacia (*Vachellia farnesiana* [L.] Wight & Arn.)

José Luis Aguirre-Noyola, Verónica García-Iglesias, Gustavo Cuaxinque-Flores and José Daniel Chávez-González

Abstract

The genus *Sinorhizobium* (*Ensifer*) exhibits remarkable metabolic versatility and adaptive capacity, establishing effective nitrogen-fixing symbioses with diverse legume crops, including the pioneer shrub *Vachellia farnesiana*. This association is crucial for ecological restoration in arid and nutrient-deficient Mexican dryland ecosystems and for the remediation of contaminated sites. However, the prolonged preservation of specialized native strains through continuous serial subculturing risks genomic instability and the loss of extrachromosomal replicons harboring essential nod, nif, and heavy metal tolerance determinants. Here, we describe an optimized, reproducible, and effective glycerol-based cryopreservation protocol specifically adapted for *Sinorhizobium* strains associated with *V. farnesiana*. The protocol is based on the use of a glycerol solution for immediate ultra-low-temperature freezing at −80°C. Strain reactivation via a frozen stock-scraping technique minimized the physiological shock associated with freeze–thaw cycles. Post-preservation validation confirmed the typical *Sinorhizobium* morphology and, critically, demonstrated the retention of symbiotic competence through in vitro root nodulation assays with *V. farnesiana* seedlings. This optimized procedure ensures the viable and functionally intact preservation of microbial genetic resources for potential large-scale reforestation and restoration applications.

Keywords Biological nitrogen fixation, Bacterial preservation, Root nodulation, Bioremediation

1 Introduction

Rhizobia are a diverse polyphyletic group of soil bacteria with the capacity for nitrogen fixation, root nodule formation, and the establishment of specialized symbioses with plants in the family Fabaceae [1]. Among the rhizobial genera studied, the genus *Sinorhizobium* (*Ensifer*) has been widely recognized for its metabolic versatility and adaptive capacity across diverse environments [2]. *Sinorhizobium* species form highly effective symbioses with various legumes, inducing the formation of specialized root structures known as nodules,

Marco A. Ramírez-Mosqueda et al. (eds.), *Methods for The Conservation of Microorganisms*, Methods in Molecular Biology, vol. 3090, https://doi.org/10.1007/978-1-0716-5623-5_5,

where biological nitrogen fixation occurs. This physiological process enables plants to establish symbiotic associations and thrive in diverse environments, particularly in nitrogen-deficient soils [3]. The genes and molecular mechanisms involved in *Sinorhizobium* adaptation to distinct stress conditions are critical for the establishment of a functional symbiotic relationship. These mechanisms involve the transcription of genes associated with nodulation, nitrogen fixation, exopolysaccharide biosynthesis, nitrogenase production, and protein function [4]. The primary role of *Sinorhizobium* strains in association with legume crops is the modulation of plant development, nutrient acquisition, and tolerance to environmental stressors [5]. Consequently, the study of *Sinorhizobium* species has emerged as a crucial strategy for optimizing productivity in nutrient-deficient agricultural systems and for the ecological remediation of contaminated soils [5, 6].

In this context, several legume crops have been proposed as important species for the phytoremediation of metal-contaminated soils due to their capacity to establish symbiotic associations with *Sinorhizobium* strains. Specifically, the genus *Vachellia* (commonly known as "huizache") within the family Fabaceae is widely distributed throughout Mexican dryland ecosystems, with notable prevalence in contaminated soils (Fig. 1) [7, 8]. Furthermore, the symbiotic association between *Vachellia* and *Sinorhizobium* has been extensively reported in arid and semi-arid regions. Accordingly, the study of this relationship is focused on identifying and proposing suitable native plant candidates for the reforestation and restoration of degraded soils through inoculation with *Sinorhizobium* strains for large-scale applications [9–11].

The successful ecological application of these specialized *Sinorhizobium* strains requires the preservation of their genetic integrity. Prolonged preservation through consecutive subculturing on solid media frequently results in genetic drift, reduced symbiotic efficiency, or the loss of extrachromosomal replicons harboring crucial

Fig. 1 Symbiotic and nitrogen-fixing nodules of sweet acacia (*Vachellia farnesiana*) growing in mine tailings from Mexico. (**A**) "Huizache" plants and (**B**) root nodules with rhizobia

nod, *nif*, and heavy metal tolerance genes. From this perspective, the use of cryoprotectants such as glycerol represents a highly effective methodology for the stabilization of microbial genetic resources [12]. This optimized method prevents intracellular ice formation, stabilizes lipid bilayers and native protein conformations, and strictly controls metabolic activity while preventing the lethal osmotic shock associated with ultra-low temperatures [13].

This optimized protocol offers a reproducible and cost-effective solution for the long-term preservation of *Sinorhizobium* (*Ensifer*) genetic resources. The glycerol-based cryopreservation procedure was specifically adapted for *Sinorhizobium* strains capable of nodulating *V. farnesiana*, with the aim of ensuring the viability and functional traits of the strains after thawing.

2 Materials

2.1 Bacterial Strains

The *Sinorhizobium* strains (*S. fredii* and *S. americanus*) employed in this protocol correspond to native strains, recovered from functional root nodules of the pioneer sweet *V. farnesiana* colonizing mine tailings (Taxco, Guerrero, Mexico). Taxonomic results of the isolates were established through 16S rRNA gene amplification and sequencing.

2.2 Culture Media

1. Tryptone–Yeast Extract (TY) solid agar medium was used due to its nutritional formulation, which optimally supports the characteristic fast-growing kinetics of *Sinorhizobium*. Crucially, supplementation with $CaCl_2$ reinforces the bacterial cellular envelope, providing structural stabilization that significantly enhances biological resilience against subsequent cryogenic stress.

2.3 Solutions

1. Cryoprotectant agent: 40% (v/v) glycerol solution, prepared in distilled water and strictly sterilized by autoclaving or by sterile filtration (0.22 μm).
2. Acid-base titrants: 1 N HCl and 1 N NaOH for pH adjustment calibration.
3. Calcium supplement stock solution: 100× $CaCl_2 \cdot 2\,H_2O$ stock solution (87 g/L), sterilized via 0.22 μm membrane microfiltration to prevent heat-induced precipitation.
4. Microbiological stains: Standard Gram stain kit components, including crystal violet (1%, w/v), Lugol's iodine, decolorizing agent (95% ethanol), and safranin counterstain (0.5%–1%, w/v).

2.4 Infrastructure

1. Culture room: Horizontal laminar flow clean bench, Bunsen burner or dry glass bead sterilizer, UV–Vis spectrophotometer, and a temperature-controlled orbital incubator shaker.
2. Media provision: Precision analytical balance, laboratory spatulas, high-pressure sterilizer, calibrated pH meter, magnetic hot plate stirrer, sterile microfilters (0.22 μm pore size), volume micropipettes (100–1000 μL), sterile polystyrene Petri dishes (100 × 15 mm), and sterile 2-mL cryopreservation vials.
3. Laboratory Glassware: Borosilicate beakers (500 mL and 1 L capacities), screw-cap media storage bottles (1 L), glass graduated cylinders (50 mL, 100 mL, and 1 L), and Erlenmeyer culture flasks (250 mL and 500 mL).

3 Methods

3.1 Culture Media Preparation

1. Prepare the Tryptone-Yeast Extract medium by dissolving 5 g of tryptone, 3 g of yeast extract, and 15 g of bacteriological agar (for solid media) in 900 mL of sterile distilled water.
2. Calibrate the pH to 7.0 utilizing 1 N NaOH or 1 N HCl as titrants.
3. Fill to final volume of 990 mL with sterile distilled water.
4. Sterilize the medium by autoclaving at 121°C for 15 min.
5. Allow the autoclaved medium to cool to approximately 50°C. Under strict aseptic conditions, supplement the medium with 10 mL of a sterilized calcium supplement stock solution (100×) to achieve a final working concentration of 0.87 g/L and mix gently.
6. Aliquot approximately 25 mL of the supplemented TY medium into sterile Petri dishes within a laminar flow, allowing the matrix to polymerize completely.

3.2 Bacterial Growth Conditions

1. Under sterile conditions within a laminar flow cabinet, inoculate the native *Sinorhizobium* strains in the prepared TY agar plates employing a continuous cross-streaking technique with a sterile inoculation loop to ensure optimal single-colony isolation.
2. Incubate the plates in an inverted position at a constant 30°C for 48 h.
3. Post-incubation, visually confirm the distinct, well-isolated colonies exhibiting the common morphological traits of *Sinorhizobium*: circular, translucent, mucoid colonies, convex elevation, and a pearly-white characteristic (Fig. 2).

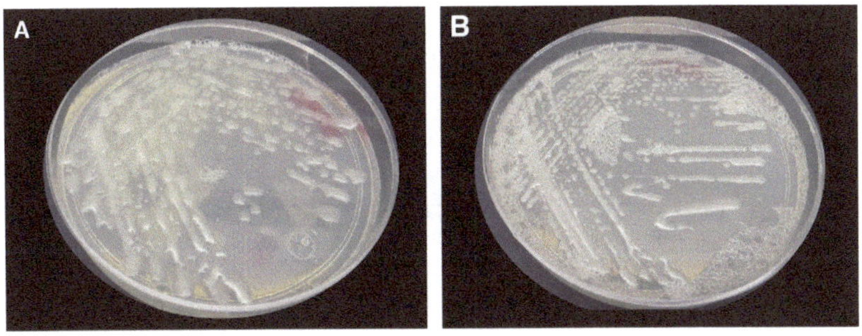

Fig. 2 Colonial morphology of symbiotic *Sinorhizobium* strains. (**A**) *Sinorhizobium fredii* (SVF1) and (**B**) *Sinorhizobium americanus* (SVF3) grown on TY medium

4. Select a representative, morphologically confirmed single colony and rigorously verify culture purity via Gram staining prior to future applications (*see* **Note 1**).

3.3 Glycerol-Based Cryopreservation of Sinorhizobium Strains

1. Under sterile conditions in the laminar flow, inoculate a singlet, and fresh colony into 50 mL of TY broth within a 150-mL Erlenmeyer flask.
2. Cultivate the suspension in an orbital incubator shaker at 30°C and 150–180 rpm for a period of 24 to 36 h.
3. Supervise cellular growth via UV–Vis spectrophotometry at an optical density of 600 nm (OD_{600}). Biomass harvesting should proceed exclusively when the OD_{600} reaches approximately 0.8–1.2, marking the critical transition from the late-exponential to the early-stationary growth phase.
4. Homogenize the liquid culture with an equal volume of a sterilized glycerol solution to achieve a final cryoprotectant concentration of 20% (v/v).
5. Aliquot the stabilized bacterial suspension into sterile 2-mL cryovials and subject them to immediate ultra-low temperature freezing (−80°C) for long-term preservation.

3.4 Reactivation of Sinorhizobium Strains Post-Cryopreservation

1. Cryovial from −80°C should be handled in a rack with ice and transfer it strictly under a laminar flow.
2. Uncap the cryovial and excise a small fragment of the frozen biomass utilizing a sterile inoculation loop (*see* **Note 2**).
3. Immediately streak the excised ice crystal onto a fresh TY agar plate employing the continuous cross-streaking technique.
4. Reseal the cryovial and return it to the −80°C freezer, rigorously avoiding any partial or complete thawing of the main stock.
5. Incubate the inoculated plates at 30°C for 48 h. Post-incubation, evaluate the isolates for characteristic growth kinetics, expected morphology, and structural purity via

Gram staining. Additional assessments of the functionality of plant-microorganism interactions should also be conducted (*see* **Note 3**).

4 Notes

1. *Sinorhizobium* cells typically exhibit a Gram-negative phenotype, appearing pink after Gram staining, and present as straight or slightly curved rods occurring singly or in pairs. The presence of Gram-positive bacteria or abnormal morphologies may indicate culture contamination. Microscopic observation should be performed using a 100× oil-immersion objective.
2. You must work rapidly about 10–15 s. The frozen stock-scraping technique facilitates the reactivation of microbial cultures, avoiding the complete thawing of the cryovial, thereby mitigating the severe physiological shock induced by recurrent unfreezing cycles. Nevertheless, this method must be executed with extreme rapidity and its frequency strictly limited per vial.
3. The functionality of symbiotic traits should be validated through an in vitro plant–microbe interaction assay using *Vachellia farnesiana* plants [7]. The biological nitrogen fixation capacity of the inoculated strain must be confirmed through effective root nodulation (Fig. 3).

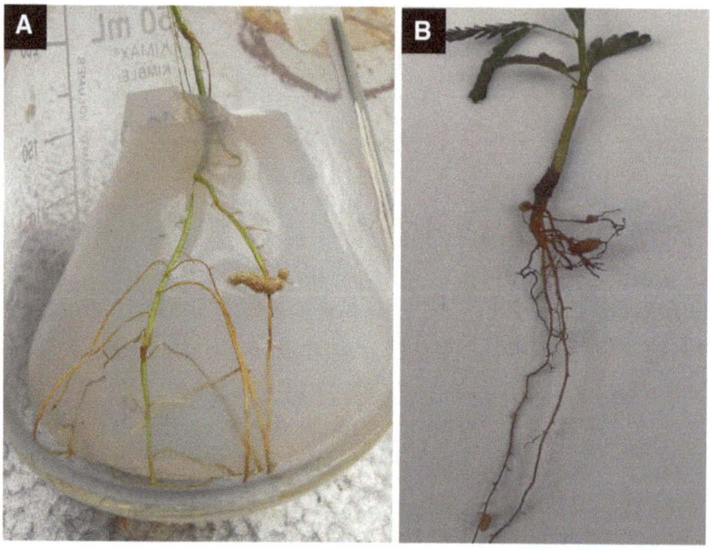

Fig. 3 Root nodules of *Vachellia farnesiana* induced by *Sinorhizobium* strains following glycerol-based cryopreservation. (**A**) In vitro plant–microbe symbiotic growth assay. (**B**) Root nodule formation in *V. farnesiana* after 60 days

Conflict of Interest

The authors have no conflicts of interest to declare that are relevant to the content of this chapter.

References

1. Poole P, Ramachandran V, Terpolilli J (2018) Rhizobia: from saprophytes to endosymbionts. Nat Rev Microbiol 16:291–303. https://doi.org/10.1038/nrmicro.2017.171
2. Kearsley JVS, Sather LM, Finan TM (2024) *Sinorhizobium* (*Ensifer*) *meliloti*. Trends Microbiol 32:516–518. https://doi.org/10.1016/j.tim.2024.03.004
3. Fagorzi C et al. (2018) Harnessing rhizobia to improve heavy-metal phytoremediation by legumes. Genes (Basel) 9:542. https://doi.org/10.3390/genes9110542
4. Jach ME, Sajnaga E, Ziaja M (2022) Utilization of legume-nodule bacterial symbiosis in phytoremediation of heavy metal-contaminated soils. Biology (Basel) 11:676. https://doi.org/10.3390/biology11050676
5. Chen W et al (2025) The role of symbiotic nitrogen-fixing bacteria, *Rhizobium* and *Sinorhizobium*, as "bridges" in the rhizosphere of legumes after fomesafen application. Appl Soil Ecol 209:106013. https://doi.org/10.1016/j.apsoil.2025.106013
6. Li P et al (2024) Different responses of *Sinorhizobium* sp. upon Pb and Zn exposure: mineralization versus complexation. Environ Pollut 343:123260. https://doi.org/10.1016/j.envpol.2023.123260
7. Aguirre-Noyola JL, Garcia-Iglesias V, Chavez-Gonzalez JD (2026) Short-term conservation and exploitation of sweet acacia (*Vachellia farnesiana* L. Wight & Arn.) through an efficient scarification-based germination protocol. Methods Mol Biol 3011:81–90. https://doi.org/10.1007/978-1-0716-5130-8_6
8. Romero MF et al (2021) Metagenomics of mine tailing rhizospheric communities and its selection for plant establishment towards bioremediation. Microbiol Res 247:126732. https://doi.org/10.1016/j.micres.2021.126732
9. Bouhnik O et al (2024) Characterization of *Sinorhizobium* strains isolated from arid and semi-arid areas of Morocco promoting the growth of *Vachellia gummifera*. Biocatal Agric Biotechnol 61:103369. https://doi.org/10.1016/j.bcab.2024.103369
10. Garcidueñas-Piña C et al (2022) *Prosopis laevigata* and *Acacia farnesiana* in association with *Rhizobium* sp. remove copper from soil. S Afr J Bot 151:417–424. https://doi.org/10.1016/j.sajb.2022.09.019
11. Li Z et al (2024) Natural nodulation and nitrogen fixation of *Acacia auriculiformis* grown in technosol eco-engineered from Fe ore tailings. Plant Soil 497:25–41. https://doi.org/10.1007/s11104-023-06342-7
12. Gomez-Godinez LJ et al (2026) Preservation of microorganisms and microbiomes: methods, impacts, and future prospects. Biopreserv Biobank. https://doi.org/10.1177/19475535251401795
13. Jiang P, Li Q, Liu B, Liang W (2023) Effect of cryoprotectant-induced intracellular ice formation and crystallinity on bacteria during cryopreservation. Cryobiology 113:104786. https://doi.org/10.1016/j.cryobiol.2023.104786

Check for updates

Chapter 6

Long-Term Cryopreservation of *Pseudomonas* spp. for Biobanking and Biotechnology Applications

José Luis Aguirre-Noyola, José Daniel Chávez-González, Jeiry Toribio-Jiménez, Lorena Jacqueline Gómez-Godínez and Gustavo Cuaxinque-Flores

Abstract

The genus *Pseudomonas* encompasses one of the most metabolically versatile and ecologically diverse groups among Gram-negative bacteria, comprising over 300 validly described species distributed across virtually all ecosystems. Members of this genus are of considerable relevance in medicine, environmental science, and industrial biotechnology, owing to their capacity to produce a wide array of bioactive compounds, as well as their involvement in biodegradation of xenobiotics and plant growth promotion. Among the most extensively studied representatives are *P. aeruginosa* and *P. putida*, which serve as contrasting model strains in pathogenicity research and environmental biotechnology, respectively. Improper storage conditions may compromise key physiological traits of these strains, including pigment production and biosurfactant-synthesizing capacity, thereby limiting their utility in downstream applications. The long-term preservation of *Pseudomonas* strains in biobanks is therefore essential to safeguard biological material of scientific, biotechnological, and biomedical value. In this chapter, we describe a standardized glycerol-based cryopreservation protocol at −80°C for the long-term preservation of *Pseudomonas* strains, along with the evaluation of their functional stability following reactivation. The method relies on glycerol as a penetrating cryoprotectant, which minimizes cellular damage associated with ice crystal formation and osmotic stress while preserving cell viability, genetic integrity, and metabolic functionality. Post-reactivation validation includes the assessment of fluorescence emission and the capacity to produce biosurfactants. This protocol provides a reliable and reproducible tool for the preservation of biotechnologically relevant microbial genetic resources.

Keywords Glycerol, *Pseudomonas aeruginosa*, *Pseudomonas putida*, Biosurfactants

1 Introduction

The genus *Pseudomonas* (family *Pseudomonadaceae*) constitutes one of the most diverse and metabolically versatile groups among Gram-negative bacteria [1]. Currently, more than 300 validly described species are recognized; however, comparative genomics studies suggest that its actual diversity is considerably greater [2]. The remarkable adaptive

Marco A. Ramírez-Mosqueda et al. (eds.), *Methods for The Conservation of Microorganisms*, Methods in Molecular Biology, vol. 3090, https://doi.org/10.1007/978-1-0716-5623-5_6,

capacity of *Pseudomonas* is associated with high genetic plasticity, the presence of mobile genetic elements, horizontal gene transfer, and a highly flexible metabolism that enables the colonization of a wide variety of environments [3]. *Pseudomonas* spp. exhibit a cosmopolitan distribution and have been isolated from virtually all ecosystems studied to date, including agricultural soils, the rhizosphere, and plant tissues; marine, freshwater, and sediment environments; polar and volcanic regions; animal microbiomes; wastewater, mines, as well as hospital settings [4, 5]. Furthermore, numerous species possess significant biotechnological relevance due to their ability to produce phenazines, biosurfactants, enzymes, polyhydroxyalkanoates, antimicrobial compounds, and specialized catabolic pathways involved in xenobiotic degradation, which confer an important role in industrial biotechnology and bioremediation [6, 7].

Among the most extensively studied groups are the *Pseudomonas aeruginosa* and *Pseudomonas putida* complexes, which represent contrasting models of ecological adaptation, pathogenicity, and biotechnological exploitation. *Pseudomonas aeruginosa* is recognized for its clinical relevance as a member of the ESKAPE pathogens and for exhibiting multidrug resistance (MDR), in addition to its ability to persist on hospital surfaces, aquatic systems, and medical devices [8]. In contrast, *Pseudomonas putida* is distinguished by its high metabolic versatility and its association with soil and plant-related environments, being widely used in bioremediation processes for contaminant degradation and in agriculture as a plant growth-promoting bacterium [9]. Due to their broad genetic and functional diversity, *Pseudomonas* strains constitute strategic microbial genetic resources whose preservation in biobanks is essential to ensure the future availability of biological material with scientific, environmental, agricultural, industrial, and biomedical value.

Cryopreservation represents one of the most effective methods for maintaining the viability, genetic stability, and physiological functionality of bacterial strains over prolonged periods [10]. Among the cryoprotectants used in microbiology, glycerol has become the most widely employed compound due to its high efficiency, low cost, relatively low toxicity, and compatibility with a broad diversity of Gram-positive and Gram-negative bacteria [11]. Its main advantage lies in its high cryoprotective capacity against freezing-induced damage, allowing the maintenance of high levels of cellular viability after long-term storage. In addition, glycerol exhibits lower toxicity compared to other permeable cryoprotectants, such as dimethyl sulfoxide (DMSO), ethylene glycol (EG), and propylene glycol (PG) [12]. Furthermore, glycerol is compatible with automated biobank systems and the large-scale storage of microbial collections for years and even decades at −80°C [13].

The present chapter describes a dual cryopreservation protocol for pathogenic and environmental *Pseudomonas* strains using glycerol as a cryoprotective agent. For this purpose, *Pseudomonas aeruginosa* and *Pseudomonas putida* will be employed as model strains.

2 Material

2.1 Bacterial Strains

1. The bacterial strains used in this protocol correspond to *Pseudomonas aeruginosa* PAO1 and *Pseudomonas putida* KT2440, two model strains widely employed in microbiology, bacterial physiology, and biotechnological research. *P. aeruginosa* PAO1 represents a clinically relevant reference strain due to its capacity for biofilm formation, pigment production, and biosurfactant synthesis [14]. *P. putida* KT2440 is recognized for its high metabolic versatility and its potential in biodegradation processes and environmental biotechnology [15].

2.2 Culture Media

1. King B Agar (*Pseudomonas* Agar F). This medium was used for the isolation, maintenance, and phenotypic evaluation of *Pseudomonas* strains, particularly to promote the production of fluorescent pigments characteristic of the genus, such as fluorescein and pyoverdine [16]. The formulation includes proteose peptone (20 g/L), K_2HPO_4 (1.5 g/L), $MgSO_4 \cdot 7H_2O$ (1.5 g/L), glycerol (10 mL/L), and bacteriological agar (15 g/L).
2. PPGAS Broth (*Pseudomonas* Phosphate Glucose Ammonium Sulfate). PPGAS broth was used for liquid growth prior to cryopreservation and for physiological validation following strain reactivation, as it provides suitable conditions for metabolic development and the production of pigments and biosurfactants in *Pseudomonas* species [17]. The medium composition consists of glucose (10 g/L), NH_4Cl (1.1 g/L), KCl (0.7 g/L), K_2HPO_4 (1.4 g/L), KH_2PO_4 (1.4 g/L), and $MgSO_4 \cdot 7H_2O$ (0.2 g/L).

2.3 Solutions

1. Cryoprotective solution: 40% (v/v) glycerol prepared in distilled water and sterilized by autoclaving or 0.22 μm sterile filtration.
2. pH adjustment solutions: 1 N NaOH and 1 N HCl used for calibration of culture medium pH.
3. Sterile magnesium sulfate solution: 10 mM $MgSO_4$ used for cell handling, preparation of bacterial suspensions, and maintenance of osmotic stability during strain processing.

4. Gram staining reagents: crystal violet (1%, w/v), Gram's iodine (Lugol's solution), 95% ethanol-based decolorizer, and safranin (0.5%–1%, w/v).
5. Sudan Black B-stained mineral oil: prepare a 0.1% (w/v) solution of Sudan Black B in mineral oil. Mix thoroughly until a homogeneous dark coloration is obtained. Filter through Whatman No. 1 filter paper to remove undissolved particles.

2.4 Infrastructure and Equipment

1. Microbiological work area: horizontal laminar flow hood, Bunsen/Fisher burner, UV–Vis spectrophotometer, UV transilluminator, and orbital incubator with temperature control.
2. Media preparation and sterilization area: analytical balance, autoclave, calibrated pH meter, magnetic stirrer with heating plate, automatic micropipettes, sterile Petri dishes, and 0.22 μm sterile membrane filters.
3. Glassware: beakers, graduated cylinders, storage bottles, and Erlenmeyer flasks of various capacities.
4. Preservation system: 2 mL cryovials and −80°C ultra-low temperature freezer designated for long-term storage of cryopreserved bacterial strains.

3 Methods

3.1 Culture Media Preparation

King B Agar

1. Dissolve 20 g of proteose peptone, 1.5 g of K_2HPO_4, 1.5 g of $MgSO_4 \cdot 7H_2O$, and 15 g of bacteriological agar in 900 mL of distilled water.
2. Add 10 mL of glycerol and mix thoroughly until fully homogenized (*see* **Note 1**).
3. Adjust the pH of the medium to 7.2 using 1 N NaOH or 1 N HCl.
4. Bring the final volume to 1 L with distilled water.
5. Sterilize by autoclaving at 121°C for 15 min.
6. Allow the medium to cool to approximately 50°C, then dispense approximately 25 mL per sterile Petri dish under aseptic conditions.

PPGAS Broth

1. Prepare the basal solution by dissolving 10 g of glucose, 1.1 g of NH_4Cl, and 0.7 g of KCl in approximately 700 mL of distilled water.
2. Prepare a separate phosphate solution containing 1.4 g of K_2HPO_4 and 1.4 g of KH_2PO_4 in 200 mL of distilled water.

3. Prepare an independent $MgSO_4 \cdot 7\,H_2O$ solution by dissolving 0.2 g in 100 mL of distilled water; sterilize using a 0.22 μm membrane filter.
4. Adjust the pH of the basal solution to 7.0 using 1 N NaOH or 1 N HCl.
5. Sterilize the basal solution and the phosphate solution by autoclaving at 121 °C for 15 min.
6. Allow the autoclaved solutions to reach room temperature.
7. Combine the basal solution, phosphate solution, and sulfate solution to obtain a final volume of 1 L (*see* **Note 2**).
8. Store the prepared medium at 45 °C until use.

Bacterial Growth Conditions

1. Under aseptic conditions, inoculate a single isolated colony of *P. aeruginosa* PAO1 or *P. putida* KT2440 onto King B agar plates using a continuous cross-streak technique with a sterile inoculation loop.
2. Incubate the plates in an inverted position at 30 °C for 48 h (*see* **Note 3**).
3. Visually confirm characteristic strain growth, including colonial morphology and pigment production (Fig. 1).
4. Select isolated colonies and verify purity by Gram staining prior to proceeding with the cryopreservation protocol (*see* **Note 4**).

Cryopreservation of Pseudomonas Strains Using Glycerol

1. Inoculate a fresh colony of each strain into 50 mL of PPGAS liquid medium contained in an Erlenmeyer flask.

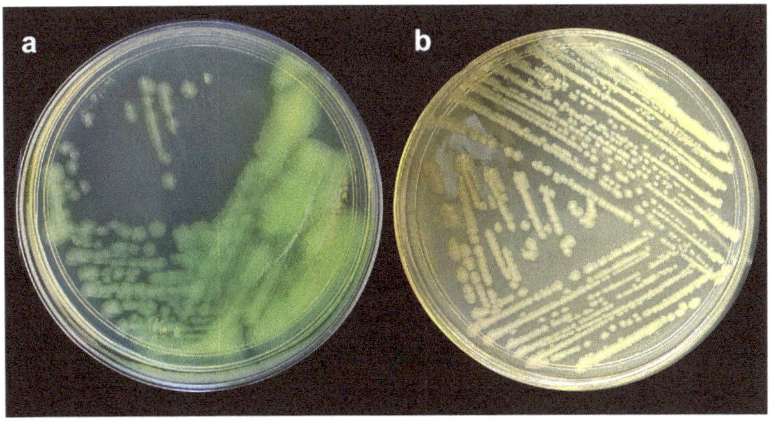

Fig. 1 Colonial morphology of *Pseudomonas* spp. strains on King B agar. (**A**) *P. aeruginosa* PAO1 produces colonies with irregular to slightly undulate margins, flat to slightly convex elevation, and a diffusible blue-green pigmentation due to pyocyanin production, which impregnates the surrounding medium. (**B**) *P. putida* KT2440 produces colonies with circular to slightly irregular morphology, entire margins, convex elevation, and a smooth, opaque surface

2. Incubate cultures at 30°C with constant orbital shaking at 150 rpm for 18–24 h.
3. Monitor bacterial growth by spectrophotometric readings at OD_{600}. Proceed with cryopreservation when cultures reach OD_{600} values of approximately 0.8 to 1.0.
4. Thoroughly mix the bacterial culture with an equal volume of sterile 40% glycerol solution to obtain a final glycerol concentration of 20% (v/v).
5. Dispense the suspension into sterile 2 mL cryovials under aseptic conditions.
6. Immediately transfer the cryovials to a −80°C ultra-low temperature freezer for long-term storage.

3.2 Reactivation of Cryopreserved Strains

1. Remove cryovials from the ultra-low temperature freezer and keep them on ice during handling.
2. Under laminar flow, remove a small portion of frozen biomass using a sterile inoculation loop.
3. Immediately streak the material onto fresh King B agar plates.
4. Promptly return the cryovial to the ultra-low temperature freezer to avoid repeated freeze–thaw cycles of the primary stock.
5. Incubate the inoculated plates at 30°C for 48 h.
6. Expose the cultures to UV light at 365 nm to verify that the reactivated strains retain their characteristic fluorescence emission (*see* **Note 5**, Fig. 2).

3.3 Post-Reactivation Validation of Biosurfactant Production

1. Select a single isolated colony of *P. aeruginosa* PAO1 or *P. putida* KT2440 obtained after reactivation and inoculate it into 50 mL of PPGAS liquid medium in an Erlenmeyer flask.
2. Incubate cultures at 30°C with constant orbital shaking at 150 rpm for 48–72 h.
3. Remove the flask from the orbital shaker and observe the formation of stable foam at the air–liquid interface. Foam persistence for more than 30 min after agitation is considered indicative of biosurfactant activity (*see* **Note 6**).
4. To evaluate surface tension disruption, perform the drop collapse assay: place a 10 µL droplet of culture supernatant onto a hydrophobic surface (e.g., the inner surface of a Parafilm strip). A collapsed or flattened droplet indicates active surface tension reduction, whereas a stable, spherical droplet indicates a negative test result.
5. To evaluate oil dispersal activity, add 40 mL of distilled water to a Petri dish and allow the surface to stabilize. Carefully place 20 µL of Sudan Black B-stained mineral oil onto the water surface using a micropipette, forming a uniform dark-colored

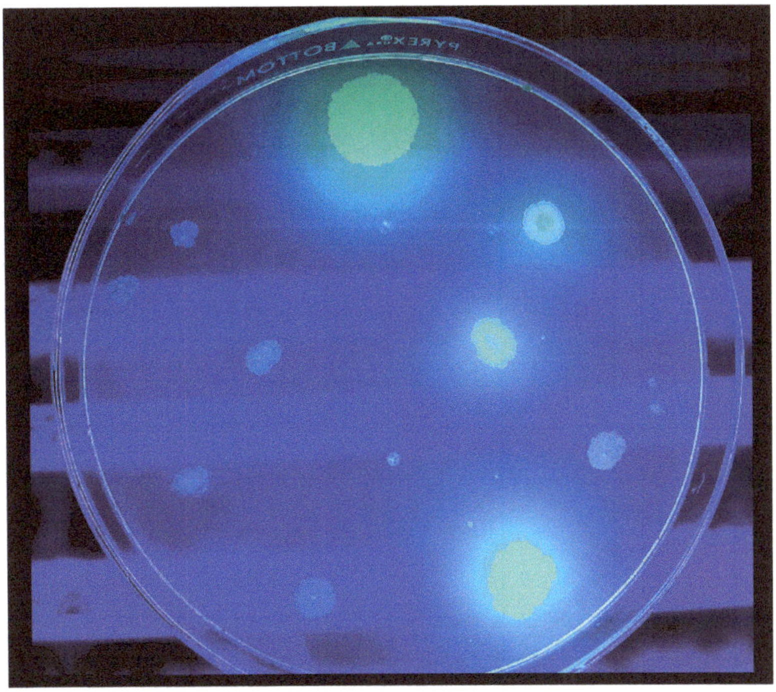

Fig. 2 Post-reactivation screening of *Pseudomonas* spp. colonies grown on King B agar under UV light at 365 nm. Fluorescent colonies (blue-green, indicated as positive) retain pyoverdine-producing capacity following cryopreservation, whereas non-fluorescent colonies (opaque, indicated as negative) indicate loss of this physiological trait. Fluorescence serves as a rapid and reliable indicator of post-reactivation functional stability

oil film. Then, gently add 10 μL of culture supernatant to the center of the oil film. The diameter of the clear zone formed by oil displacement shall be measured in millimeters after 30 s using a ruler.

6. To assess emulsification capacity, combine 2 mL of culture supernatant with 2 mL of mineral oil in a sterile glass tube. Vortex the mixture at maximum speed for 2 minutes, then allow it to stand undisturbed for 24 h at room temperature. Calculate the emulsification index (E_{24}) as follows:

$$E_{24}(\%) = [\text{height of emulsified layer}/\text{total height of liquid.phase}] \times 100$$

An E_{24} value ≥40% is considered indicative of effective emulsification activity.

7. Record all observations, including foam height (mm), drop collapse result (positive/negative), oil dispersal activity (mm), and E_{24} values (%) (Fig. 3), and compare results between reactivated strains and fresh reference cultures to confirm that

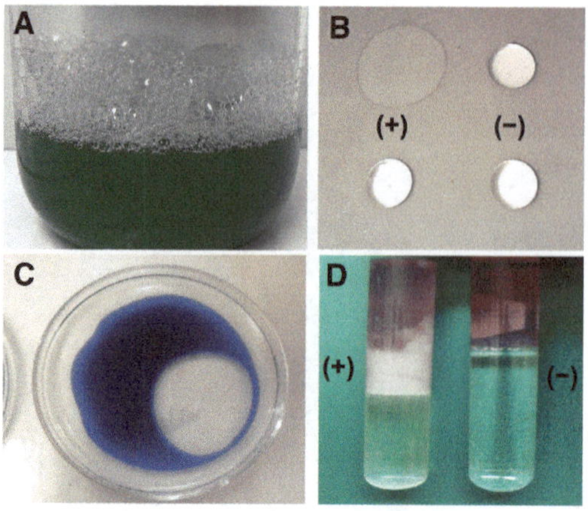

Fig. 3 Assays for the detection of biosurfactant production by *Pseudomonas* spp. following post-reactivation. (**A**) stable foam formation, (**B**) drop collapse assay, (**C**) oil dispersal activity, and (**D**) emulsification index (E_{24}). (+) positive result; (–) negative result

cryopreservation did not compromise biosurfactant-producing capacity.

4 Notes

1. Glycerol functions as a highly assimilable alternative carbon source for *Pseudomonas* species. Its metabolism activates central carbon pathways associated with both gluconeogenic and glycolytic processes, thereby promoting the availability of metabolic precursors and energy required for secondary metabolite biosynthesis.
2. The basal solution, phosphate solution, and sulfate solution for PPGAS broth must be prepared and sterilized separately to prevent salt precipitation during autoclaving. Precipitate formation reduces nutrient bioavailability and may adversely affect bacterial growth as well as the production of characteristic *Pseudomonas* pigments. It is recommended to combine the solutions only after they have reached room temperature and under strict sterile conditions.
3. *Pseudomonas aeruginosa* strains may be incubated at 37°C to promote pyoverdine production and other virulence-associated metabolites. Additionally, a characteristic grape-like odor may be perceived during growth, associated with the release of 2-aminoacetophenone.

4. *Pseudomonas* cells stained by the Gram technique appear as thin, Gram-negative bacilli that are straight or slightly curved, generally found as isolated cells or arranged in pairs, exhibiting characteristic pink staining.
5. Exposure of King B agar plates to UV light at 365 nm is a useful approach to verify that reactivated strains retain their capacity to produce fluorescent compounds, particularly siderophores such as pyoverdine.
6. *Pseudomonas* strains produce a biosurfactant known as rhamnolipid (primarily mono- or di-rhamnolipids), a surface-active agent capable of reducing surface tension. These compounds are classified as anionic glycolipid-type surfactants and possess multiple functional properties, including their use as wetting, emulsifying, and foaming agents.

References

1. Hesse C et al. (2018) Genome-based evolutionary history of Pseudomonas spp. Environ Microbiol 20:2142–2159. https://doi.org/10.1111/1462-2920.14130
2. Girard L et al. (2021) The ever-expanding Pseudomonas genus: description of 43 new species and partition of the Pseudomonas putida group. Microorganisms 9:1766. https://doi.org/10.3390/microorganisms9081766
3. Gross H, Loper JE (2009) Genomics of secondary metabolite production by Pseudomonas spp. Nat Prod Rep 26:1408–1446. https://doi.org/10.1039/b817075b
4. Alattas H, Glick BR, Murphy DV, Scott C (2024) Harnessing Pseudomonas spp. for sustainable plant crop protection. Front Microbiol 15:1485197. https://doi.org/10.3389/fmicb.2024.1485197
5. Rossi C et al. (2018) Biofilm formation, pigment production and motility in Pseudomonas spp. isolated from the dairy industry. Food Control 86:241–248. https://doi.org/10.1016/j.foodcont.2017.11.018
6. Orellana-Saez M et al. (2019) In-depth genomic and phenotypic characterization of the Antarctic psychrotolerant strain Pseudomonas sp. MPC6 reveals unique metabolic features, plasticity, and biotechnological potential. Front Microbiol 10:1154. https://doi.org/10.3389/fmicb.2019.01154
7. Chauhan M, Kimothi A, Sharma A, Pandey A (2023) Cold adapted Pseudomonas: ecology to biotechnology. Front Microbiol 14:1218708. https://doi.org/10.3389/fmicb.2023.1218708
8. Kamurai B, Mombeshora M, Mukanganyama S (2020) Repurposing of drugs for antibacterial activities on selected ESKAPE bacteria Staphylococcus aureus and Pseudomonas aeruginosa. Int J Microbiol 2020:8885338. https://doi.org/10.1155/2020/8885338
9. Weimer A, Kohlstedt M, Volke DC, Nikel PI, Wittmann C (2020) Industrial biotechnology of Pseudomonas putida: advances and prospects. Appl Microbiol Biotechnol 104:7745–7766. https://doi.org/10.1007/s00253-020-10811-9
10. Gomez-Godinez LJ et al. (2026) Preservation of microorganisms and microbiomes: methods, impacts, and future prospects. Biopreserv Biobank https://doi.org/10.1177/19475535251401795
11. Bircher L, Geirnaert A, Hammes F, Lacroix C, Schwab C (2018) Effect of cryopreservation and lyophilization on viability and growth of strict anaerobic human gut microbes. Microb Biotechnol 11:721–733. https://doi.org/10.1111/1751-7915.13265
12. Jungare KA, Radha R, Sreekanth D (2022) Cryopreservation of biological samples – a short review. Mater Today Proc 51:1637–1641. https://doi.org/10.1016/j.matpr.2021.11.203
13. De Paoli P (2005) Bio-banking in microbiology: from sample collection to epidemiology, diagnosis and research. FEMS Microbiol Rev 29:897–910. https://doi.org/10.1016/j.femsre.2005.01.005
14. Grace A, Sahu R, Owen DR, Dennis VA (2022) *Pseudomonas aeruginosa* reference strains PAO1 and PA14: a genomic, phenotypic, and therapeutic review. Front Microbiol 13:1023523
15. Dos Santos VM, Heim S, Moore ERB, Strätz M, Timmis KN (2004) Insights into the genomic basis of niche specificity of Pseudomonas putida KT2440. Environ Microbiol 6(12):1264–1286

16. King EO, Ward MK, Raney DE (1954) Two simple media for the demonstration of pyocyanin and fluorescin. J Lab Clin Med 44(2):301–307

17. Wild M, Caro AD, Hernández AL, Miller RM, Soberón-Chávez G (1997) Selection and partial characterization of a Pseudomonas aeruginosa mono-rhamnolipid deficient mutant. FEMS Microbiol Lett 153(2):279–285

Check for updates

Chapter 7

Lyophilization Protocol for Long-Term Preservation of Ureolytic *Sporosarcina* Strains with Biotechnological Applications

Gustavo Cuaxinque-Flores, José Luis Aguirre-Noyola, Esperanza Martínez-Romero and Oscar Talavera-Mendoza

Abstract

The long-term preservation of biotechnologically relevant bacterial genetic resources is essential, particularly for *Sporosarcina* spp. strains capable of microbially induced carbonate precipitation (MICP) for toxic metal immobilization. Standard glycerol-based cryopreservation requires an ultra-low-temperature cold chain (−80°C), thereby increasing operational costs and limiting field-scale inoculum transport. To address this bottleneck, lyophilization offers a robust, cost-effective alternative compatible with standard freezing temperatures (−4°C to −20°C). Here, we standardize and validate an optimized lyophilization protocol for *Sporosarcina* spp. using 20% skim milk as a cryoprotectant. Following a 2-year storage period at −20°C, the lyophilized cells demonstrated high viability and fully retained the urease enzymatic activity required for MICP. These findings demonstrate that this protocol preserves the phenotypic stability and bioremediation potential of *Sporosarcina* strains without the need for specialized ultra-cold equipment. Consequently, this method provides a highly scalable and viable framework for the formulation of bacterial inocula tailored for in situ field applications.

Keywords *Sporosarcina*, Lyophilization, Cryopreservation, Microbially induced calcium carbonate precipitation

1 Introduction

The genus *Sporosarcina* was first described by Kluyver and Van Niel [1]. This group of bacteria belongs to the phylum Bacillota (formerly Firmutes), class Bacilli, order Caryophanales, and family Planococcaceae. The taxonomy of the genus is dynamic, and as of January 2026, 22 validly published taxa have been described, following the consolidation of several heterotypic synonyms [2]. These ubiquitous bacteria inhabit diverse environments, ranging from soil, water, and host-associated microbiomes to extreme ecological niches such as sulfide-bearing tailings [3, 4].

Marco A. Ramírez-Mosqueda et al. (eds.), *Methods for The Conservation of Microorganisms*, Methods in Molecular Biology, vol. 3090, https://doi.org/10.1007/978-1-0716-5623-5_7,

Morphologically, the genus comprises Gram-positive, motile, endospore-forming bacteria that exhibit notable phenotypic plasticity, appearing either as elongated bacilli or spherical cells [3, 5, 6]. Their capacity to form endospores grants them high tolerance to abiotic stress [7]. Chemotaxonomically, they are distinguished by a menaquinone-7 (MK-7) respiratory profile and an A4α-type peptidoglycan (L-Lys-L-Ala-D-Asp), which serve as key phenotypic markers differentiating them from closely related taxa [5, 8].

The biotechnological interest in the genus *Sporosarcina* lies in its ureolytic activity, which is mediated by urease metalloenzyme (EC 3.5.1.5), a protein encoded by an operon comprising structural subunits (UreABC) and accessory proteins (UreDEFG/H) [9]. While *S. pasteurii* is the established model for microbially induced carbonate precipitation (MICP), substantial genomic and functional heterogeneity exists across the genus, with ureolytic capacity varying significantly at the strain level [10]. MICP relies on the enzymatic hydrolysis of urea, which drives environmental alkalization and shifts the inorganic carbon equilibrium toward carbonate (CO_3^{2-}) formation (Fig. 1). This process facilitates the precipitation of stable mineral phases, enabling both the immobilization of toxic metals and the structural restoration of materials via biocementation [11].

Recent bioprospecting of indigenous strains has highlighted *S. pasteurii* UB10 and, particularly, *S. luteola* UB5 for their exceptional ureolytic capabilities. Notably, strain UB5 features a unique genomic architecture containing two functional copies of the urease operon, conferring enhanced enzymatic activity. Consequently, UB5 has demonstrated superior efficiency in immobilizing metals within acid mine drainage (AMD) environments through the formation of thermodynamically stable carbonates [10]. Given the high biotechnological value of these functional traits, it is imperative to implement robust preservation strategies. Conventional deep-freezing methods are equipment-dependent and logistically constrained. Lyophilization emerges as a critical alternative to safeguard these genetic resources, ensuring long-term phenotypic stability and cellular viability without reliance on ultra-low -temperature cold chains.

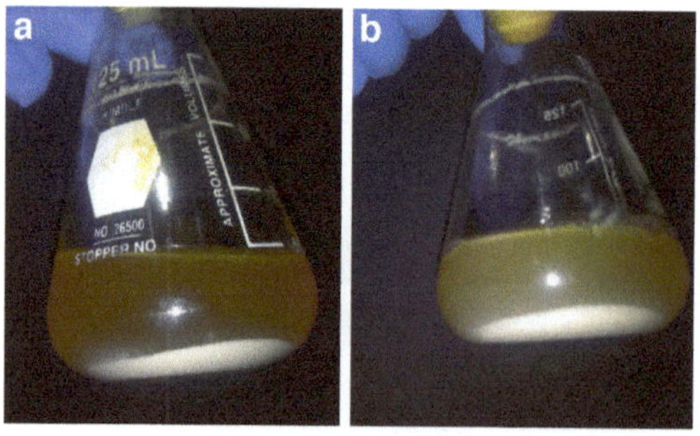

Fig. 1 Induced calcium carbonate precipitation by *Sporosarcina luteola* UB5 (**a**) and *Sporosarcina pasteurii* UB10 (**b**)

This work describes an optimized freeze-drying protocol for the preservation of ureolytic *Sporosarcina* strains. We evaluate the efficacy of this method in maintaining metabolic integrity and urease capacity, providing a scalable and cost-effective framework for the long-term management of microbial resources destined for biotechnological applications.

2 Materials

2.1 Bacterial Strains and Culture Media

1. The *Sporosarcina* strains used in this protocol were *Sporosarcina luteola* UB5 and *Sporosarcina pasteurii* UB10, isolated from sulfide-bearing tailings in the mining region of Taxco de Alarcón, Guerrero. These strains were analyzed and identified through *16S rRNA* gene and whole-genome sequencing [1, 2].
2. Nutrient broth + Urea (NBU) was used to culture the *Sporosarcina* strains, as it provides the ideal nutrients and the necessary nitrogen source to activate the enzymatic mechanism of urease. This medium contains (g/L) 10 of peptone, 4 of yeast extract, 10 of glucose, and 16 of agar, adjusted to a pH of 7.0.

2.2 Solutions

1. A urea solution (20%, v/v) was prepared in distilled water. The solution was sterilized by filtration using Syringe filters (0.22 μm).
2. Solutions for adjusting pH (0.1 N): hydrochloric acid (HCl) and sodium hydroxide (NaOH).
3. Prepare a cryoprotective solution using 20% skim milk. Weigh 20 g of skim milk for every 100 mL of distilled water, adjust to pH of 7.0, and sterilize in an autoclave for 5 min at 121°C and 15 PSI, maintaining this duration to prevent thermal denaturation.

2.3 Equipment

1. Incubation Room: Horizontal laminar flow cabinet; Bunsen burner or glass bead sterilizer; Shaking incubator.
2. Analytical Instrument Room: Benchtop pH-meter; Spectrophotometer; Benchtop lyophilizer
3. Media preparation room: Standard Analytical Balance; Spatula, Autoclave; Hot plate stirrer; Syringe filter 0.22 μm; micropipettes 100–1000 μL; Petri dishes (100 mm × 15 mm); Cryogenic tubes (2 mL).
4. Glassware: Beakers (100–1000 mL); Borosilicate Glass Reagent Bottles with Polypropylene Cap (250–1000 mL); Graduated

cylinders (100–1000 mL); Erlenmeyer flasks (125–500 mL); Glass tubes for freeze-drying (9 × 200 mm), Pasteur pipette.

5. Others: Sealing Film; Magnetic Stir Bar; Marker Pen

3 Methods

3.1 Preparation of Culture Medium

1. Prepare NBU media by dissolving peptone (10 g), yeast extract (4 g), glucose (10 g) in 800 mL of distilled water. Add agar (16 g) for solid media.
2. Adjust the pH to 7.0 using 1 N NaOH or 1 N HCl.
3. Adjust the final volume to 1000 mL
4. Sterilize by autoclaving 15–25 min at 121°C and 15 PSI.
5. After cooling the medium to approximately 40°C, aseptically add 100 mL of sterile urea solution, to obtain a final concentration of 2% urea and homogenize gently (*see* **Note 1**).
6. For solid media, dispense approximately 25 mL of NBU medium into sterile Petri dishes under a laminar flow cabinet and allow the medium to solidify.

3.2 Strain Recovery and Pure Culture Preparation

1. Under aseptic conditions in a laminar flow cabinet, streak the *Sporosarcina* UB5 and UB10 strains onto NBU agar plates using a sterile inoculation loop and the cross-streak method
2. Incubate the plates at 30°C for 48 h.
3. After incubation, examine the isolated colonies: UB5 colonies are small and cream-yellow-colored, while UB10 colonies are larger, convex, and white in appearance (Fig. 2).

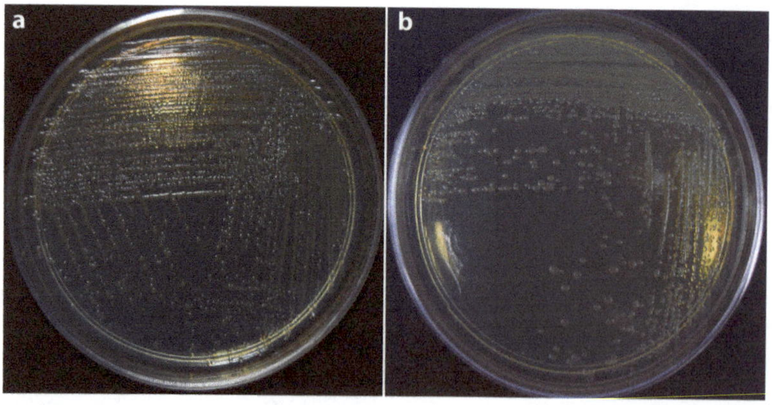

Fig. 2 Colony morphology of *S. luteola* UB5 (**a**) and S. *pasteurii* UB10 (**b**) strains recovered from lyophilized samples

4. Select a well-isolated colony and inoculate it into liquid NBU medium for the freeze-drying process (cryopreservation method).

3.3 Preparation of Sporosarcina Biomass for Lyophilization

1. Before beginning the freeze-drying process, sterilize the freeze-drying tubes (9 × 200 mm) with lids and label them properly.
2. Under axenic conditions, inoculate a fresh single colony into 150 mL Erlenmeyer flasks containing 50 mL of NBU broth
3. Incubate at 30°C with shaking at 250 rpm for 24 h.
4. Monitor bacterial growth by measuring the optical density at 600 nm (OD_{600}) using a spectrophotometer. Proceed when OD_{600} reaches approximately 0.8–1.2 corresponding to late exponential or early stationary phase.
5. Transfer 10 mL of the fresh culture to 15-mL conical tubes and centrifuge at 4000–8000 × *g* for 10 min at 4°C.
6. Carefully decant the supernatant and retain the cell pellet
7. Resuspend the cell pellet in 3 mL of fresh NBU broth
8. Mix in a 1:1 ratio (fresh culture: cryoprotective solution); in this case, 3 mL of the culture from each strain and 3 mL of the cryoprotective solution were used. Gently homogenize, taking care to avoid foaming or mechanical damage (*see* **Note 2**).
9. Dispense 3 mL of the mixture into each pre-sterilized freeze-drying tube using a sterile Pasteur pipette, avoiding contact with the tube walls to maintain sterility.
10. Close the tubes with loose caps or sterile cotton plugs to allow vapor escape during lyophilization. Do not seal hermetically at this stage.
11. Pre-freeze the samples at −80°C for at least 12 h (*see* **Note 3**), ensuring complete solidification before lyophilization.

3.4 Freeze-Drying Protocol

1. Transfer the frozen tubes quickly to the benchtop lyophilizer to avoid thawing. Connect the tubes one by one, allowing the system vacuum to re-stabilize between each connection.
2. Primary drying (sublimation phase): Maintain shelf temperature between −40°C and −20°C and pressure below 0.1–0.2 mbar for 12–24 h, or until visible ice is completely removed.
3. Secondary drying (desorption phase): Gradually increase the temperature to 20°C–25°C under continued vacuum and maintain for an additional 6–12 h to remove residual bound water.

3.5 Sealing and Storage

1. Once the samples are completely dry, seal the vials immediately under vacuum or an inert atmosphere to prevent moisture uptake and ensure long-term stability.
2. Store the sealed lyophilized samples at 4°C for short- to medium-term preservation or at −20°C for long-term storage (Fig. 3).

3.6 Recovery of *Sporosarcina* Strains after Lyophilization

1. Retrieve the vial stored at 4°C or −20°C and allow it to equilibrate to room temperature before opening, to prevent condensation inside the vial. Transfer it to a laminar flow cabinet.
2. Open the vial aseptically. Add 0.5–1 mL of sterile NBU broth directly to the vial to rehydrate the lyophilized material. Allow it to stand for 5–10 min at room temperature.
3. Gently resuspend the cells by slow pipetting or mild swirling to obtain a homogeneous suspension.

Fig. 3 Lyophilized samples of *S. luteola* UB5 (**a**, **b**) and S. *pasteurii* UB10 (**c**, **d**), stored at −20°C for over 2 years

4. Streak an aliquot of the rehydrated suspension onto NBU agar plates and incubate at 30°C for 48 h to restore metabolic activity and obtain isolated colonies.
5. Perform serial dilutions of the remaining suspension, plate onto NBU agar, and determine the colony-forming unit (CFU) count. Compare this value with that of the pre-lyophilization culture to estimate survival rate.

3.7 Evaluation of Ureolytic Activity After Lyophilization

1. Prepare plates with NBU medium supplemented with phenol red (0.012 g/L) as a pH indicator.
2. Inoculate the medium by cross-streaking with an isolated colony obtained from the recovery of *Sporosarcina* strains after lyophilization.
3. Incubate the cultures at 30°C for 24 to 48 h.
4. Monitor ureolytic activity by observing the colorimetric shift of the medium to an intense pink color, which indicates environmental alkalinization driven by active urea hydrolysis (Fig. 4).
5. The ability of the strain to precipitate calcium carbonate ($CaCO_3$) can be confirmed (*see* **Note 4**).

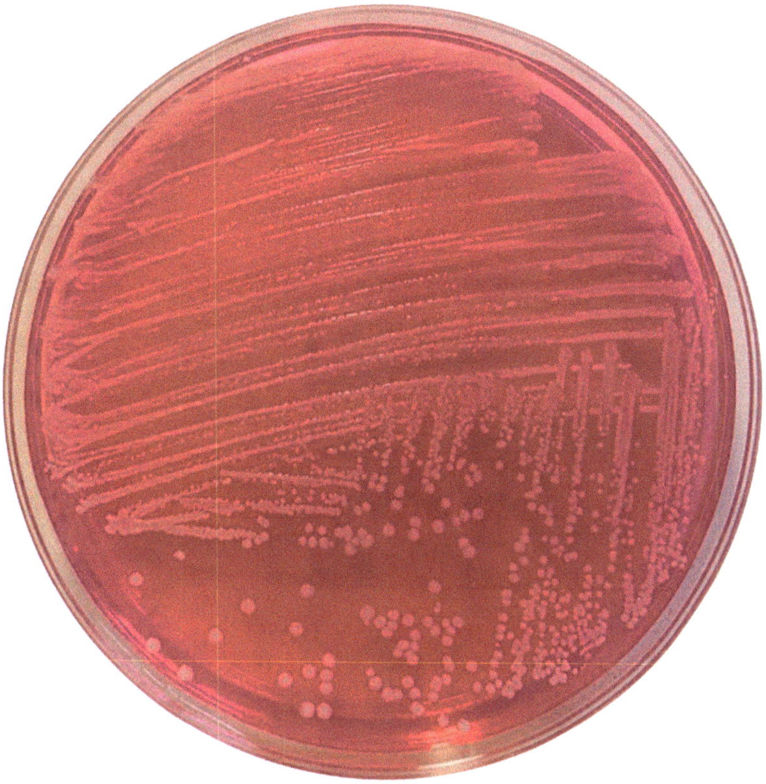

Fig. 4 Representative ureolytic activity of *Sporosarcina* strains after recovery from lyophilization, assessed in NBU medium supplemented with phenol red

4 Notes

1. Urea is thermolabile and must not be autoclaved.
2. A disaccharide (sucrose or trehalose, 5%–10% w/v) can be added to enhance cellular protection during the drying process.
3. Alternatively, rapid freezing can be performed using an ethanol bath with dry ice.
4. Although this colorimetric assay confirms the restoration of ureolytic activity after lyophilization, applications such as biocementation may require precise identification of the precipitated calcium carbonate formed in liquid culture. In such cases, structural characterization by X-ray diffraction is recommended.

Acknowledgments

The authors thank Luis Israel Tapia Serrano for providing the photographs of the freeze-dried powder.

Conflict of Interest Statement The authors have no conflicts of interest to declare that are relevant to the content of this chapter.

References

1. Kluyver AJ, Van Niel CB (1936) Prospects for a natural system of classification of bacteria. Zentralblatt für Bakteriologie, Parasitenkunde, Infektionskrankheiten und Hygiene, Abteilung II 94:369–403
2. Freese HM, Meier-Kolthoff JP, Sardà Carbasse J, Afolayan AO, Göker M (2026) TYGS and LPSN in 2025: a Global Core Biodata Resource for genome-based classification and nomenclature of prokaryotes within DSMZ Digital Diversity. Nucleic Acids Res 54(D1):D884–D891. https://doi.org/10.1093/nar/gkaf1110
3. Oliver A, Kay M, Cooper KK (2018) Comparative genomics of cocci-shaped Sporosarcina strains with diverse spatial isolation. BMC Genomics 19(1):310. https://doi.org/10.1186/s12864-018-4635-8
4. Cuaxinque-Flores G, Aguirre-Noyola JL, Hernández-Flores G, Martínez-Romero E, Romero-Ramírez Y, Talavera-Mendoza O (2020) Bioimmobilization of toxic metals by precipitation of carbonates using Sporosarcina luteola: an *in vitro* study and application to sulfide-bearing tailings. Sci Total Environ 724:138124. https://doi.org/10.1016/j.scitotenv.2020.138124
5. Yoon JH, Lee KC, Weiss N, Kho YH, Kang KH, Park YH (2001) Sporosarcina aquimarina sp. nov., a bacterium isolated from seawater in Korea, and transfer of Bacillus globisporus (Larkin and Stokes 1967), Bacillus psychrophilus (Nakamura 1984) and Bacillus pasteurii (Chester 1898) to the genus Sporosarcina as Sporosarcina globispora comb. nov., Sporosarcina psychrophila comb. nov. and Sporosarcina pasteurii comb. nov., and emended description of th. Int J Syst Evol Microbiol 51(3):1079–1086. https://doi.org/10.1099/00207713-51-3-1079
6. De Vos P, Garrity GM, Jones D, Krieg NR, Ludwig W, Rainey FA, Schleifer K-H, Whitman WB (2009) Bergey's manual of systematic bacteriology: volume three: the firmicutes. Springer. https://doi.org/10.1007/b92997
7. Kwon S-W, Kim B-Y, Song J, Weon H-Y, Schumann P, Tindall BJ, Stackebrandt E, Fritze D (2007) Sporosarcina koreensis sp. nov. and Sporosarcina soli sp. nov., isolated from soil in Korea. Int J Syst

Evol Microbiol 57(8):1694–1698. https://doi.org/10.1099/ijs.0.64352-0
8. An S-Y, Haga T, Kasai H, Goto K, Yokota A (2007) Sporosarcina saromensis sp. nov., an aerobic endospore-forming bacterium. Int J Syst Evol Microbiol 57(8):1868–1871. https://doi.org/10.1099/ijs.0.64962-0
9. Pei D, Liu Z, Hu B (2025) A novel urease gene structure of Sporosarcina pasteurii with double operons. Mol Genet Genomics 300(1):25. https://doi.org/10.1007/s00438-025-02236-8
10. Cuaxinque-Flores G, Talavera-Mendoza O, Aguirre-Noyola JL, Hernández-Flores G, Martínez-Miranda V, Rosas-Guerrero V, Martínez-Romero E (2024) Molecular and geochemical basis of microbially induced carbonate precipitation for treating acid mine drainage: the case of a novel Sporosarcina genomospecies from mine tailings. J Hazard Mater 476:135005. https://doi.org/10.1016/j.jhazmat.2024.135005
11. Murugan R, Suraishkumar GK, Mukherjee A, Dhami NK (2021) Influence of native ureolytic microbial community on biocementation potential of Sporosarcina pasteurii. Sci Rep 11(1):20856. https://doi.org/10.1038/s41598-021-00315-5

Chapter 8

Protocol for Cryopreservation of *Bacillus* Strains with Biotechnological Potential

Alejandro Bolaños-Dircio, Augusto Rojas-Aparicio and Yanet Romero-Ramírez

Abstract

Cryopreservation is an important method in microbiology that guarantees the genetic stability and long-term viability of microbial strains of biotechnological significance, including heat-resistant bacteria of the *Bacillus* genus. The most prevalent and efficient method for bacteria is the use of glycerol as a cryoprotective agent at low temperatures, typically −80°C. This protocol delineates a technique for preserving three thermoresistant strains (*B. licheniformis* M2-7, *B. subtilis* ANF2, and *B. tequilensis* M2) with great biotechnological potential using glycerol. This is a low-cost, dependable, and standard method for preserving the integrity of bacterial strains in research laboratories, thereby increasing their utility for future investigations.

Keywords Bioremediation, Biofertilization, Biocontrol, Biotechnological potential, Cryopreservation

1 Introduction

Microorganisms are important elements in research and biotechnology because they produce metabolites, fuels, and food and are tools for generating knowledge. Microorganisms are cultivated in various different culture media and are used for numerous research and industrial processes. However, they exhibit a high level of genetic variation and a high rate of mutation in reproduction across species; therefore, it is imperative to preserve microorganisms *in vitro* in *ex situ* collections. In this way, microbial collections in a laboratory provide biological material to analyze conservation strategies and ensure genetic resources in the face of important future scenarios, for example, in industry, agriculture, and bioremediation, among others [1].

There are numerous methods for the preservation of bacterial strains, such as periodic subculturing, freeze-drying, drying, and cryopreservation [2]. The main objective of these methods is to

Marco A. Ramírez-Mosqueda et al. (eds.), *Methods for The Conservation of Microorganisms*, Methods in Molecular Biology, vol. 3090, https://doi.org/10.1007/978-1-0716-5623-5_8,

maintain bacterial strains in a viable state without morphological, genetic, or physiological changes [3]. Glycerol is employed as a cryoprotectant due to its ability to disrupt the hydrogen bonds between water molecules, which prevents the formation of ice crystals during freezing and maintains viability for future reseeding, as it has a low biological mutation rate and a wide range of applications [4]. Among the advantages of cryopreservation over other preservation methods, it is that it can be applied to most microbial genera, reduces the risk of contamination, and can be easily performed in any laboratory [5].

The Laboratory of Molecular Microbiology and Environmental Biotechnology works with bacteria of the genus *Bacillus* has evaluated the *Bacillus licheniformis* M2-7 strain in the degradation of complex hydrocarbons such as gasoline, where it has been observed that it has the capacity to degrade up to 100 ppm of this product [6–8]. In the particular case of polycyclic aromatic hydrocarbons, such as phenanthrene, anthracene, pyrene, and benzo[a]pyrene, the strain has been reported to possess the capacity to biotransform them into less toxic metabolites [9]. It also promotes plant growth (biofertilization and phytostimulation), improving the growth, development, and yield of crops such as chili peppers (*Capsicum annuum* L.) [10], lettuce (*Lactuca sativa* var longifolia), epazote (*Dysphania ambrosioides* L.), celery (*Apium graveolens* var. Secalinum) [11], and radish (*Rhapanus sativus* L.) [12]. In addition, this strain exhibits antagonistic characteristics against a variety of phytopathogenic fungi, including *Colletotrichum* sp., *Aspergillus oryzae*, and *A. niger* [13]. Recent research has shown that *B. subtilis* ANF2 and *B. tequilensis* M2 possess biofertilization and phytostimulation properties in the growth of tomato (*Solanum lycopersicum* L.) [14]. Additionally, they have an antagonistic capacity against ochratoxigenic fungi (*Fusarium odoratisimum* and *Curvularia lunata*) by producing compounds that diffuse into the environment and volatile organic compounds (VOCs). The metabolic characteristics of these strains render them a valuable biological resource; therefore, it is essential to optimize conservation to guarantee the quality and viability of these species following extended storage.

2 Materials

2.1 Bacterial Strains

1. Three thermoresistant strains (*B. licheniformis* M2-7, *B. subtilis* ANF2, and *B. tequilensis* M2) with great biotechnological potential.

2.2 Culture Medium

Prepare culture media Luria-Bertani Medium (LB) and solutions using distilled water and analytical-grade reagents.

1. Add the media components to a flask containing approximately 80% of the total volume of distilled water, g/L: 10.0 casein peptone, 5.0 yeast extract, and 10.0 NaCl (*see* **Note 1**).
2. Mix with a magnetic stirrer until the reagents are completely dissolved.
3. Add the remaining water to reach the desired final volume.
4. Adjust the pH to 7.3 using NaOH or HCl solutions (*see* **Note 2**).
5. Sterilize at 121°C for 15 min.

2.3 Preparation of 40% Glycerol

1. Dilute 40 mL of absolute glycerol ($C_3H_5(OH_3)$) in 40 mL of distilled water in a volumetric flask.
2. Cover the volumetric flask, shake, and mix well until the liquids are evenly mixed.
3. Make up to 100 mL with distilled water, pour into an Erlenmeyer flask, and sterilize at 121°C for 15 min.

3 Methods

3.1 Reactivation of Bacterial Strains

1. Select well-isolated colonies from the LB plates and confirm the purity of each bacterial strain by Gram staining and evaluation of colony morphology (Fig. 1).
2. Using a sterile inoculating loop, collect a single colony from each purified strain. Transfer the inoculum into an Erlenmeyer flask containing liquid LB medium (*see* **Note 1**).
3. Incubate the cultures at 37°C and 120 rpm for 24 h to obtain exponentially growing cells for cryopreservation.

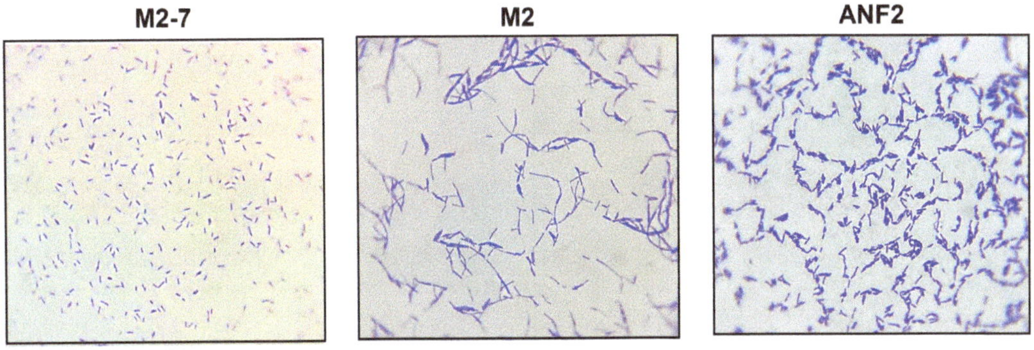

Fig. 1 Gram staining of *Bacillus licheniformis* M2-7, *B. subtilis* M2 and *B. tequilensis* ANF2 after 24 h of incubation at 37°C on LB plates

3.2 Cryopreservation of Strains and Storage

1. Dispense 500 μL aliquots of sterile 40% (v/v) glycerol into 1.5 mL microcentrifuge tubes. Prepare the tubes in triplicate or in a higher number according to the required quantity of cryopreserved stocks.
2. Add 500 μL of the bacterial culture grown in the Erlenmeyer flask to each tube containing the glycerol solution.
3. Mix thoroughly by pipetting up and down using a sterile 1 mL pipette tip until a homogeneous suspension is obtained.
4. Label each tube with the strain name and preparation date, and store at −80°C in the corresponding storage rack (Fig. 2) (*see* **Note 3**).

3.3 Feasibility Test

1. Remove the cryopreserved stocks from the −80°C freezer and allow them to thaw at room temperature.
2. Using a sterile inoculating loop, collect a small aliquot of each glycerol stock and streak onto solid LB medium using the streak plate technique (*see* **Note 2**).
3. Incubate the plates at 37°C for 24 h.
4. After incubation, verify that the colony morphology and purity of each recovered strain are consistent with the original culture characteristics (Fig. 3).

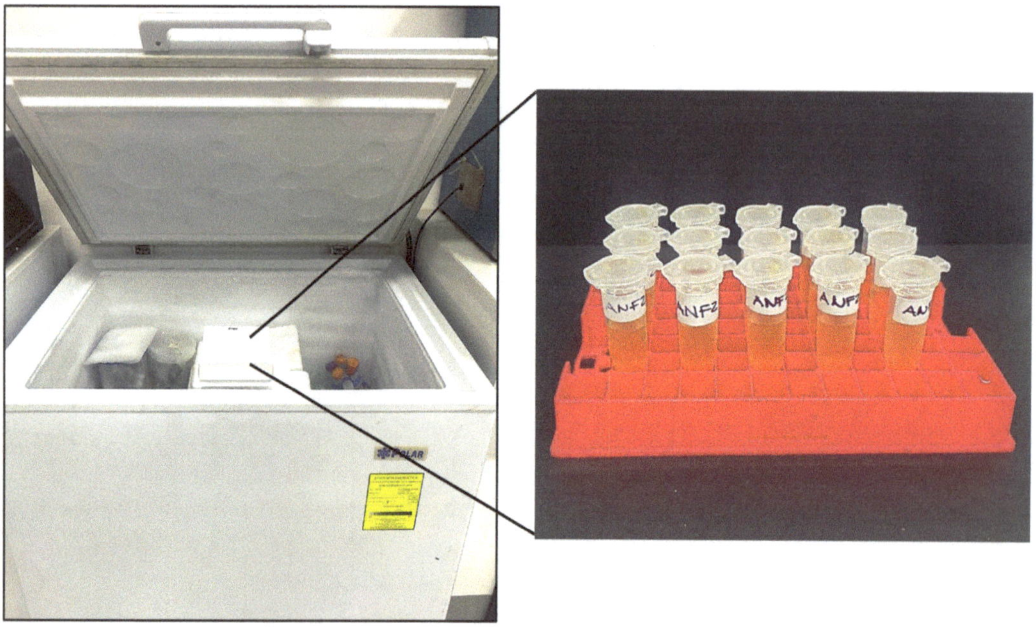

Fig. 2 Freezer storage of vials of each strain at −80°C

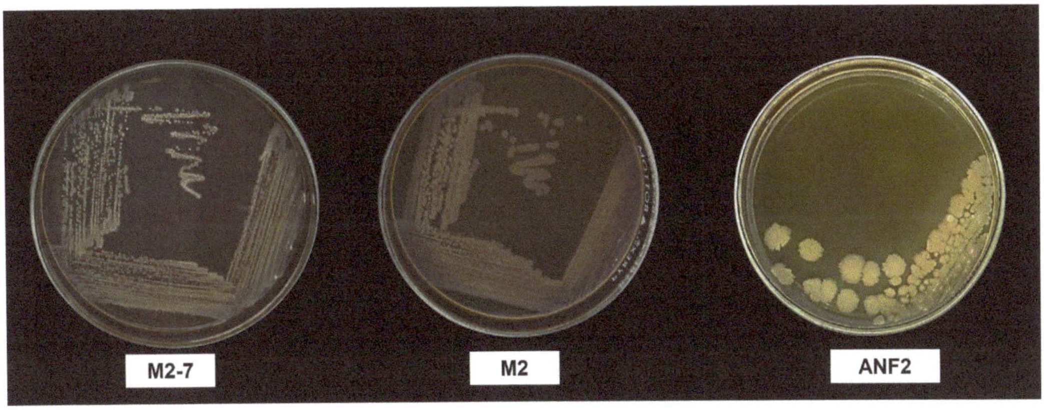

Fig. 3 Feasibility test of *Bacillus licheniformis* M2-7, *B. subtilis* M2, and *B. tequilensis* ANF2

4 Notes

1. For solid LB medium, add 17.0 g/L of bacteriological agar. The entire procedure must be performed under aseptic conditions to avoid contamination.
2. Adjusting the pH of the LB medium is important to ensure optimal growth of *Bacillus* strains, as it provides an adequate concentration of H^+ ions for proper enzymatic and metabolic activity.
3. The final ratio should be 1:1, consisting of 500 μL of bacterial culture and 500 μL of 40% glycerol solution.

Conflict of Interest

The authors have no conflicts of interest to declare that are relevant to the content of this chapter.

References

1. Stromberg PM, Dedeurwaerdere T, Pascual U (2013) The heterogeneity of public *ex situ* collections of microorganisms: empirical evidence about conservation practices, industry spillovers and public goods. Environ Sci Policy 33:19–27
2. Wei L, Meixia W, Baolin L (2017) The effect of cryopreservation on biospercimens and biomacromolecules. Chin J Biomed Eng 36:615–621
3. Montes de Oca N, Gonzáles RA, Riverón Y, Nuñez A, Villoch A, Rodríguez N (2008) Establecimiento y desarrollo de la colección de cultivos del CENSA. Revista de Salud. Animal 30:17–24
4. Wang Y, Jin Z, Wu L (2007) Preservation and management of microbial strains. Shangahai J Prev Med 19:87–88
5. Pegg DE (2015) Principles of cryopreservation. Methods Mol Biol (Clifton, N.J.) 1257:3–19
6. Rojas-Aparicio A, Hernández-Eligio JA, Toribio-Jiménez J, Rodríguez-Barrera M, Castellanos-Escamilla M, Romero-Ramírez Y (2018) Genetic expression of poba and fabHB

in Bacillus licheniformis M2-7 in the presence of benzo[a]pyrene. Genet Mol Res 17:4–6.

7. Ángel LIS, Segura D, Jiménez JT, Barrera MÁR, Pineda CO, Ramírez YR (2020) Carbon storage regulator A (*csrA*) gene regulates motility and growth of *Bacillus licheniformis* in the presence of hydrocarbons. Microbiol Biotechnol Lett 48:185–192
8. Morales-Blancas EJA, Ortuño PC, Toribio JJ, Rodríguez BMA, Toledo HE, Romero RY (2023) The *catE* gene of *Bacillus licheniformis* M2-7 isessential for growth in Benzopyrene and its expression is regulated by the Csr system. World J Microbiol Biotechnol 39:1–13
9. Guevara-Luna J, Serrano-Ángel LI, Rodrígez-Barrera MÁ, Hernández-Flores G, Toribio-Jiménez J, Toledo-Hernández E, Romero-Ramírez Y (2022) Isolated bacteria from hot springs able to use hydrocarbons as carbon source. Rev Int Contam Ambient 38:68–77
10. Bolaños-Dircio A, Toribio-Jiménez J, Rodríguez-Barrera MA, Hernández-Flores G, Toledo-Hernández E, Palemón-Alberto F, Romero-Ramírez Y (2021) *Bacillus licheniformis* M2-7 improves the growth and performance of *Capsicum annuum* L. Agrociencias 55:227–242
11. Castro-Don Juan AB, Romero-Ramírez Y, Rodríguez-Barrera MA, Bolaños-Dircio A, Toribio-Jiménez J (2024) Production of vegetables with biofertilizers based on native microorganisms: a case study in Tixtla, Guerrero, México. Rev Int Contam Ambient 40:13–25
12. Sánchez-Marcelo J, Ramírez-Cruz A, Toledo-Hernández E, Návez -González D, Romero-Ramírez Y, Bolaños-Dircio A (2023) Evaluación del efecto de *Bacillus licheniformis* M2-7 como biofertilizante en cultivo *de Raphanus sativus* L. (Brassicales: Brassicaceae). Acta Agrícola y Pecuaria 9:6
13. Albarrán-de la Luz L, Rodríguez-Barrera M, Hernández-Flores G, Lopezaraiza-Mikel M, Alemán-Figueroa L, Toribio-Jiménez J, Romero-Ramírez Y (2022) Antagonism of *Bacillus licheniformis* M2-7 against phytopathogen fungi of *Mangifera indica* L. Rev Int Contam Ambient 38:1–10
14. Flores-Higuera H, Bello-Gatica B, Basilio-Santana NA, Rios-Meza S, Bolaños-Dircio A, Romero-Ramírez Y (2025) Evaluación de bacterias del género Bacillus como control biológico de patógenos vegetales. Ambiente Biotecnol Camino Sustent 2:708–714.

Chapter 9

Isolation of Rhizospheric Bacteroidota Using Oligotrophic Supplemented Culture Media

Sergio Montoya-Montoya, Idolina Flores-Cortez and Eduardo Valencia-Cantero

Abstract

Isolation of Bacteroidota is crucial in microbial ecology laboratories because these organisms play important roles in the environment, including relevant interactions with plants and animals. Therefore, effective methods are needed for their successful isolation. Here, we describe modified Soil Extract (SE), Tryptic Soy (TS), and Reasoner's 2A (R2A) culture media that facilitate the isolation of Bacteroidota, primarily by addressing their auxotrophies and specific growth requirements. This approach enables both morphological and molecular identification of the isolates.

Keywords Cyanocobalamin, Agar, Gellan gum, Soil extract, Tryptic soy, Reasoner's 2A

1 Introduction

Phylum Bacteroidetes [1], recently renamed Bacteroidota [2], comprises Gram-negative, bacillary, nonendospore-forming bacteria [1]. Members of this phylum are found in a wide range of environments and perform important ecological functions, including maintaining intestinal health, antagonizing plant pathogens, and promoting plant growth [3]. The complexity of establishing optimal growth conditions makes certain Bacteroidota species particularly challenging to isolate and study. In this context, the predominance of fastidious bacteria and the need to induce a culturable state may contribute to this challenge [4].

Bacteroidota, as fastidious bacteria, refers to the difficulty of these organisms to grow in laboratory conditions due to diverse factors such as nondisponible nutrients because of complexes formed during processes like autoclave sterilization that induce Maillard reaction products (MRPs also known as Amadori products) formed between reducing sugars and amino acids during cooking forming inaccessible nutrients complexes [5, 6], the difference in growth velocity

Marco A. Ramírez-Mosqueda et al. (eds.), *Methods for The Conservation of Microorganisms*, Methods in Molecular Biology, vol. 3090, https://doi.org/10.1007/978-1-0716-5623-5_9,

between bacterial species is a constant additional problem to resolve for a successfully isolation, and additionally the auxotrophy in bacteria represents an important challenge to be solved through culture media supplementation, among the main Bacteroidota auxotrophies reported are biotin, thiamine (vitamin B1), and cyanocobalamin (vitamin B12) auxotrophy [8].

In this context, it is necessary to develop effective methods for successfully isolating fastidious Bacteroidota, enabling their study and advancing our understanding of their ecology and environmental interactions. Several straightforward methods for preparing culture media have been developed to facilitate the broad isolation of bacterial groups from environmental samples. These include using gellan gum as a gelling agent instead of agar [9] and separately autoclaving phosphate and agar [10]. Although both methods are simple, they have been shown to increase the success of isolating diverse bacteria from various environmental sources, such as soil, sediment, and freshwater. Moreover, numerous studies have demonstrated that these techniques are also effective for culturing recalcitrant bacteria in a range of environments [11–15]. Despite these advances, these cultivation methods have not yet been widely adopted for obtaining axenic cultures of rhizospheric bacteria from plants.

2 Materials

2.1 Culture Media

Prepare all culture media, the vitamin-amino acid solution, and the phosphate buffer using deionized water.

1. Soil Extract (SE): yeast extract 0.1 g/L and 40 mL/L of soil extract (*see* **Note 1**).
2. Tryptic Soy (TS): casein peptone 17 g/L, soy peptone 3 g/L, NaCl 5 g/L, K_2HPO_4 2.5 g/L, and dextrose 2.5 g/L (*see* **Notes 2 and 3**).
3. Reasoner's 2A (R2A): yeast extract 0.5 g/L, proteose peptone 0.5 g/L, casamino acids 0.5 g/L, glucose 0.5 g/L, starch 0.5 g/L, sodium pyruvate 0.3 g/L, K_2HPO_4 0.3 g/L, and $MgSO_4$ 0.024 g/L (*see* **Notes 2 and 3**).
4. Gelling agent: Use bacteriological agar or gellan gum (e.g., Phytagel) as gelling agents at concentrations of 15 and 25 g/L, respectively (*see* **Note 3**). Culture media SE, TS, and R2A, solidified with agar, are designated as SEA, TSA, and R2AA, respectively, and solidified with gellan gum are SEP, TSP, and R2AP.
5. Vitamin-amino acids solution: thiamine hydrochloride 10 mg/L, riboflavin 10 mg/L, 4-aminobenzoic acid 40 mg/L, D(+)-biotin 10 mg/L, calcium D(+)-pantothenate 100 mg/L,

nicotinic acid 50 mg/L, pyridoxine hydrochloride 100 mg/L, folic acid 30 mg/L, mg/L, cyanocobalamin 50 mg/L, tyrosine 1.159 g/L, tryptophan 0.25 g/L, and leucine 2.5 g/L.

6. Phosphate buffer: NaCl 7.597 g/L, NaH_2PO_4 0.411 g/L, and Na_2HPO_4 0.993 g/L. Adjust pH to 7.

2.2 Biological Sample

1. One gram of rhizosphere originated from the plant of interest.

2.3 Genomic DNA (gDNA) Extraction

1. SET lysis buffer: NaCl 7.5 mM, EDTA 25 mM, Tris-HCl 20 mM, adjust pH 7.5.
2. SDS (sodium dodecyl sulfate) 10%.
3. NaCl 5 M.
4. Lysozyme (20 mg/mL).
5. Chloroform.
6. Isopropanol.
7. Ethanol 70%.
8. Nuclease-free water.
9. RNase A.

3 Methods

3.1 Preparation and Sterilization of Culture Media

1. Culture media and gelling agents must be sterilized separately using the wheat heat method, autoclave standard conditions (temperature 121°C, pressure 1 atm) for 15 min (*see* **Note 3**). The phosphate buffer must be sterilized through this strategy.
2. Vitamin-amino acids must be sterilized by filtration (using a PES membrane with a pore size of 0.22 μm).
3. Once sterilized and tempered, both volumes must be homogenized and supplemented with vitamin-amino acids solution (1 μL per mL).
4. Pour off culture media homogenized and supplemented on Petri dishes (100 × 15 mm) inside a laminar flow hood. Once solidified, set apart until use.

3.2 Obtaining the Rhizospheric Inoculum

1. Stock dilution: weigh 1 g of rhizosphere, place each sample in a sterile 50 mL tube, add 10 mL of phosphate buffer, and shake at 200 rpm for 5 min to promote liberation of rhizosphere bacteria.
2. Once the stock dilution is prepared, prepare serial 1:10 dilutions five times in sterile phosphate buffer (10^{-1}–10^{-5}).

3.3 Bacterial Inoculation and Isolation

1. One hundred microliters of each dilution (from 10^{-1} to 10^{-5}), including stock dilutions, will be spread onto Petri dishes containing SEA, R2AA, and TSA (Fig. 1a–c), SEP, R2AP, and R2AP (Fig. 1d–f).
2. The plates must be incubated at 30°C for 7–30 days with constant monitoring.
3. Single colonies of bacteria, which are morphologically different from one another, will be reisolated on new SEA, SEP, TSA, TSP, R2AA, or R2AP dishes to obtain axenic cultures (*see* **Notes 4, 5, and 6**).
4. Long-term store in glycerol 85% complemented with nutritive broth 15% (v/v) at −4°C.

3.4 Morphological Bacterial Identification

1. The presence of transparent, viscous white colonies (Fig. 2a, c) and yellow or orange colonies (Fig. 2b) exhibiting indefinite, irregular growth (Fig. 2e, f) is characteristic of the colonial morphology of certain Bacteroidota species. Additionally, some of these species can hydrolyze gellan gum (Fig. 2d).

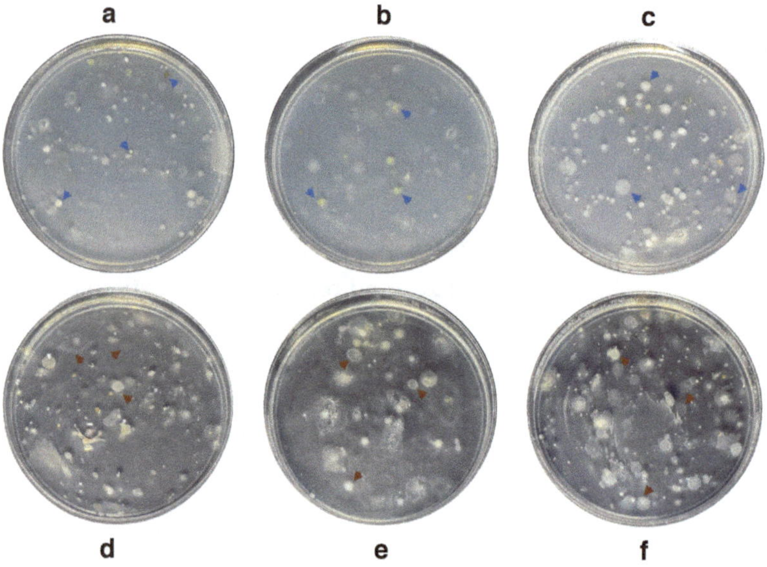

Fig. 1 Bacterial growth and morphology originating from the rhizosphere of *Triticum durum*. Stock dilution was plated on plates, and bacterial growth is demonstrated on the following culture media: (**a**) SEA, (**b**) R2AA, and (**c**) TSA. The dark blue arrowheads indicate orange and pale, bright yellow colonies (**a** and **b**), as well as viscous, opaque, white, and transparent colonies (**c**). Growth and colonial morphology observed on (**d**) SEP, (**e**) R2AP, and (**f**) TSP show viscous, transparent white colonies with irregular edges. Photographs were taken 72 h after incubation at 30°C

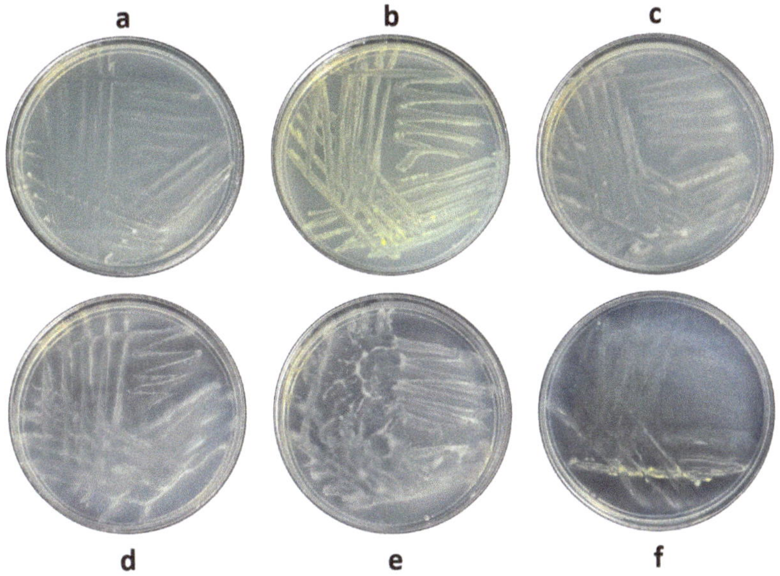

Fig. 2 Bacteroidota species isolated from the rhizosphere of *T. durum*. Several rhizospheric Bacteroidota species were isolated using our protocol: (**a**) *Pedobacter ginsengisoli* UMSMT45, (**b**) *Flavobacterium humidisoli* UMSMT60, (**c**) *Flavobacterium panacis* UMSMT84, isolated on SEA, R2AA, and TSA culture media, respectively; (**d**) *Pedobacter nutrimenti* UMSMT43, (**e**) *Sphingobacterium* sp. UMSMT76, and (**f**) *Chryseobacterium candidae* UMBP116, isolated on SEP, R2AP, and TSP culture media, respectively. Photographs were taken 72 h after incubation at °C

3.5 Molecular Identification

gDNA Extraction: Once Axenic Cultures are Obtained, gDNA Must be Extracted from the Isolated Bacteria as Follows

1. Take a loopful of culture and place it in a 1.5 mL microtube.
2. Add 200 μL of SET lysis buffer.
3. Add 1 μL of lysozyme stock solution (20 mg/mL) to 19 μL of sample to obtain a final concentration of 1 mg/mL. Vortex and incubate at 37°C for 1 h with frequent agitation (500 rpm).
4. Add 20 μL (1/10 vol) of SDS 10% and incubate at 55°C for 2 h with frequent agitation (500 rpm).
5. Add 74 μL (1/3 vol) of NaCl 5 M and 222 μL (1 vol) of chloroform, and incubate at room temperature for 30 min with frequent agitation (150 rpm).
6. Centrifuge at 12,000 rpm for 15 min, collect the upper phase, and transfer it to a clean tube.
7. Precipitate DNA by adding 1 vol of isopropanol and incubating overnight at −20°C.
8. Centrifuge at 12,000 rpm for 10 min, then carefully discard the supernatant without losing the pellet.

9. Wash the pellet with 200 μL of ethanol 70%, resuspend the pellet, and centrifuge at 12,000 rpm for 3 min.
10. Air-dry the pellet in a laminar flow hood, then resuspend in 50 μL of nuclease-free water.
11. Treat with 1 μL of RNase A and incubate at room temperature for 15 min.
12. Store at −20°C until use (*see* **Note 5**).

Amplification of 16S rRNA Gene Using the Primers 27F-1495R [16]

1. The PCR reaction must be prepared according to the Master mix manufacturer's instructions, and the resulting amplicons are visualized on a 1% agarose gel in 0.5× TAE buffer (20 mM Tris-acetate, 0.5 mM EDTA, pH 8) stained with fluorescent nucleic acid dye.
2. The amplicons must be sequenced using techniques such as the Sanger method.

Bioinformatic Analysis of 16S rRNA Gene Sequences

1. The sequences must be compared with those deposited in the NCBI database using BLASTn (https://blast.ncbi.nlm.nih.gov/Blast.cgi) against the rRNA databases, excluding sequences from uncultured or environmental samples and prioritizing sequences from type material. The frequency of obtaining Bacteroidota isolates varies across sample sources. In rhizospheric soil, a proportion of 6%–12% is common.

4 Notes

1. Prepare a soil extract using 300 g of soil, then sterilize it in an autoclave for 1 h. Subsequently, filter the extract through a Whatman Grade 1 filter paper (11 μm) and sterilize the filtrate again in an autoclave for 1 h. The final concentration of the soil extract is 10× (100 mL/L); however, in this protocol, 40 mL (a 1:25 dilution) is used to prepare the SE culture medium.
2. Dissolve culture media and gelling agents in 100 mL and 900 mL of deionized water, respectively. For the soil extract, homogenize it with yeast extract and bring it up to 100 mL with deionized water.
3. Axenic cultures are produced on the original culture media used to isolate Bacteroidota.
4. Optimal growth of the isolated Bacteroidota species occurs between 24 and 72 h after inoculation at 30°C.
5. gDNA must be quantified using a Nanodrop, and for PCR reactions, prepare dilutions to 50 ng/μL.

Acknowledgments

This work was supported by Valencia-Macías Foundation (México, Grant 15.1). S.M-M was funded by SECIHTI-México.

Conflict of Interest The authors have no conflicts of interest to declare that are relevant to the content of this chapter.

References

1. Krieg NR, Ludwig W, Euzéby J, Whitman WB (2010) Phylum XIV. Bacteroidetes phyl. nov. In: Krieg NR, Ludwig W, Whitman WB, Hedlund BP, Paster BJ, Staley JT, Ward N, Brown DR (eds) Bergey's manual of systematic bacteriology, 2nd edn, vol 4. Springer, New York, NY, pp 25–469. https://doi.org/10.1007/978-0-387-68572-4_3
2. Oren A, Garrity GM (2021) Valid publication of the names of forty-two phyla of prokaryotes. Int J Syst Evol Microbiol 71:005056. https://doi.org/10.1099/ijsem.0.005056
3. Figueiredo G, Osório H, Mendes MV, Mendo S (2025) A review on the expanding biotechnological frontier of *Pedobacter*. Biotechnol Adv 82:108588. https://doi.org/10.1016/j.biotechadv.2025.108588
4. Tanaka Y, Tozawa E, Iwashita T, Morishita Y, Tamaki H, Toyama T, Morikawa M, Kamagata Y, Mori K (2024) Successful isolation of diverse *Verrucomicrobiota* strains through the duckweed-microbes co-cultivation method. Microbe Environ 39:ME24019. https://doi.org/10.1264/jsme2.ME24019
5. Dyer DG, Dunn JA, Thorpe SR, Bailie KE, Lyons TJ, McCance DR, Baynes JW (1993) Accumulation of Maillard reaction products in skin collagen in diabetes and aging. J Clin Invest 91:2463–2469. https://doi.org/10.1172/JCI116481
6. ALjahdali N, Carbonero F (2019) Impact of Maillard reaction products on nutrition and health: current knowledge and need to understand their fate in the human digestive system. Crit Rev Food Sci Nutr 59:474–487. https://doi.org/10.1080/10408398.2017.1378865
7. Gorski L, Shariat NW, Richards AK, Siceloff AT, Noriega AA, Harhay DM (2024) Growth assessment of *Salmonella enterica* multi-serovar populations in poultry rinsates with commonly used enrichment and plating media. Food Microbiol 119:104431. https://doi.org/10.1016/j.fm.2023.104431
8. Yang JA, Yang SH, Kim J, Kwon KK, Oh HM (2017) Comparative genome analysis of the *Flavobacteriales bacterium* strain UJ101, isolated from the gut of *Atergatis reticulatus*. J Microbiol 55:583–591. https://doi.org/10.1007/s12275-017-7172-2
9. Tamaki H, Sekiguchi Y, Hanada S, Nakamura K, Nomura N, Matsumura M, Kamagata Y (2005) Comparative analysis of bacterial diversity in freshwater sediment of a shallow eutrophic lake by molecular and improved cultivation-based techniques. Appl Environ Microbiol 71:2162–2169. https://doi.org/10.1128/AEM.71.4.2162-2169.2005
10. Tanaka T, Kawasaki K, Daimon S, Kitagawa W, Yamamoto K, Tamaki H, Tanaka M, Nakatsu CH, Kamagata Y (2014) A hidden pitfall in the preparation of agar media undermines microorganism cultivability. Appl Environ Microbiol 80:7659–7666. https://doi.org/10.1128/AEM.02741-14
11. Kato S, Yamagishi A, Daimon S, Kawasaki K, Tamaki H, Kitagawa W, Abe A, Tanaka M, Sone T, Asano K, Kamagata Y (2018) Isolation of previously uncultured slow-growing bacteria by using a simple modification in the preparation of agar media. Appl Environ Microbiol 84:e00807–e00818. https://doi.org/10.1128/AEM.00807-18
12. Tamaki H, Hanada S, Sekiguchi Y, Tanaka Y, Kamagata Y (2009) Effect of gelling agent on colony formation in solid cultivation of microbial community in lake sediment. Environ Microbiol 11:1827–1834. https://doi.org/10.1111/j.1462-2920.2009.01907.x
13. Nishioka T, Elsharkawy MM, Suga H, Kageyama K, Hyakumachi M, Shimizu M (2016) Development of culture medium for the isolation of *Flavobacterium* and *Chryseobacterium* from rhizosphere soil. Microbe Environ 31:104–110. https://doi.org/10.1264/jsme2.ME15144
14. Kato S, Terashima M, Yama A, Sato M, Kitagawa W, Kawasaki K, Kamagata Y (2020) Improved isolation of uncultured anaerobic bacteria using medium prepared with separate sterilization of agar and phosphate. Microbe Environ 35:ME19060. https://doi.org/10.1264/jsme2.ME19060

15. Nishioka T, Tamaki H (2022) Improved cultivation and isolation of diverse endophytic bacteria inhabiting *Dendrobium* roots by using simply modified agar media. Microbiol Spectr 10:e0223822. https://doi.org/10.1128/spectrum.02238-22

16. Frank JA, Reich CI, Sharma S, Weisbaum JS, Wilson BA, Olsen GJ (2008) Critical evaluation of two primers commonly used for amplification of bacterial 16S rRNA genes. Appl Environ Microbiol 74:2461–2470. https://doi.org/10.1128/AEM.02272-07

Check for updates

Chapter 10

Methodological Approaches for the Preservation of *Akkermansia muciniphila* as Microbial Genetic Resource

Haydee Eliza Romero-Luna, Audry Peredo-Lovillo, Guadalupe Luna-Solano, Carlos Alberto Cruz-Cruz and Carlos Eduardo Lara-Hidalgo

Abstract

Akkermansia muciniphila is part of the intestinal microbiota and uses mucin as its primary source of carbon and nitrogen. It is considered a promising next-generation probiotic, with numerous beneficial functions. It regulates metabolic functions and immune response and improves the intestinal barrier, to name a few. This translates into the prevention of diseases such as obesity and type 2 diabetes, among others. However, despite its diverse benefits, *A. muciniphila* presents technical challenges in its cultivation and application, such as the need for a medium with a specific composition, the bacteria's high sensitivity to oxygen, a slow growth rate and stability, as well as differences between strains. In this sense, this protocol reflects the conditions for the preservation and maintenance of *A. muciniphila*.

Keywords Periodic transfer protocol, Promising probiotic, Mucin, Short-chain fatty acids, Homeostasis, Gut microbiota

1 Introduction

Akkermansia muciniphila, a member of the Verrucomicrobia phylum, is a Gram-negative, oval-shaped, nonmotile, nonspore-forming, and strictly anaerobic bacterium. It was first isolated by Derrien et al. [1] from the human gut as *Akkermansia muciniphila* (type strain Muc^T; ATCC BAA-835). This bacterium, which degrades mucin, has been named as a next-generation probiotic due to its diverse health benefits, including strengthening the intestinal barrier, regulating metabolic functions, and modulating immune responses [2, 3]. For the above, its relevance as a microbial therapeutic agent has increased. So, preserving *A. muciniphila* as a microbial genetic resource has become a key area of interest. However, its strict anaerobic nature and sensitivity to environmental conditions pose significant challenges for long-term storage, transport, and formulation. Therefore, developing practical methodological approaches to maintain

Marco A. Ramírez-Mosqueda et al. (eds.), *Methods for The Conservation of Microorganisms*, Methods in Molecular Biology, vol. 3090, https://doi.org/10.1007/978-1-0716-5623-5_10,

its viability, genetic integrity, and functional properties is essential for research and potential clinical applications.

1.1 Importance of Akkermansia muciniphila in the Gut Microbiota

The intestinal microbiota is critical to supporting the nutrition and health of humans. *A. muciniphila* is part of the gut microbiota, mainly consisting of bacteria that colonize the mucus layer, as it inhabits the mucosal layer of the distal sections of the large and small intestines. Its presence accounts for approximately 3%–5% of the total bacteria of the gut microbiota [4]. For this bacterium, mucin serves as both the primary source of energy and a source of amino acids and sugars. In turn, this breakdown promotes colonization and the production of short-chain fatty acids, which in turn provide energy to the host [5]. Furthermore, the generation of these short-chain fatty acids increases the number of mucin-producing goblet cells, resulting in enhanced mucin production and, consequently, the restoration of mucosal thickness and intestinal barrier integrity [6]. Additionally, its presence helps maintain intestinal homeostasis, which, through various mechanisms, aids in preventing diseases such as obesity, insulin resistance, and neuropsychiatric disorders, among others. This emphasizes its importance within the gut microbiota [7–9].

1.2 Potential Nutritional and Metabolic Benefits

A. muciniphila is being regarded as a next-generation probiotic with nutritional and metabolic advantages. This bacterium can modulate the gut microbiota and as mentioned above, strengthen the intestinal barrier by affecting metabolic pathways, thereby helping improve metabolic health [10–12], e.g., *A. muciniphila* contribute to the regulation of metabolic disorders such as obesity and type 2 diabetes, through increased insulin sensitivity, reduced body fat, and improved glucose tolerance [13, 14]. Furthermore, as mentioned, this microorganism enhances the integrity of the intestinal barrier by increasing the thickness of the intestinal mucus layer and strengthening the gut epithelial barrier. This reduces intestinal permeability and prevents the infiltration of harmful bacteria into the bloodstream, known as endotoxemia, which is associated with inflammation and metabolic diseases [14, 15]. Another benefit is that it generates short-chain fatty acids, among other compounds that offer various benefits, such as promoting the growth of beneficial gut microorganisms and facilitating the secretion of GLP-1. Additionally, the short-chain fatty acids produced by *A. muciniphila* have been shown to reduce inflammation [15]. This microorganism has also been shown to affect lipid metabolism, which contributes to maintaining cardiovascular health and preventing metabolic syndrome [16].

All these benefits increase interest in *A. muciniphila* as a therapeutic agent. However, these benefits can only be realized after overcoming the various technical challenges involved in cultivating and maintaining this microorganism.

1.3 Technical Challenges in Its Cultivation and Maintenance

The cultivation and maintenance of *A. muciniphila* face several technical challenges (Fig. 1), including: (1) the specific composition of the medium, (2) sensitivity to oxygen, (3) growth rate, (4) stability, and (5) differences between strains.

Media Composition

As mentioned above, *A. muciniphila* uses mucin as its primary source of carbon and nitrogen. This ingredient is expensive and difficult to purify. Because it is derived from animal sources, it is unsuitable for large-scale production. For this reason, alternatives are under development. Indeed, the use of N-acetylglucosamine, glucose, and other prebiotics is being studied; however, this could affect the physiology and functionality of the microorganism compared to mucin-based cultures [17].

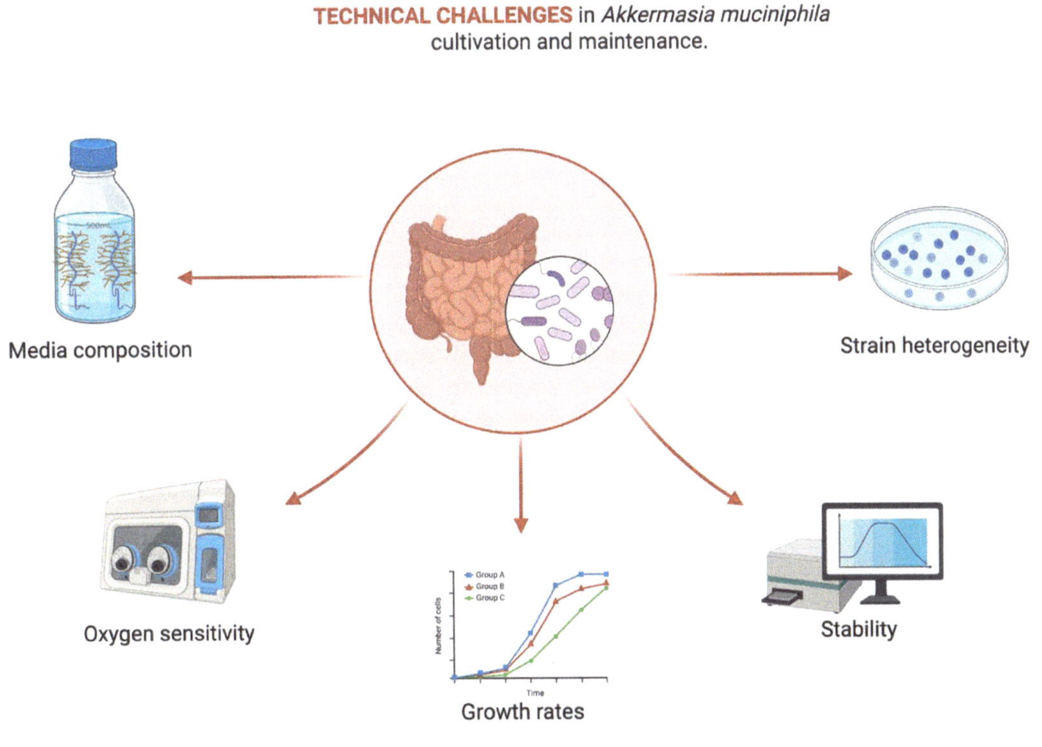

Fig. 1 Technical challenges in the cultivation and maintenance of *Akkermansia muciniphila*. Schematic diagram showing the main technical challenges involved in cultivating and maintaining *A. muciniphila*. At the center, there is an illustration of the human intestine, accompanied by a magnified circle that highlights purple, rod-shaped bacterial cells. From this central image, five red arrows branch out, each referring to a key factor: media composition, represented by a flask with a bubbling culture medium; strain heterogeneity, shown as a Petri dish containing bacterial colonies; oxygen sensitivity, indicated by an anaerobic chamber; growth rates, depicted with a typical bacterial growth curve; and stability, represented by a computer screen displaying an analytical trace. Overall, the figure provides a visual summary of the experimental difficulties commonly faced when handling and propagating this strictly anaerobic intestinal microorganism

Oxygen Sensitivity

For proper growth and maintenance, *A. muciniphila* requires strictly anaerobic environments, which are difficult to achieve on an industrial scale [18].

Growth Rates

Akkermansia muciniphila exhibits slow cell growth and a slow growth rate in mucin-based media. For this reason, various strategies have been studied to optimize yield, such as pH control, carbon–nitrogen ratios, and supplementation with growth promoters (galactose, sialic acid, lactose, and chitosan). However, a study focused on the cost, reproducibility, and functional properties of other strains is needed [19, 20].

Stability

A. muciniphila is considered unstable, as its viability is rapidly affected under storage conditions, and therefore, its functionality as well. In this regard, techniques such as pasteurization and the production of postbiotics have been used to improve its stability during storage [18].

Strain Heterogeneity

Although *A. muciniphila* is considered a promising probiotic bacterium, different strains and phylogroups have been studied, and all of them present variations concerning their sensitivity to oxygen, growth, and health benefits [21].

With this in mind, a periodic transfer protocol for *A. muciniphila* would offer several technical and experimental advantages. Nutrient depletion and the accumulation of toxic metabolites impair viability; however, this can be remedied through a periodic transfer protocol. It can also prevent phenotypic variations due to the stress generated by the depleted medium. It would also allow the bacteria to maintain their ability to degrade mucin [1]. Its experimental advantages include reproducible results by maintaining similar physiological conditions in each experiment; it also allows for control of the growth phase and constant availability, saving time and reducing variability [22].

2 Materials

2.1 Bacterial Strains and Its Sources: Description of the Strains

1. The strain *Akkermansia muciniphila* ATCC BAA-835 is used in this protocol procedure (*see* **Note 1**). Cryopreserved vials of the strain stored at −80°C are activated and used to initiate subcultivation.

2.2 Culture Media and Supplements

1. Basal mucin medium (BMM) is used as culture media [23] (*see* **Note 2**). The composition is detailed in Table 1. The primary carbon and nitrogen source to *A. muciniphila* ATCC BAA-835 is mucin (porcine gastric type II), and the medium must

Table 1
Composition of BMM for *A. muciniphila*

Component	Concentration
Porcine gastric mucin (Type II)	10 g/L
Yeast extract	2 g/L
Peptone	2 g/L
NaCl	4 g/L
K_2HPO_4	0.5 g/L
KH_2PO_4	0.5 g/L
$MgSO_4 \cdot 7H_2O$	0.2 g/L
$CaCl_2 \cdot 2H_2O$	0.09 g/L
L-cysteine	0.1 g/L
Vitamin K1-Hemin solution	10 mL/L
Resazurin	1 mg/L
Distilled water	1 L

be adjusted to pH 6.8 and prereduced with L-cysteine [24] for subsequent use.

2.3 Equipment and Conditions

1. Laboratory equipment: autoclave, refrigerator, ultra-freezer, potentiometer, heating plate, analytical balance, optical microscope, incubator 37°C, and spectrophotometer.
2. Laminar flow hood for aseptic conditions.
3. Membrane filter (0.22 μm) for sterilization of sensitive substances.
4. Anaerobic jars with sachets and anaerobic indicator (Fig. 2).
5. Anaerobic gas cylinder with mixture (80% N_2, 10% CO_2, 10% H_2).
6. Crystal material: hungate tubes, Schott bottles, anaerobic Petri dishes, Erlenmeyer flasks, beakers, graduated cylinder, and micropipettes.
7. Other materials: sterile syringes and needles, Falcon tubes, quartz microcuvette, timer, butyl rubber stoppers, and aluminum seals.

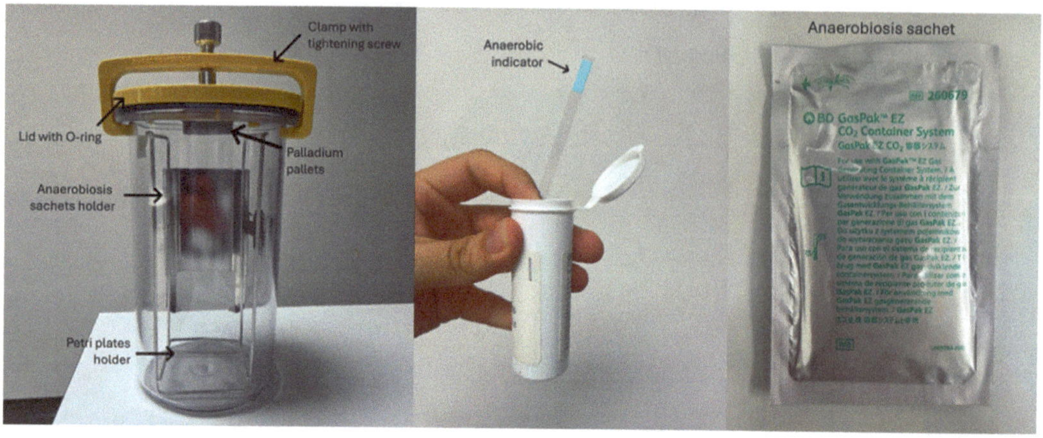

Fig. 2 Components of an anaerobic jar system used for the cultivation of obligatory anaerobic microorganisms. This illustration shows the main parts of an anaerobic jar system used to grow bacteria that cannot tolerate oxygen. The jar is cylindrical and tightly sealed with a lid to keep the interior airtight. Inside, several Petri dishes with bacterial cultures are placed on the bottom. A small gas-generating envelope is attached to the inner wall, releasing hydrogen and carbon dioxide once activated. Near the top, a palladium catalyst helps remove any remaining traces of oxygen, while an anaerobic indicator strip changes color to confirm that the atmosphere inside is oxygen-free. Together, these components create and maintain the conditions required for the cultivation of strict anaerobes

3 Methods

3.1 Preparation of the Media

1. Weigh all the components listed in Table 1, except L-cysteine and resazurin, and dissolve in the corresponding volume of distilled water with gentle stirring (*see* **Note 3**).
2. Adjust the pH to 6.8 using HCl or NaOH 1 N.
3. Autoclave at 121°C for 15 min.
4. Prepare a stock solution of L-cysteine and resazurin, which should be filtered using a sterile 0.2 μm syringe filter. Avoid heat sterilization, as this amino acid can be degraded.
5. L-cysteine as a reducing agent with the purpose of degrading the dissolved oxygen in the media, and resazurin as a redox indicator, which must turn around colorless in anaerobic conditions.
6. When the media is autoclaved and cooled to 40°C, add the corresponding volume of the stock solution of L-cysteine and resazurin.
7. Distribute the volume into Hungate tubes (10 mL per tube) and cap the tubes with butyl rubber stoppers and aluminum seals.

8. Flush the tubes headspace with an anaerobic gas mixture (80% N_2, 10% CO_2, 10% H_2) for 10 min [17, 25].
9. Incubate the media overnight in the anaerobic chamber before use. That is due to media prereducing.

3.2 Inoculation and Culture Conditions: Anaerobic Environment, Temperature, Incubation Time, Growth Monitoring, Special Conditions

1. With a sterile syringe, a volume of 100 μL Cryopreserved vials of the *A. muciniphila* ATCC BAA-835 is inoculated in each Hungate tube containing 10 mL of BMM (*see* **Notes 4** and **5**).
2. Incubate at 37°C in an anaerobic chamber with an anaerobic indicator and without shaking.
3. Monitor growth by OD_{600nm} each 6 h for 24 h.
4. Using a sterile syringe, take an aliquot of fermented BMM and place it in a quartz microcuvette and measure absorption at OD_{600nm} in a spectrophotometer.
5. Typically, the absorbance reaches a value of 1 at 24 h.

3.3 Periodic Transfer

1. Every 72 h, inside a laminar flow hood, extract 5 mL of the growth phase culture using a sterile syringe.
2. Transfer the extracted volume to an Erlenmeyer flask containing 45 mL of fresh BMM, previously reduced.
3. Incubate the culture under the conditions specified in the section above.
4. Record the transfer number, date and time, and absorption at OD_{600nm} before and after the transfer.
5. Maintain a record of the history of each subculture, as well as the occurrences during the fermentation process.
6. Every ten subculture passes, perform a cryopreservation cycle with sterile 20% glycerol to maintain the integrity of the strain [26] (*see* **Note 6**).

3.4 Monitoring Microbial Culture Viability

1. Take 0.1 mL of the growth phase culture using a micropipette and make serial dilutions in 0.9 mL of sterile saline solution (10^{-1}–10^{-8}).
2. Plate each serial dilutions on BMM agar and incubate at 37°C for 48 h inside an anaerobic chamber.
3. Verify viability through microbial counting on plates.
4. Perform a Gram stain and check the morphological characteristics under a microscope.
5. *A. muciniphila* is a Gram-negative bacterium with an ellipsoid or oval shape, measuring 0.7–1.0 μm long and 0.4–0.5 μm wide [20].
6. Discard any culture with unexpected morphology.

4 Notes

1. The protocol is optimized for *A. muciniphila* ATCC BAA-835. Strain-specific variations may occur and should be considered.
2. Mucin is a fundamental component for the growth of *A. muciniphila*. Therefore, it must be highly pure and completely dissolved in the culture medium.
3. Resazurin must always remain colorless in the culture medium. If it changes to a pink hue, this indicates the presence of oxygen and invalidates anaerobic conditions.
4. The use of a gas-tight syringe and needle is essential when transferring the subculture, as it prevents oxygen contamination.
5. At no time should the culture be exposed to air. All manipulations must be carried out in an anaerobic chamber or using anaerobic techniques.
6. It is recommended that strain identity be verified by 16S rRNA sequencing after multiple subculture passages.

Conflict of Interest

The authors have no conflicts of interest to declare that are relevant to the content of this chapter.

References

1. Derrien M, Vaughan EE, Plugge CM et al (2004) *Akkermansia muciniphila* gen. nov., sp. nov., a human intestinal mucin-degrading bacterium. Int J Syst Evol Microbiol 54(Pt 5):1469–1476
2. Zhao Q, Yu J, Hao Y et al (2022) *Akkermansia muciniphila* plays critical roles in host health. Crit Rev Microbiol 49(1):82–100
3. Ghotaslou R, Nabizadeh E, Memar MY et al (2023) The metabolic, protective, and immune functions of *Akkermansia muciniphila*. Microbiol Res 266:127245
4. Si J, Kang H, You HJ et al (2022) Revisiting the role of *Akkermansia muciniphila* as a therapeutic bacterium. Gut Microbes 14(1):2078619
5. Liu MJ, Yang JY, Yan ZH et al (2022) Recent findings in *Akkermansia muciniphila*-regulated metabolism and its role in intestinal diseases. Clin Nutr 41(10):2333–2344
6. Cheng D, Xie MZ (2021) A review of a potential and promising probiotic candidate—*Akkermansia muciniphila*. J Appl Microbiol 130 (6):1813–1822
7. Aggarwal V, Sunder S, Verma SR (2022) Disease-associated dysbiosis and potential therapeutic role of *Akkermansia muciniphila*, a mucus-degrading bacterium of gut microbiome. Folia Microbiol 67(6):811–824
8. Misera A, Marlicz W, Podkówka A, Łoniewski I, Skonieczna-Żydecka K (2024) Possible application of *Akkermansia muciniphila* in stress management. Microbiome Res Rep 3(4):48
9. Zeng Z, Chen M, Liu Y et al (2025) Role of *Akkermansia muciniphila* in insulin resistance. J Gastroenterol Hepatol 40(1):19–32
10. Wu W, Kaicen W, Bian X et al (2023) *Akkermansia muciniphila* alleviates high-fat-diet-related metabolic-associated fatty liver disease by modulating gut microbiota and bile acids. Microb Biotechnol 16(10):1924–1939
11. Xu W, Zhang S, Yang Y et al (2024) Therapeutic potential of dietary nutrients and medicinal foods against metabolic disorders: targeting *Akkermansia muciniphila*. Food Front 5(2):329–349

12. Zhang M, Wang Y, Gan Y (2024) The potential role of *Akkermansia muciniphila* in liver health. Future Microbiol 19(12):1081–1096
13. Zhao S, Liu W, Wang J et al (2017) *Akkermansia muciniphila* improves metabolic profiles by reducing inflammation in chow diet-fed mice. J Mol Endocrinol 58(1):1–14
14. Niu H, Zhou M, Zogona D et al (2024) *Akkermansia muciniphila*: a potential candidate for ameliorating metabolic diseases. Front Immunol 15:1370658
15. Gao F, Cheng C, Li R et al (2025) The role of *Akkermansia muciniphila* in maintaining health: a bibliometric study. Front Med 12:1484656
16. Yan J, Sheng L, Li H (2021) *Akkermansia muciniphila*: is it the Holy Grail for ameliorating metabolic diseases? Gut Microbes 13(1):1984104
17. Geerlings SY, van der Ark K, Nijsse B et al (2024) Omics-based analysis of *Akkermansia muciniphila* cultivation in food-grade media. Microbiome Res Rep 3(3):36
18. Mruk-Mazurkiewicz H, Kulaszyńska M, Czarnecka W et al (2024) Insights into the mechanisms of action of *Akkermansia muciniphila* in the treatment of non-communicable diseases. Nutrients 16(11):1695
19. Meng X, Xv C, Lv J et al (2024) Optimizing *Akkermansia muciniphila* isolation and cultivation: insights into gut microbiota composition and potential growth promoters in a Chinese cohort. Microorganisms 12(5):881
20. Wu H, Qi S, Yang R et al (2024) Strategies for high-cell-density cultivation of *Akkermansia muciniphila* and its potential metabolism. Microbiol Spectr 12(1):e02386-23
21. Luo Y, Lan C, Li H et al (2022) Rational consideration of *Akkermansia muciniphila* targeting intestinal health: advantages and challenges. NPJ Biofilms Microbiomes 8(1):81
22. Lagier JC, Edouard S, Pagnier I et al (2015) Current and past strategies for bacterial culture in clinical microbiology. Clin Microbiol Rev 28(1):208–236
23. Liu J, Liu H, Liu H et al (2023) Live and pasteurized *Akkermansia muciniphila* decrease susceptibility to *Salmonella* Typhimurium infection in mice. J Adv Res 52:89–102
24. Gu HY, Fan JX, Bao P et al (2024) Probiotic biofilm-modified 3D-printing scaffolds for improving chemo-immunotherapy of bone tumor and promoting osteogenesis. Adv Funct Mater 34(9):2311015
25. Li ZT, Zhu L, Hu GA et al (2021) *Akkermansia muciniphila* fermentation culture based on a novel bionic large intestine dynamic digestion model. Food Biosci 43:101260
26. Ouwerkerk JP, Aalvink S, Belzer C et al (2017) Preparation and preservation of viable *Akkermansia muciniphila* cells for therapeutic interventions. Benef Microbes 8(2):163–170

Chapter 11

Emerging Strategies for the Preservation of Microorganisms of Biomedical and Environmental Importance

Jasmín Salazar-Mendoza, Janette Betzabeth Villarruel-Muñoz, Lorena Jacqueline Gómez-Godínez, José Luis Aguirre-Noyola, Hugo Ernesto Flores-López and Ignacio Valenzuela-Chavira

Abstract

Microorganisms represent the largest component of global biodiversity and play essential roles in biogeochemical cycling, biomedical innovation, and environmental resilience. However, the majority of microbial diversity remains unexplored or inadequately conserved, which poses a threat to scientific reproducibility, public health, and sustainable bioeconomic development. This review evaluates emerging strategies for conserving microorganisms of biomedical and environmental relevance, emphasizing the preservation of viability, genetic stability, and functional integrity across diverse taxa and microbial communities. Traditional methods, such as cryopreservation and lyophilization, are foundational; they present limitations for extremophiles, strict anaerobes, and complex microbiomes. Recent developments incorporate omics-guided preservation, dry-state technologies, and synthetic biology to improve long-term stability and functional fidelity. Innovations such as next-generation freeze-drying, microbiome-level conservation, and engineered microbial systems enhance storage, transport, and reproducibility while reducing dependence on cold-chain infrastructure. The review further addresses regulatory, ethical, and infrastructural challenges related to global biobanking and equitable access to microbial resources. New frameworks prioritize the conservation of functional microbial assemblages over isolated strains, supporting applications in antimicrobial discovery, advanced biotherapeutics, agriculture, and environmental remediation. Advancing microbial conservation is therefore essential for protecting planetary health, advancing biomedical research, and supporting a resilient, sustainable bioeconomy amid global environmental change.

Keywords Microbial conservation, Synthetic biology, Precision engineering, Stress resistance, Resuscitation-promoting factors

1 Introduction

The limitations of traditional preservation methodologies, particularly their inability to maintain the functional integrity of complex consortia and the viability of extremely sensitive taxa, have necessitated a paradigm shift in microbial conservation [1,2]. We are currently

Marco A. Ramírez-Mosqueda et al. (eds.), *Methods for The Conservation of Microorganisms*, Methods in Molecular Biology, vol. 3090, https://doi.org/10.1007/978-1-0716-5623-5_11,

transitioning from an era of "passive dormancy" to one of "precision engineering," where microorganisms are no longer merely archived but strategically redesigned to withstand environmental and industrial rigors [3]. This chapter explores the convergence of synthetic biology, omics-guided strategies, and advanced materials science as the new foundation for safeguarding the Earth's microbial heritage [4,5].

At the core of this transformation is the application of synthetic biology to rewire microbial chassis for enhanced robustness. By installing multigene modules derived from polyextremophiles such as the DNA-binding protein HU and chaperones like DnaJ, researchers can now confer multistress resistance to industrial and biomedical strains, ensuring they remain viable under fluctuating thermal and desiccation conditions [6,7]. Furthermore, the development of "cell factories" that utilize abiotic-stress-inducible promoters allows for the synchronization of protective metabolites with environmental cues, effectively decoupling biological performance from chronological time [8,9].

Furthermore, the development of "cell factories" that utilize abiotic-stress-inducible promoters allows for the synchronization of protective metabolites with environmental cues, effectively decoupling biological performance from chronological time [8,9].

A critical advancement discussed in this chapter is the active management of microbial dormancy and resuscitation. Rather than relying on stochastic recovery, the use of Resuscitation-promoting factors and the modulation of energy supply allows for the precise "wake-up" of viable but nonculturable populations, ensuring that preserved resources can immediately execute their ecological or therapeutic roles upon reactivation [10,11]. Moreover, as we move toward conserving entire "functional assemblages" rather than isolated strains, the focus shifts to maintaining the interaction networks and spatial structures that drive ecosystem resilience [12,13].

Finally, this chapter addresses the imperative of "Safety by Design." The deployment of engineered microbes for environmental remediation or advanced biotherapeutics requires robust biocontainment modules, such as CRISPR-based kill switches and "expiry-date" circuits, to prevent unintended genomic persistence in the biosphere [14–16]. By integrating these emerging technologies with global initiatives like the Microbiota Vault, we establish a sustainable framework for a resilient bioeconomy in which microbial diversity is actively managed to mitigate the most pressing global challenges [1,4].

2 Need for Emerging Approaches

Traditional methods have notable limitations. Advances in systems biology and microbial ecology highlight the need for new strategies [2]. Current practices still focus on single-species cultures, yet these do not suffice for preserving complex microbial groups that depend on interactions that are lost when isolated [17]. Maintaining community structure and function, therefore, demands the development of alternative methods [18].

The use of omics technologies requires preservation steps that maintain cell viability and ensure genetic and transcriptomic stability, thereby reducing noise in gene expression after recovery [19]. However, several challenges persist, among them sustainability and biosafety concerns. Cryopreservation, for instance, requires high energy inputs. In response, alternatives like improved lyophilization, deep eutectic solvents, and storage in polymers or microfluidic matrices have emerged [2]. These options enable room-temperature storage and broaden access to biobanking [13].

The limitations of current preservation strategies, together with advancements in systems biology and microbial ecology, have created a gap between expanding knowledge of microbial diversity and the ability to accurately preserve it [2]. Bridging this disparity requires preservation methods that integrate new developments in cryobiology, materials science, and molecular biology, aligning preservation practices with scientific progress [13].

2.1 Advanced Cryopreservation Strategies

Contemporary cryobiology has evolved beyond the paradigm of slow freezing toward approaches that precisely modulate the thermodynamics of intracellular and extracellular water [20]. Traditional methods primarily aim to reduce ice nucleation through controlled cooling rates. In contrast, advanced strategies aim to induce vitrification, the transition of water into an amorphous, glass-like state without crystallization [20]. This transition preserves the structural and molecular integrity of cells and biomaterials, minimizing mechanical and osmotic damage associated with conventional freezing. Innovation in cryopreservation increasingly integrates principles from molecular biology, supramolecular chemistry, and materials science to enhance the stability of complex biological systems [9].

Next-Generation Cryoprotectants and Biomimetic Approaches

A significant challenge in cryopreservation is the toxicity of conventional cryoprotectants, such as dimethyl sulfoxide and glycerol [21]. These chemicals protect cells from freezing-induced damage but can also cause osmotic stress, membrane disruption, and cytotoxicity, ultimately compromising cell viability. To overcome these limitations, next-generation cryoprotectants inspired by biological systems have been developed to reduce toxicity and enhance molecular stability during cooling.

Among these alternatives, natural deep eutectic systems have emerged as particularly promising. Composed of naturally occurring metabolites such as sugars, amino acids, and polyols, NADES form highly viscous liquids that establish extensive hydrogen-bond networks with water [22]. These systems lower the glass transition temperature, stabilize membranes and proteins, and enable vitrification at lower concentrations than those required by conventional cryoprotectants [23]. Due to their reduced cytotoxicity and enhanced biocompatibility, NADES are suitable for both clinical and environmental applications [24].

Additionally, antifreeze proteins and synthetic peptides that inhibit ice recrystallization have been developed. Modeled after adaptations found in polar organisms, these molecules bind to the surfaces of forming ice crystals and inhibit their growth through thermal hysteresis [25]. Synthetic analogues and small-molecule mimetics (antifreeze peptides) can provide similar protective effects while reducing immunogenicity and enhancing their potential for biomedical applications [21,26].

Vitrification and Ultra-Rapid Freezing

Vitrification represents one of the most significant advances in cryopreservation. Unlike conventional freezing, this approach aims to completely avoid ice formation by combining extremely rapid cooling rates with highly viscous cryoprotectant solutions. The process typically involves controlled osmotic dehydration, brief exposure to high cryoprotectant concentrations, and immediate immersion in liquid nitrogen, achieving cooling rates exceeding 10,000°C/min and producing a stable glassy state [27].

Recent advancements include 3D and microfluidic vitrification techniques for preserving structures such as hydrogels, organoids, and microbial consortia [28]. DMSO-free mixtures containing ampholytic polymers and modified alginate matrices offer cellular protection during rapid cooling. However, challenges remain in uniformly heating large volumes and preventing devitrification during warming. Approaches such as magnetic nanoparticles (nanowarming) and laser heating have been developed to achieve rapid, uniform rewarming [29,30].

Cellular Stress Control and Limitations

Despite recent advancements, cryopreservation continues to induce cellular stress that requires careful management [2]. Elevated concentrations of cryoprotectants can lead to osmotic imbalance and toxicity, while freezing and thawing processes generate oxidative stress and cellular damage [31]. Gradual, stepwise addition of cryoprotectants allows cells to adapt more effectively [2]. Furthermore, combining different cryoprotectants can reduce toxicity while maintaining protective efficacy [9].

Oxidative stress is a significant limitation, especially during thawing, when reoxygenation can generate reactive oxygen species. Incorporating antioxidants into preservation media, including mitochondria-targeted compounds (e.g., Mito-TEMPO or MitoPBN), enhances post-thaw recovery and maintains cellular functionality [31]. In conjunction with optimizing the physiological state of the inoculum and precisely controlling cooling and warming rates, these strategies are critical for efficient recovery [2].

Biomedical and Environmental Applications

Advanced cryopreservation techniques are transforming both clinical and environmental biobanking. In medical contexts, these methods enable long-term storage of organoids, tissues, and three-dimensional constructs, supporting regenerative medicine and scientific research [28]. Meanwhile, the high-fidelity preservation of clinical and probiotic strains supports disease surveillance and advances the development of microbiome-based therapies [1].

In environmental applications, biomimetic cryoprotectants and vitrification facilitate the preservation of microbial groups and species that are otherwise difficult to conserve. This capability is essential for maintaining soil and aquatic microbiomes and safeguarding biological resources for bioremediation, bioenergy production, and environmental biotechnology [18]. Next-generation cryopreservation is increasingly adopting integrated strategies to protect microbial and cellular diversity across medicine, biotechnology, and sustainability initiatives [1].

2.2 Emerging Dry-State Preservation Technologies

Dry-state preservation represents a promising and sustainable alternative for conserving biological resources. Drawing inspiration from anhydrobiosis, the process by which certain organisms survive with minimal water, this approach involves controlled dehydration to form amorphous glassy matrices at room temperature [1]. Unlike cryopreservation, which inhibits chemical reactions through low temperatures, dry-state preservation immobilizes biomolecules in a stable glassy phase, thereby restricting molecular mobility [32]. This method aligns with the principles of Green Biobanking, which aims to reduce energy consumption and the carbon footprint associated with cold storage while ensuring safe and accessible preservation of biological materials [1].

Next-Generation Freeze-Drying

Conventional freeze-drying has long served as the gold standard for preserving microbial strains in a dry state. However, its effectiveness is limited for sensitive microorganisms due to random ice nucleation and desiccation-induced stress [2]. To address these limitations, next-generation freeze-drying technologies employ controlled nucleation systems, such as rapid depressurization or ultrasound, to synchronize ice crystal formation and produce more uniform porous structures. This technology reduces intracellular mechanical damage and improves preservation outcomes [33].

The design of lyoprotective formulations has been optimized based on the "water replacement hypothesis," which holds that disaccharides such as trehalose and sucrose replace water interactions with cell membranes, forming hydrogen bonds. This mechanism preserves the membrane structure during dehydration. Molecular modeling predicts critical parameters, such as collapse and glass transition temperatures, which help optimize drying cycles and increase efficiency without affecting cell viability.

Spray Drying and Hybrid Technologies

Spray drying offers a continuous, scalable alternative to batch lyophilization, especially for industrial-scale production of probiotics and biofertilizers. This technique converts liquid solutions into dry powders via atomization in heated air streams but has historically been limited by thermal stress. Recent advancements, including low-temperature atomizers and ultrasonic nozzles, have substantially improved microbial survival rates [34].

Hybrid technologies, such as spray-freeze-drying, combine atomization in liquid nitrogen with subsequent lyophilization to produce highly porous particles that rehydrate rapidly [35]. This method combines the structural protection of cryopreservation with the logistical advantages of dry storage. Recent studies indicate that spray-freeze-drying can encapsulate anaerobic microbial consortia, with viability rates exceeding 90% after extended storage at room temperature.

Encapsulation and Protective Matrices

For sensitive microorganisms or those necessitating controlled release, encapsulation within polymeric matrices is critical. Techniques such as electrospraying use high-voltage electric fields to generate microcapsules without heat or toxic solvents, thereby preserving cell viability and biomolecular stability [36].

Hydrogels composed of alginate, pectin, or chitosan can be engineered to respond to environmental stimuli such as pH changes or enzymatic activity, thereby enabling targeted release. The layer-by-layer technique involves coating individual cells with nanometric polyelectrolyte layers, which protect against oxidation and adverse conditions while preserving metabolic activity following rehydration [37].

Advantages for Stability, Transport, and Sustainability

Dry-state preservation technologies are transforming biobank logistics by providing thermal stability at ambient temperatures. When the glass transition temperature of the preservation matrix exceeds room temperature, biological materials can be stored for extended periods without refrigeration, thereby minimizing the risks associated with power failures or accidental thawing [2].

Transport is streamlined because lyophilized or encapsulated powders eliminate the need for dry ice or liquid nitrogen, thereby reducing shipping costs and facilitating compliance with international transport regulations. Environmentally, decreasing dependence on ultra-low-temperature freezers significantly reduces the carbon footprint of microbial collections, aligning preservation practices with global sustainability objectives and the principles of green biobanking [1]. Consequently, dry-state preservation is emerging as a central strategy for the long-term conservation of microorganisms and other biological systems, integrating energy efficiency, functional stability, and global accessibility within a sustainable scientific framework.

2.3 Omics-Guided and Molecular-Based Strategies

The transition toward functional stability is increasingly characterized by the implementation of high-resolution multiomic assessments that bypass the inherent limitations of traditional genomic abundance. While DNA-based methods identify potential targets, they are insufficient for measuring actively expressed genes or confirming the metabolic status of community members [38,39]. Experimental evidence from metaproteomic studies has demonstrated that the technical reproducibility of taxonomic and functional profiles is strictly dependent on immediate stabilization; for instance, flash-freezing human fecal samples at −196°C or −80°C has been validated as the only method to maintain identification rates for key groups like Actinobacteria and preserve the "snapshot" of the biological system [38,39]. Furthermore, primary research on stool metatranscriptomics indicates that transcript levels do not necessarily correlate with protein levels, and the inherent instability of mRNA makes isolation and storage conditions particularly critical for maintaining a map of active gene expression [38,40].

Experimental results reveal that immediate cryopreservation is essential because protein and transcript profiles are highly sensitive to rapid shifts in environmental conditions, such as air exposure and temperature changes [38,39]. In a controlled trial, transcriptomes of samples preserved in RNA later remained unchanged for up to 6 days at room temperature, whereas stabilizers such as RNAProtect were ineffective beyond 24 h [40]. However, quantitative analysis of metabolic pathways has shown that certain preservation methods can introduce significant biases, such as the depletion of transcripts associated with carbohydrate transport and metabolism, potentially leading to an underestimation of the community's functional capacity [40]. Furthermore, long-term studies have shown that ultralow-temperature cryopreservation has a minimal impact on microbiota diversity compared to fresh cultures. That biobanked microbiota can maintain their functional responses to prebiotics for up to 12 months when stored in deoxygenated glycerol buffers [41]. Collectively, these experimental findings underscore that the functional integrity of biobanked fecal microbiota is highly sensitive to the initial handling and storage parameters, requiring purpose-optimized collection strategies to ensure data comparability in clinical and environmental research [38,39,42].

2.4 Bioengineering and Synthetic Biology Approaches

The rational redesign of microbial chassis to withstand the physical rigors of desiccation and thermal fluctuations represents a definitive departure from traditional empirical stabilization. Synthetic biology facilitates the strategic rewiring of regulatory networks and the installation of multigene modules derived from polyextremophiles, such as the DNA-binding protein HU and the chaperone DnaJ, which have been experimentally validated to confer multistress resistance in industrial strains [6,7]. These "cell factories" use dynamic feedback regulation and abiotic-stress-inducible promoters to synchronize the expression of protective metabolites with environmental cues, thereby ensuring high performance under fluctuating conditions [8,9]. Notably, the orchestration of metabolic regulation and induced dormancy provides an elegant mechanism for long-term stabilization, moving beyond passive survival toward active persistency [43]. Experimental investigations demonstrate that dormancy can be actively induced by modulating energy supply and extracellular signaling, allowing cells to persist at near-zero metabolic rates, focused solely on biomolecule repair [43, 44]. The subsequent resuscitation is precisely controlled by the application of Resuscitation-promoting factors, peptidoglycan-active enzymes that terminate dormancy and facilitate the recovery of viable but nonculturable populations [10,11,45].

Extending these synthetic architectures to open environments requires implementing genetic controllers that mitigate the evolutionary burden of gene expression and maintain circuit longevity [46,47]. Experimental designs with rhizobacteria have successfully used synthetic promoters to enable "on-demand" functionality for soil restoration and carbon sequestration [48]. Concomitantly, the environmental deployment of these organisms is strictly governed by "safety by design" principles and robust biocontainment modules [3]. Genetically stable CRISPR-based kill switches and "expiry-date" circuits, such as the "demon and angel" expression constructs, have been validated to achieve escape rates below −10°C while ensuring effective bioremediation within defined temporal windows [14–16]. These breakthroughs necessitate a transparent ethical framework that balances innovation with anticipatory regulation to manage the persistence of engineered material in the biosphere [3,49].

2.5 Conservation of Microbial Consortial and Microbiomes

The conservation of microbial consortia and microbiomes requires strategies that preserve the complex interaction networks and spatial structures of the community. Experimental evaluations using shotgun metagenomics have demonstrated that the technical and compositional reproducibility of human fecal microbiome samples is best maintained by room-temperature stabilization methods such as FLOQSwab-ADT or immediate flash-freezing [50]. Long-term studies on human fecal microbiota stored at −196°C in liquid nitrogen have shown that ultralow-temperature cryopreservation has a minimal impact on diversity and community composition compared to fresh cultures [41]. Similarly, research on whole microbial cheese communities has experimentally demonstrated that freeze-drying with maltodextrin preserves the community's metabolic maturation patterns, enabling its successful reuse in subsequent fermentation cycles [51].

A central focus is ecological function-based conservation, which prioritizes the preservation of mutualistic interactions and biogeochemical capabilities. Experimental design of synthetic microbial communities has shown that stability is driven by narrow-spectrum resource-utilizing bacteria that enhance metabolic interactions [13]. Simplified, functional consortia derived from the *Brachypodium* rhizosphere have exhibited consistent reproducibility and revivability from glycerol stocks, maintaining their original microbial composition and taxonomic richness after multiple transfers [12]. Furthermore, studies on the maintenance of microbial structures using cost-effective preservation buffers have shown that human fecal and saliva samples can remain stable at room temperature for up to 4 weeks, even under temperature fluctuations, which is essential for large-scale environmental sampling where freezing facilities are unavailable [52].

The standardization of emerging strategies is vital to advancing global microbiome research. Inter-laboratory trials have validated standardized operating procedures for DNA isolation and purification from complex biological matrices, showing high reproducibility across different research units [53]. The development of whole-cell reference reagents, such as the WC-Gut RR, enables researchers to evaluate the performance of various DNA extraction kits and bioinformatic pipelines, thereby distinguishing their ability to accurately reconstruct taxonomic profiles [41]. Additionally, the introduction of technical reporting standards, such as the STREAMS guidelines, promotes the FAIR data principles by providing a clear framework for managing and reporting environmental and host-associated microbiome data [54]. These standardized protocols ensure that microbiome datasets, including those from preserved museum specimens, can be effectively integrated and compared across studies to understand global microbial diversity [55].

2.6 Integration into Modern Biobanks

Modernizing microbial biobanking requires transitioning to integrated, high-throughput protocols that prioritize functional fidelity alongside taxonomic archiving. Experimental validations have substantiated that the combined use of deoxygenated glycerol buffers and ultra-low-temperature cryopreservation significantly enhances the recovery of fastidious anaerobic pathways [30,41]. These approaches are now coupled with rapid ex vivo assays to empirically confirm that biobanked communities retain their original fermentative responses, such as the production of short-chain fatty acids, upon resuscitation [30]. This shift toward "next-generation biobanking" is underpinned by the routine use of whole-cell standards to calibrate preservation cycles, ensuring that biological material remains functionally consistent across decentralized laboratory networks [56].

The governance of these collections is increasingly defined by digitalization and molecular traceability. The adoption of the STREAMS guidelines provides a robust technical reporting standard for managing the complex metadata associated with environmental microbiomes [54]. Digital traceability has been further revolutionized by shotgun metagenomics, which captures substantially higher microbial diversity and taxonomic resolution than traditional 16 S rRNA sequencing, particularly for archival field specimens [55]. Furthermore, quality control now involves inter-laboratory validation of SOPs, effectively minimizing technical bias in functional biomarker identification and ensuring that postpreservation data are statistically reproducible [13,53].

Finally, the standardization of these emerging frameworks is spearheaded by pivotal global initiatives like the UK Crop Microbiome Cryobank and the Microbiota Vault. These projects have established experimental benchmarks for the long-term archiving of agricultural and clinical diversity, utilizing reproducible synthetic soil communities to validate the resilience of microbial networks over successive storage cycles [57,58]. By defining rigorous community standards for the assembly and resuscitation of synthetic consortia, researchers ensure that engineered microbial products are industrially scalable [57,59]. This structured approach, grounded in experimental data on taxonomic evenness and ecological resilience, provides the essential foundation for the ethical and systematic utilization of global microbial heritages [60,61].

3 Impact of Emerging Strategies on Microbial Functionality

A fundamental paradigm shift in preservation science is the necessary decoupling of cellular viability from biological performance, recognizing that high survival rates do not inherently guarantee metabolic potency. Recent experimental frameworks for viability enumeration emphasize that "potency" must be quantified alongside strain-level functionality to ensure that probiotics and industrial starters retain their health-promoting or fermentative capabilities [62]. Multiomic investigations have elucidated that while lyophilized Staphylococcus and Brevibacterium may exhibit high culturability, their ability to express critical carbohydrate-active enzymes can be transiently suppressed, necessitating a strategic "resuscitation window" to restore full metabolic performance [13,30]. This observation underscores that the "post-thaw lag phase" is not a passive recovery period but a stage of intense metabolic reconfiguration that must be finely tuned to ensure preserved microorganisms can immediately execute their ecological or industrial roles [63].

The implications of these functional insights are profound for both biomedical and environmental applications, particularly in advancing live biotherapeutic products and climate-resilient agriculture. In clinical contexts, the high-fidelity preservation of fecal microbiota has been shown to maintain the community's dynamic response to prebiotic stimuli—a critical factor for the therapeutic efficacy of microbiota transplants in the management of metabolic and gastrointestinal disorders [30,41]. From an environmental perspective, the implementation of synthetic "expiry-date" circuits in soil-remediating bacteria allows for the precise, temporal control of bioremediation; these engineered strains can detoxify pollutants like phenol within a targeted window before self-destructing, thus preventing long-term genomic persistence in the biosphere [3,16].

Such precision depends entirely on the ability to predict and control the microbe's functional state after the transition from dry-state dormancy back to active metabolism [37].

The ultimate benchmark for industrial-scale preservation remains the reproducibility of these complex functional outcomes over extended storage periods. Experimental trials on whole microbial cheese communities have validated that lyophilization with maltodextrin effectively "locks" the metabolic maturation patterns of the consortium, facilitating its functional reuse across successive fermentation cycles without loss of profile [51]. Furthermore, the development of simplified, functional consortia from the *Brachypodium* rhizosphere has demonstrated that taxonomic richness and mutualistic networks can remain consistent across multiple transfers and cryopreservation events [12]. These findings suggest that by adopting "interaction-aware" preservation strategies, researchers can maintain the reproducibility of complex microbial functions, thereby accelerating the commercialization of robust microbial solutions for sustainable waste management and agriculture [18,64].

4 Current Challenges and Future Perspectives

Despite recent advances, significant technical and scaling challenges hinder the widespread adoption of next-generation preservation technologies. The transition from lab-scale optimization to industrial-scale implementation often involves a loss of precision in vacuum and temperature management. For example, while Flash Freeze-Drying can reduce processing time by over 60% in controlled trials, maintaining the uniform sublimation front required for 89.94% viability in large-scale batches remains technically difficult [33,65]. Additionally, the metabolic burden of synthetic gene circuits can reduce the evolutionary longevity of engineered strains, as nonfunctional mutants may outcompete functional ones during the large-scale expansion required before preservation [46,66]. Overcoming these bottlenecks requires the development of more robust genetic controllers and optimized cycle pressures that can be maintained across larger industrial volumes [65,67].

Furthermore, successful technology transfer and the associated costs of these methods are primary concerns for global biobanking. While vacuum-sealed bacterial tablets and optimized lyophilization pressures (0.3–1.0 mbar) offer superior survival rates, the initial capital investment in high-sensitivity vacuum systems and real-time monitoring equipment is substantial [68,69]. However, experimental data suggest that these costs can be offset by the significant reduction in total processing time and energy consumption achieved through cyclic pressure variations and optimized formulations [33]. Facilitating technology transfer to developing regions is a core mission of global initiatives, ensuring that the biological heritage of the most biodiverse areas on Earth is not lost due to a lack of technical infrastructure [1,61].

Looking forward, the future of microbial conservation will likely be defined by the integration of planetary-scale biobanking with synthetic biology. The Microbiota Vault Initiative represents a proactive, global effort to safeguard Earth's microbial diversity against the threats of climate change and habitat loss [1]. Future research will also increasingly explore the engineering of "induced dormancy" and the use of resuscitation-promoting factors to achieve absolute control over the storage and reactivation of microbial life [11,70]. These technologies, combined with the engineering of soil microbiomes for carbon sequestration and ecosystem resilience, offer a transformative path toward a sustainable bioeconomy where microbial functional diversity is not merely archived but actively managed to mitigate global environmental and biomedical challenges [4,71].

5 Conclusions

The transition from descriptive microbiology to precision engineering marks the onset of an era in which preservation is no longer confined to passive dormancy. Still, it entails the active stabilization of complex functional states. Evidence synthesized in this review suggests that metrics for success in modern biobanking have evolved. At the same time, cellular viability remains foundational; it is insufficient without guaranteeing the recovery of metabolic interaction networks and the functional potency of the consortium. Although technical challenges related to industrial scaling and the metabolic burden of synthetic circuits persist, the convergence of omics approaches with "safety-by-design" architectures offers a robust pathway for the responsible implementation of microbial technologies in open environments.

Looking ahead, the success of these emerging strategies will depend on consolidating a global ecosystem of bioinformation and technical standardization. Large-scale initiatives, such as the Microbiota Vault, not only serve as safeguards against biodiversity loss but also establish the infrastructure needed to ensure Earth's microbial heritage remains interoperable and reproducible amid global change. By prioritizing the conservation of functional assemblages over isolated strains, these molecular and technological advancements provide the architectural foundation for a sustainable bioeconomy, where microbial diversity is actively managed to mitigate the most pressing biomedical and environmental challenges of our time.

Conflict of Interest

The authors have no conflicts of interest to declare that are relevant to the content of this chapter.

References

1. Dominguez-Bello MG, Steiger D, Fankhauser M et al (2025) The microbiota vault initiative: safeguarding Earth's microbial heritage for future generations. Nat Commun 16(1):5373
2. Senaratne W, Jayaweera J (2024) Comparison of microbial preservation methods: a narrative review. Germs 14(4):375–386. https://doi.org/10.18683/germs.2024.1447
3. Chemla Y, Sweeney CJ, Wozniak CA et al (2025) Design and regulation of engineered bacteria for environmental release. Nat Microbiol 10(2):281–300. https://doi.org/10.1038/s41564-024-01918-0
4. Jones EM, Marken JP, Silver PA (2024) Synthetic microbiology in sustainability applications. Nat Rev Microbiol 22(6):345–359. https://doi.org/10.1038/s41579-023-01007-9
5. Mariam I, Rova U, Christakopoulos P et al (2025) Data-driven synthetic microbes for sustainable future. NPJ Syst Biol Appl 11(1):74. https://doi.org/10.1038/s41540-025-00556-4
6. Huanca-Juarez J, Nascimento-Silva EA, Silva NH et al (2023) Identification and functional analysis of novel protein-encoding sequences related to stress-resistance. Front Microbiol 14:1268315. https://doi.org/10.3389/fmicb.2023.1268315
7. Juarez JH, Nascimento Silva E, Silva NH et al (2023) Identification and functional analysis of novel stress-resistance genes from metagenomes of extreme environments. bioRxiv 2023:544099. https://doi.org/10.1101/2023.06.07.544099
8. Xu P, Lin NQ, Zhang ZQ et al (2024) Strategies to increase the robustness of microbial cell factories. Adv Biotechnol (Singap) 2(1):9
9. Yang Y, Shao Y, Chaffin TA et al (2022) Performance of abiotic stress-inducible synthetic promoters in genetically engineered hybrid poplar (*Populus tremula* × *Populus alba*). Front Plant Sci 13:1011939. https://doi.org/10.3389/fpls.2022.1011939
10. Guzman J, Raval D, Hauck D, Titz A, Poehlein A, Degenkolb T, ... , Vilcinskas A (2023) The resuscitation-promoting factor (Rpf) from *Micrococcus luteus* and its putative reaction product 1,6-anhydro-MurNAc increase culturability of environmental bacteria. Access Microbiol 5(9):000647-v4
11. Lennon JT, Lehmkuhl BK, Chen L et al (2025) Resuscitation-promoting factor (Rpf) terminates dormancy among diverse soil bacteria. mSystems 10(5):e01517–e01524
12. Chen M, Acharya SM, Yee MO et al (2024) Developing stable, simplified, functional

consortia from *Brachypodium* rhizosphere for microbial application in sustainable agriculture. Front Microbiol 15:1401794. https://doi.org/10.3389/fmicb.2024.1401794
13. Wang W, Xia Y, Zhang P et al (2025) Narrow-spectrum resource-utilizing bacteria drive the stability of synthetic communities through enhancing metabolic interactions. Nat Commun 16(1):6088. https://doi.org/10.1038/s41467-025-61432-7
14. Kato Y, Mori H (2024) Genetically stable kill-switch using "demon and angel" expression construct of essential genes. Front Bioeng Biotechnol 12:1365870
15. Rottinghaus AG, Ferreiro A, Fishbein SRS et al (2022) Genetically stable CRISPR-based kill switches for engineered microbes. Nat Commun 13(1):672
16. Tran KM, Nong NT, Ren J et al (2025) Genetic "expiry-date" circuits control lifespan of synthetic scavenger bacteria for safe bioremediation. Nucleic Acids Res 53:14
17. Kapoore RV, Padmaperuma G, Maneein S et al (2022) Co-culturing microbial consortia: approaches for applications in biomanufacturing and bioprocessing. Crit Rev Biotechnol 42(1):46–72. https://doi.org/10.1080/07388551.2021.1921691
18. Giangeri G, Campanaro S, Kyrpides NC, Angelidaki I (2025) Unlocking the potential of designed microbial consortia: a breakthrough for sustainable waste management and climate resilience. Environ Sci Ecotechnol 25:100558
19. Duong VA, Pan E, Dabral P et al (2025) Metaproteomic analysis to assess the impact of storage media on human gut microbiome in fecal samples. Proteomics 25(19):39–49. https://doi.org/10.1002/pmic.70035
20. Abraham A, Mtewa AG, Chiutula C et al (2025) Prevalence of antibiotic resistance bacteria in manure, soil, and vegetables in urban Blantyre, Malawi, from a farm-to-fork perspective. Int J Environ Res Public Health 22(8):1273. https://doi.org/10.3390/ijerph22081273
21. Xia B, Wang J, Chen H et al (2024) Recent advances in antifreeze peptide preparation: a review. Molecules 29(20):4913. https://doi.org/10.3390/molecules29204913
22. Mallya AS, Yadav P, Zakhia S et al (2025) Low-temperature characterization of sugar alcohol-based type V deep eutectic solvents for anti-icing and cryopreservation applications. Sustain Chem Pharm 48:102256. https://doi.org/10.1016/j.scp.2025.102256
23. Jesus AR, Duarte ARC, Paiva A (2022) Use of natural deep eutectic systems as new cryoprotectant agents in the vitrification of mammalian cells. Sci Rep 12(1):8095. https://doi.org/10.1038/s41598-022-12365-4
24. Bedair HM, Samir TM, Mansour FR (2024) Antibacterial and antifungal activities of natural deep eutectic solvents. Appl Microbiol Biotechnol 108(1):198. https://doi.org/10.1007/s00253-024-13044-2
25. Ghalamara S, Silva S, Brazinha C et al (2022) Structural diversity of marine anti-freezing proteins, properties and potential applications: a review. Bioresour Bioprocess 9(1):5. https://doi.org/10.1186/s40643-022-00494-7
26. Wang Y, Zhang F, Niu X et al (2026) Modular antifreeze peptides: sequence-driven control of ice recrystallization and hydration dynamics. ACS Appl Mater Interfaces https://doi.org/10.1021/acsami.5c17507
27. Joshi P, Rabin Y (2023) Thermomechanical stress analyses of nanowarming-assisted recovery from cryopreservation by vitrification in human heart and rat heart models. PLoS One 18(8):e0290063. https://doi.org/10.1371/journal.pone.0290063
28. de Barros NR, Harb SV, da Silva Horinouchi CD et al (2025) Advances in 3D bioprinting and microfluidics for organ-on-a-chip platforms. Polymers (Basel) 17(22):3078. https://doi.org/10.3390/polym17223078
29. Liu B, Shen Z, Zhou Q et al (2025) Antibiotic resistance genes in the animal manure-amended soil-plant system: occurrence, transmission, bacterial hosts, and human health risks. Ecotoxicol Environ Saf 303:118890. https://doi.org/10.1016/j.ecoenv.2025.118890
30. Zhang X, Walker K, Mayne J et al (2022) Evaluating live microbiota biobanking using an ex vivo microbiome assay and metaproteomics. Gut Microbes 14(1):2035658. https://doi.org/10.1080/19490976.2022.2035658
31. Masoudi R, Asadzadeh N, Sharafi M (2021) Effects of freezing extender supplementation with mitochondria-targeted antioxidant Mito-TEMPO on frozen-thawed rooster semen quality and reproductive performance. Anim Reprod Sci 225:106671. https://doi.org/10.1016/j.anireprosci.2020.106671
32. Senaratne W, Jayaweera J (2024) Comparison of microbial preservation methods: a narrative review. Germs 14(4):375–386. https://doi.org/10.18683/germs.2024.1447
33. Acosta-Piantini E, Villarán MC, Martínez Á et al (2024) Examining the effect of freezing temperatures on the survival rate of micro-encapsulated probiotic Lactobacillus acidophilus LA5 using the flash freeze-drying (FFD) strategy. Microorganisms 12(3):506. https://doi.org/10.3390/microorganisms12030506
34. Ramachandran RP, Nadimi M, Cenkowski S et al (2024) Advancement and innovations in drying of biopharmaceuticals, nutraceuticals, and functional foods. Food Eng Rev

16(4):540–566. https://doi.org/10.1007/s12393-024-09381-7
35. Rahmati F, Hosseini SS, Mahuti Safai S et al (2020) New insights into the role of nanotechnology in microbial food safety. 3 Biotech 10(10):425. https://doi.org/10.1007/s13205-020-02409-9
36. Farahmand A, Ghorani B, Emadzadeh B et al (2024) Double protection of probiotics in alginate hydrogel through emulsification incorporated with freeze drying and coaxial wet-electrospraying: survivability and targeted delivery. LWT 204:116459. https://doi.org/10.1016/j.lwt.2024.116459
37. Li C, Yang X (2022) Editorial: synthetic biology approaches for stress adaptation and improved metabolism in microorganisms. Front Bioeng Biotechnol 10:1013073. https://doi.org/10.3389/fbioe.2022.1013073
38. Hickl O, Heintz-Buschart A, Trautwein-Schult A, Hercog R, Bork P, Wilmes P, Becher D (2019) Sample preservation and storage significantly impact taxonomic and functional profiles in metaproteomics studies of the human gut microbiome. Microorganisms 7(9):367
39. Mordant A, Kleiner M (2021) Evaluation of sample preservation and storage methods for metaproteomics analysis of intestinal microbiomes. Microbiol Spectr 9(3):e01877-21
40. Reck M, Tomasch J, Deng Z et al (2015) Stool metatranscriptomics: a technical guideline for mRNA stabilisation and isolation. BMC Genomics 16(1):494
41. Teo Y, Lavrinienko A, Albertos Torres D, Asare PT, Ruder A, Dominguez-Bello MG, Vonaesch P (2025) Evaluating long-term stool preservation methods for maximizing the recovery of viable human fecal microbiota. Gut Microbes Reports 2(1):2594958
42. Krause JL, Engelmann B, Lallinger DJ, Rolle-Kampczyk U, von Bergen M, Chang HD (2024) Multi-omics analysis unravels the impact of stool sample logistics on metabolites and microbial composition. Microorganisms 12(10):1998
43. Mancini L, Pilizota T (2023) Environmental conditions define the energetics of bacterial dormancy and its antibiotic susceptibility. Biophys J 122(16):3207–3218
44. Vidal E, Lindsay MR, Bradley JA et al (2025) Subsurface life on Earth as a key to unlock extraterrestrial mysteries. Microb Biotechnol 18(12):e70286
45. Xie M, Li Y, Xu L et al (2021) Optimization of bacterial cytokine protein production by response surface methodology for environmental bioremediation. RSC Adv 11(57):36105–36115
46. Barajas C, Huang HH, Gibson J, Sandoval L, Del Vecchio D (2022) Feedforward growth rate control mitigates gene activation burden. Nat Commun 13(1):7054
47. Byrom DP, Darlington APS (2025) Genetic controllers for enhancing the evolutionary longevity of synthetic gene circuits in bacteria. Nat Commun 16(1):8590
48. Dundas CM, Dinneny JR (2022) Genetic circuit design in rhizobacteria. Biodes Res 2022:9858049
49. Ou Y, Guo S (2023) Safety risks and ethical governance of biomedical applications of synthetic biology. Front Bioeng Biotechnol 11:1292029. https://doi.org/10.3389/fbioe.2023.1292029
50. Pribyl AL, Parks DH, Angel NZ et al (2021) Critical evaluation of faecal microbiome preservation using metagenomic analysis. ISME Commun 1(1):14. https://doi.org/10.1038/s43705-021-00014-2
51. Cao W, Passot S, Irlinger F et al (2024) Investigation of freezing and freeze-drying for preserving and re-using a whole microbial cheese community. Foods 13(12):1809
52. Wu C, Chen T, Xu W et al (2021) The maintenance of microbial community in human fecal samples by a cost effective preservation buffer. Sci Rep 11(1):13453. https://doi.org/10.1038/s41598-021-92869-7
53. Ferrocino I, Ferrara M, Garello M et al (2025) Standardizing microbiome research: interlaboratory validation of SOPs for sample preparation and DNA extraction from food and environmental ecosystems. Environ Microbiome 21(1):14. https://doi.org/10.1186/s40793-025-00833-z
54. Kelliher JM, Mirzayi C, Bordenstein SR et al (2025) STREAMS guidelines: standards for technical reporting in environmental and host-associated microbiome studies. Nat Microbiol 10(12):3059–3068. https://doi.org/10.1038/s41564-025-02186-2
55. Madison JD, LaBumbard BC, Woodhams DC (2023) Shotgun metagenomics captures more microbial diversity than targeted 16S rRNA gene sequencing for field specimens and preserved museum specimens. PLoS One 18(9):e0291540. https://doi.org/10.1371/journal.pone.0291540
56. Sergaki C, Anwar S, Fritzsche M et al (2022) Developing whole cell standards for the microbiome field. Microbiome 10(1):123. https://doi.org/10.1186/s40168-022-01313-z
57. Coker J, Zhalnina K, Marotz C et al (2022) A reproducible and tunable synthetic soil microbial community provides new insights into microbial ecology. mSystems 7(6):e00951-22. https://doi.org/10.1128/msystems.00951-22

58. Ryan MJ, Mauchline TH, Malone JG et al (2023) The UK crop microbiome cryobank: a utility and model for supporting phytobiomes research. CABI Agric Biosci 4(1):53. https://doi.org/10.1186/s43170-023-00190-2
59. Northen TR, Kleiner M, Torres M et al (2024) Community standards and future opportunities for synthetic communities in plant-microbiota research. Nat Microbiol 9(11):2774–2784. https://doi.org/10.1038/s41564-024-01833-4
60. Clagnan E, Costanzo M, Visca A, Di Gregorio L, Tabacchioni S, Colantoni E, ... , Bevivino A (2024) Culturomics- and metagenomics-based insights into the soil microbiome preservation and application for sustainable agriculture. Front Microbiol 15:1473666
61. Corrales C, Astrin JJ (2023) Biodiversity biobanking—a handbook on protocols and practices. Pensoft Publishers, p 271
62. Shehata HR, Pane M, Buys EM et al (2024) Editorial: emerging technologies for viability enumeration of live microorganisms. Front Microbiol 15:1546438
63. Ott E, Kawaguchi Y, Özgen N, Yamagishi A, Rabbow E, Rettberg P, ... , Milojevic T (2019) Proteomic and metabolomic profiling of *Deinococcus radiodurans* recovering after exposure to simulated low earth orbit vacuum conditions. Front Microbiol 10:909
64. Soth S, Hampton JG, Alizadeh H, Wakelin SA, Mendoza-Mendoza A (2025) Microbiomes in action: multifaceted benefits and challenges across academic disciplines. Front Microbiol 16:1550749
65. Verlhac P, Vessot-Crastes S, Degobert G et al (2020) Experimental study and optimization of freeze-drying cycles of a model Casei type probiotic bacteria. Dry Technol 38(16):2120–2133. https://doi.org/10.1080/07373937.2019.1683859
66. Byrom DP, Darlington AP (2025) Genetic controllers for enhancing the evolutionary longevity of synthetic gene circuits in bacteria. Nat Commun 16(1):8590
67. Turuvekere Sadguruprasad L, Basavaraj M (2018) Statistical modelling for optimized lyophilization of *Lactobacillus acidophilus* strains for improved viability and stability using response surface methodology. AMB Express 8(1):129
68. Cui W, Zhao L, Shi Y, Feng W, Chen X, Liu H (2025) A vacuum lyophilization and bacterial tablet-based method for culture medium evaluation and potential use in probiotic products. Front Microbiol 16:1493947
69. Hale İnci Ö (2022) The effect of different lyophilisation pressures on the microbiological stability, physicochemical, microstructural, and sensorial properties of yoghurt powders. Int Dairy J 129:105347. https://doi.org/10.1016/j.idairyj.2022.105347
70. Pan H, Ren Q (2022) Wake up! Resuscitation of viable but nonculturable bacteria: mechanism and potential application. Foods 12(1):82
71. Jansson JK, McClure R, Egbert RG (2023) Soil microbiome engineering for sustainability in a changing environment. Nat Biotechnol 41(12):1716–1728. https://doi.org/10.1038/s41587-023-01932-3

Part III

Conservation of Yeast, Filamentous Fungi, and Other Microeukaryotes

Check for updates

Chapter 12

Protocol for the Long-Term Cryopreservation of Carotenoid-Producing *Rhodotorula* spp. Strains

Juan Ramos-Garza, Lily X. Zelaya-Molina, Ismael Fernando Chávez-Díaz and José Luis Aguirre-Noyola

Abstract

The genus *Rhodotorula* comprises oleaginous yeasts widely distributed across diverse environments and recognized for their ability to synthesize carotenoids, lipids, and other high-value metabolites with biotechnological relevance. The preservation of these microorganisms is critical to maintain their genetic stability, metabolic potential, and phenotypic traits over time. Among available methods, cryopreservation using glycerol is one of the most effective strategies for long-term storage, as it protects cellular structures during freezing and minimizes metabolic activity at ultra-low temperatures. In this protocol, we describe a standardized and reproducible method for the long-term cryopreservation of the carotenoid-producing strain, *Rhodotorula glutinis*. The procedure includes cultivation under controlled conditions, preparation of cell suspensions in glycerol, stabilization at 4°C, controlled-rate freezing, and storage at −80°C. Post-reactivation is achieved by direct streaking from frozen stocks, followed by incubation and evaluation of colony morphology and purity. This protocol ensures high postthaw viability and preservation of key phenotypic characteristics, providing a reliable approach for maintaining *Rhodotorula* strains from diverse origins. Its application supports the establishment of microbial collections and facilitates future research and biotechnological exploitation of these yeasts.

Keywords Oleaginous yeasts, Microbial collections, Long-term cryopreservation

1 Introduction

The genus *Rhodotorula*, belonging to the phylum Basidiomycota, comprises oleaginous yeasts widely distributed across diverse natural and anthropogenic environments [1]. These yeasts are recognized for their ability to synthesize carotenoid pigments, which confer characteristic pink, orange, or deep red coloration, and occasionally yellow hues [2]. Colonies are typically mucoid, smooth, and glossy, while at the cellular level, cells are oval to elliptical, ranging from 5 to 10 μm in size, and exhibit multilateral or polar budding patterns [3]. This fungal genus is notable for its remarkable metabolic plasticity and adaptability to fluctuating environmental conditions. Such plasticity

Marco A. Ramírez-Mosqueda et al. (eds.), *Methods for The Conservation of Microorganisms*, Methods in Molecular Biology, vol. 3090, https://doi.org/10.1007/978-1-0716-5623-5_12,

enables the utilization of a wide range of carbon sources, including glucose, glycerol, and even petroleum-derived compounds, as well as tolerance to abiotic stress factors such as extreme pH, salinity, and temperature. These properties are associated, among other mechanisms, with the protective roles of lipids and secondary metabolites [4, 5]. Consequently, *Rhodotorula* is considered ubiquitous, occurring in soils, plant tissues, aquatic systems, air, and environments associated with human activity [6-8].

Beyond their ecological relevance, these yeasts have gained increasing attention due to their biotechnological potential. Several species have been identified as efficient platforms for the biosynthesis of high-value compounds, including lipids such as fatty acids (oleic, linoleic, palmitic, and stearic acids), triacylglycerols, phospholipids, and sterols [9, 10], as well as carotenoid pigments, notably β-carotene, torulene (3′,4′-didehydro-β,γ-carotene), and torularhodin [11, 12]. Additional metabolites associated with this genus include extracellular polysaccharides (EPS), enzymes (lipases, proteases, and amylases), organic acids (citric and succinic acids), phenolic compounds, and siderophores [13]. These attributes position *Rhodotorula* as a promising resource in sectors such as bioenergy, food industry applications, and industrial biotechnology [14].

In this context, the preservation of *Rhodotorula* strains is of strategic importance, as these microorganisms represent reservoirs of genetic and functional diversity with significant application potential. Cryopreservation in glycerol has become one of the most effective approaches for long-term maintenance, as it stabilizes cellular structures during freezing and halts metabolic activity under ultra-low temperature conditions [15]. This method facilitates efficient cell recovery and significantly reduces the likelihood of genetic changes associated with practices such as continuous subculturing [16]. Consequently, the establishment and maintenance of microbial collections not only ensure the availability of strains with defined characteristics but also safeguard biological resources of strategic value whose potential remains only partially explored [17].

Accordingly, a protocol for the cryopreservation of *Rhodotorula* yeast cells using glycerol as a cryoprotectant is described below (Fig. 1). This procedure is applicable to multiple species within the genus and to strains originating from diverse environments.

2 Materials

2.1 Yeast Strain

1. The strain used in this protocol corresponds to *Rhodotorula glutinis* Y23, isolated from the phyllosphere of blue maize grains belonging to a Mexican conical-type landrace. Molecular identification was performed by amplification and sequencing of the internal transcribed spacer 1 (ITS1) region and the

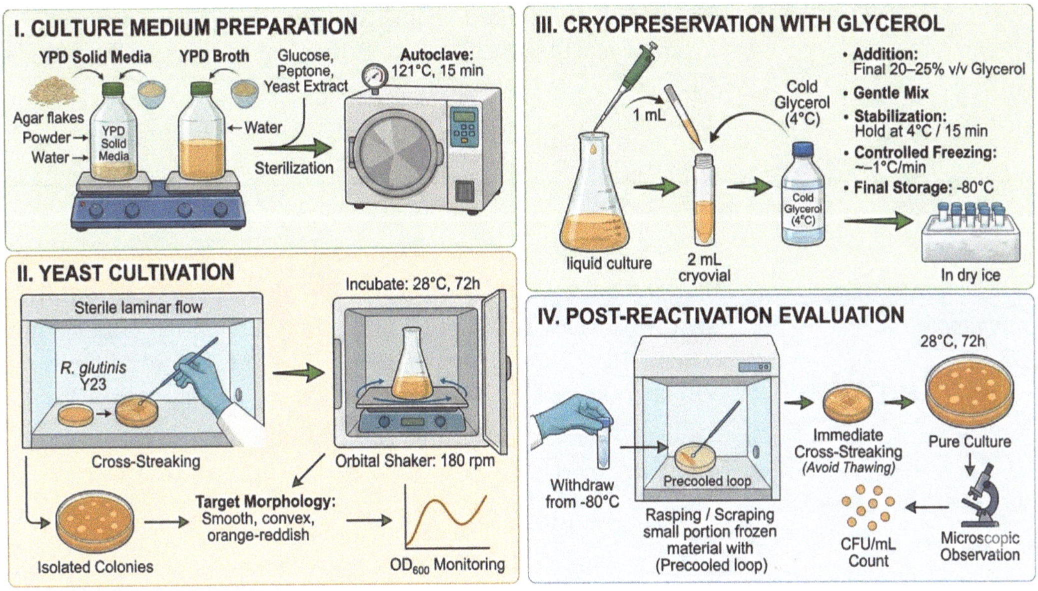

Fig. 1 Schematic workflow of *Rhodotorula glutinis* cryopreservation and post-reactivation evaluation: (I) culture medium preparation, (II) yeast cultivation, (III) cryopreservation with glycerol, and (IV) post-reactivation evaluation

D1/D2 domain of the large subunit (LSU) ribosomal DNA (rDNA). This strain was selected due to its characteristic pigmentation and its ability to promote plant growth, attributes relevant for biotechnological applications and genetic conservation [18].

2.2 Culture Medium

1. Yeast Extract–Peptone–Dextrose (YPD) medium: The nutrient composition of this medium supports rapid growth, high cell viability, and biomass production in nonconventional yeasts.
2. Agar-agar: Used as a solidifying agent.

2.3 Solutions

1. Sterile glycerol solution (40%, v/v): Used as a penetrating cryoprotectant, stabilizing membranes and proteins during freezing.
2. Staining and microscopic evaluation solutions: Lactophenol cotton blue stain, used for observing cellular morphology.

2.4 Equipment and Glassware

1. Culture room: incubator shaker, horizontal laminar flow cabinet, and Bunsen burner.
2. Media preparation room: hot plate stirrer with magnetic stir bar, analytical balance, autoclave, pH meter, spatula, 0.22 μm membrane filters, micropipettes (100–1000 μL), cryovials (2 mL), and plastic Petri dishes (100 mm × 15 mm).

3. Glassware: graduated cylinders (50 mL, 100 mL, and 1 L), Erlenmeyer flasks (250 and 500 mL), beakers (500 mL and 1 L), and 1 L screw-cap bottles.

3 Methods

3.1 Culture Medium Preparation

1. Prepare Yeast Extract–Peptone–Dextrose (YPD) medium by dissolving yeast extract (10 g/L), peptone (20 g/L), and glucose (20 g/L) in 900 mL of distilled water. For solid medium, add agar-agar (15 g/L).
2. Adjust the pH to 6.0 using 1 N NaOH or 1 N HCl.
3. Bring the final volume to 1 L with distilled water.
4. Sterilize by autoclaving at 121°C for 15 min.
5. Allow the medium to cool to approximately 50°C and gently homogenize.
6. Under laminar flow conditions, dispense 15 mL into sterile Petri dishes.
7. Allow the medium to solidify completely.

3.2 Yeast Cultivation

1. Under sterile conditions, inoculate *R. glutinis* Y23 onto YPD agar plates using the cross-streaking method with a sterile inoculation loop.
2. Incubate at 28°C for 72 h.
3. Colonies should exhibit the characteristic morphology of the genus: smooth, convex, and with orange-reddish pigmentation (Fig. 2).
4. Select a single isolated colony and perform three successive subcultures on YPD plates under the same growth conditions.

3.3 Cryopreservation with Glycerol

1. Inoculate a single isolated colony into a 250 mL Erlenmeyer flask containing 50 mL of liquid YPD medium.
2. Incubate at 28°C with orbital shaking (180 rpm) for 72 h (*see* **Note 1**).
3. Monitor growth by measuring OD_{600} until reaching a range of 0.6–1.0, corresponding to the late exponential phase.
4. Dispense 0.75 mL of culture into sterile cryovials (2 mL) and gradually add 0.75 mL of cold sterile glycerol (40%) to reach a final concentration of 20% (v/v).
5. Mix gently to minimize osmotic shock and maintain the suspension at 4°C for 15 min to allow cryoprotectant penetration and membrane stabilization.

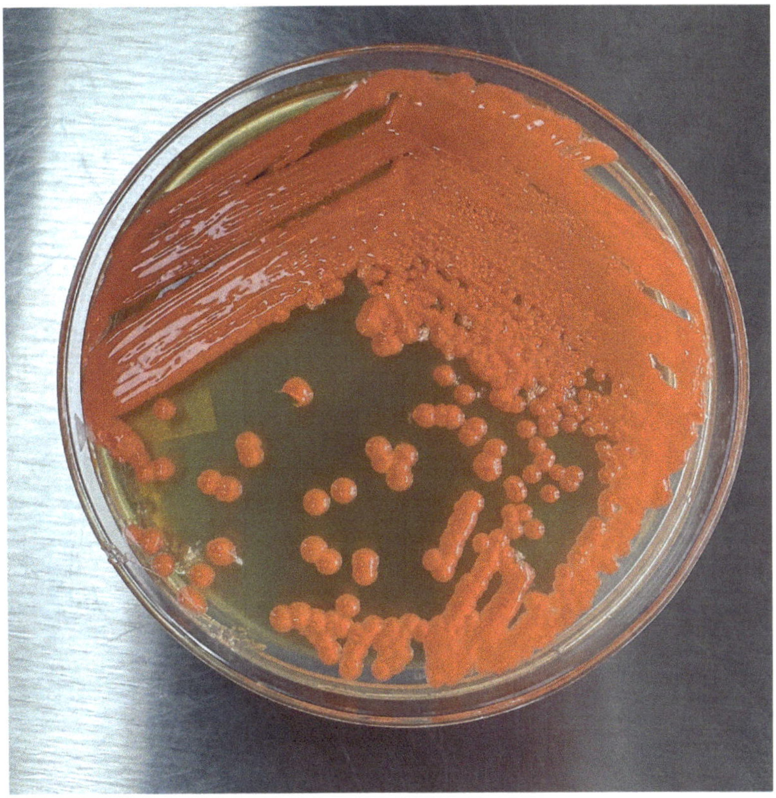

Fig. 2 Colonial morphology of *Rhodotorula glutinis* Y23, isolated from phyllosphere of maize plants. The colonies are circular and smooth, with entire margins, a shiny surface, and characteristic orange pigmentation

6. Freeze the samples under controlled cooling conditions (~−1°C/min) in an ultra-low temperature freezer prior to long-term storage at −80°C.

3.4 Post-Reactivation Protocol

1. Remove the cryovial from −80°C storage using a precooled rack or dry ice.
2. Under a laminar flow hood, open the vial and collect a small portion of the frozen material using a precooled sterile inoculation loop.
3. Immediately streak the sample onto a YPD agar plate using the cross-streaking method.
4. Close the cryovial and promptly return it to −80°C, avoiding partial or complete thawing.
5. Incubate YPD plates at 28°C for 72 h.
6. Evaluate colony morphology, pigmentation, purity, and microscopic characteristics using appropriate staining and microscopy (*see* **Note 2**).

4 Notes

1. Cells in the late exponential phase exhibit increased tolerance to osmotic stress and enhanced cryoprotective response, likely due to higher metabolic activity and membrane integrity, resulting in improved postthaw viability.
2. For microscopic observation, a small portion of a colony should be collected using a sterile inoculation loop and mounted on a glass slide with a drop of lactophenol cotton blue stain. The preparation should be gently mixed, covered with a coverslip, and observed at 10× and 40× magnification. The typical morphology of *R. glutinis* consists of spherical to ovoid cells, with no formation of pseudohyphae or true hyphae under the described culture conditions.

Conflict of Interest

The authors have no conflicts of interest to declare that are relevant to the content of this chapter.

References

1. Chentir I, Miri YB, Kot AM, Nikhanj P, Mussagy CU, Arous F, Markou G (2025) *Rhodotorula* genus: vast knowledge, limited commercialization–what's holding us back? Fungal Biol Rev 54:100459
2. Ochoa-Viñals N, Alonso-Estrada D, Pacios-Michelena S, García-Cruz A, Ramos-González R, Faife-Pérez E et al (2024) Current advances in carotenoid production by *Rhodotorula* sp. Fermentation 10(4):190
3. Khunnamwong P, Nualthaisong P, Kingphadung K, Takashima M, Sugita T, Sumerta IN, Limtong S (2025) *Rhodotorula tropicalis* sp. nov., a novel red yeast of the order Sporidiobolales isolated from Thailand, Indonesia and Japan. Int J Syst Evol Microbiol 75(3):006701
4. Benmessaoud S, Anissi J, Kara M, Assouguem A, Al-Huqail AA, Germoush MO et al (2022) Isolation and characterization of three new crude oil degrading yeast strains, *Candida parapsilosis* SK1, *Rhodotorula mucilaginosa* SK2 and SK3. Sustainability 14(6):3465
5. Sineli PE, Maza DD, Aybar MJ, Figueroa LI, Viñarta SC (2022) Bioconversion of sugarcane molasses and waste glycerol on single cell oils for biodiesel by the red yeast *Rhodotorula glutinis* R4 from Antarctica. Energy Convers Manag 16:100331
6. El Maimouni H, Aasfar A, Meftah Kadmiri I (2026) Unraveling the role of yeasts and potential endosymbionts in plant growth promotion: insights from *Rhodotorula mucilaginosa*. In: Idbella M (Ed.). The soil microbiome in agriculture: opportunities and challenges. Springer Nature Switzerland, Cham, 71–106
7. Lane DM, Valentine DL, Peng X (2023) Genomic analysis of the marine yeast *Rhodotorula sphaerocarpa* ETNP2018 reveals adaptation to the open ocean. BMC Genomics 24(1):695
8. Wykrętowicz K, Czyżewska-Dors E, Dors A, Pomorska-Mól M, Augustyniak A, Łagowski D (2025) *Rhodotorula* spp. in laboratory and veterinary clinical practice: contamination or an emerging problem? Animals 15(22):3299
9. Wu CC, Honda K, Kazuhito F (2023) Current advances in alteration of fatty acid profile in *Rhodotorula toruloides*: a mini-review. World J Microbiol Biotechnol 39(9):234
10. Zheng X, Li Y, Che C, Gao Q, Wang X, Ledesma-Amaro R et al (2025) Metabolic engineering of *Rhodotorula toruloides* for the production of linalool. J Agricult Food Chem 73(9):5395–5404

11. Li Z, Li C, Cheng P, Yu G (2022) Rhodotorula mucilaginosa—alternative sources of natural carotenoids, lipids, and enzymes for industrial use. Heliyon 8(11), e11505. https://doi.org/10.1016/j.heliyon.2022.e11505
12. Zhang TT, Wu AH, Aslam M, Song JZ, Chi ZM, Liu GL (2024) Developing *Rhodotorula* as microbial cell factories for the production of lipids and carotenoids. Green Carbon 2(4):409–420
13. Kot AM, Błażejak S, Kurcz A, Gientka I, Kieliszek M (2016) *Rhodotorula glutinis*—potential source of lipids, carotenoids, and enzymes for use in industries. Appl Microbiol Biotechnol 100(14):6103–6117
14. Zheng X, Ledesma-Amaro R, Zhao ZK, Wang Y, Zhao XQ, Yang X (2025) *Rhodotorula* yeasts as potential chassis for sustainable food biotechnology. J Agricult Food Chem 73(31):19157–19173
15. Singh SK, Baghela A (2017) Cryopreservation of microorganisms. In: Varma A, Sharma AK (Eds.) Modern tools and techniques to understand microbes. Springer International Publishing, Cham, 321–333
16. Elliott GD, Wang S, Fuller BJ (2017) Cryoprotectants: a review of the actions and applications of cryoprotective solutes that modulate cell recovery from ultra-low temperatures. Cryobiology 76:74–91
17. Gómez-Godínez LJ, Hernández-Cruz E, Guzmán-González RD, Ruiz-Sandoval G, Ruiz Rivas M, Aguirre-Noyola JL, De los Santos Villalobos S (2026) Preservation of microorganisms and microbiomes: methods, impacts, and future prospects. Biopreserv Biobank 19475535251401795
18. Ramos-Garza J, Aguirre-Noyola JL, Bustamante-Brito R, Zelaya-Molina LX, Maldonado-Hernández J, Morales-Estrada AI, Martínez-Romero E (2023) Mycobiota of Mexican maize landraces with auxin-producing yeasts that improve plant growth and root development. Plants 12(6): 1328

Chapter 13

Cryopreservation, Reactivation, and Propagation of *Candida tropicalis* (Castellani) Berkhout 750™ for Biotechnological Applications

Audry Peredo-Lovillo, Francisco Erik González-Jiménez, Haydee Eliza Romero-Luna, Carlos Eduardo Lara-Hidalgo, Guadalupe Luna-Solano and Carlos Alberto Cruz-Cruz

Abstract

Candida tropicalis is a nonconventional yeast with biotechnological applications based on its capacity to produce sweeteners, biofuels, enzymes, and polysaccharides from the assimilation of different carbon sources. The biotechnological advantages are dependent on its metabolism, which can be preserved avoiding genotypic or phenotypic changes through cryopreservation. Hence, the protocols needed to carry out the cryopreservation of *C. tropicalis* (Castellani) Berkhout 750™ obtained from the KWIK-STIK™ are described. KWIK-STIK™ provides lyophilized microorganisms useful as control strains at the laboratory level; however, these strains, once reactivated, are genetically unstable, and a series of passages compromise their metabolism. In this sense, cryopreservation favors the obtaining of pure cultures stable for prolonged periods and with the ease of obtaining large quantities of biomass from small inoculum. These advantages can be used for various purposes, either for the generation of bioactive compounds or as starters for the reproduction of new metabolically stable batches.

Keywords Nonconventional yeast, Industrial applications, KWIK-STIK™, Non-*Saccharomyces*, *Candida*, Freeze-drying, Stock culture

1 Introduction

Candida tropicalis is recognized as a nonconventional yeast with potential biotechnological applications based on its secondary metabolites, tolerance to inhibitors, and assimilation of different carbon sources, including simple carbohydrates such as glucose, xylose, galactose, sucrose, maltose, and trehalose [1]. Based on these features, *C. tropicalis* has been of interest in different industrial sectors, either to produce sweeteners, biofuels, enzyme, and polysaccharides, and also to contribute to bioremediation and medical developments [2, 3].

Marco A. Ramírez-Mosqueda et al. (eds.), *Methods for The Conservation of Microorganisms*, Methods in Molecular Biology, vol. 3090, https://doi.org/10.1007/978-1-0716-5623-5_13,

Despite being considered an opportunistic pathogen in the medical field, *C. tropicalis* stands out as a yeast with desirable characteristics for its incorporation and potential scalability in industrial processes related to the biotransformation; an interesting approach considering that current scaled processes have *Saccharomyces cerevisiae* strains as the only biotechnologically viable microorganisms, with exception in tequila, kraft beers and wine, where *Kluyveromyces marxianus* and *Pichia kluyveri*, as well as *Saccharomyces eubayanus* and *Torulaspora delbruekii* are used, respectively [3]. Nevertheless, it has been proven that nonconventional yeasts, such as *C. tropicalis*, can act as an alternative catalyst with high potential for the bioremediation of environmental pollutants, converting them into various biomolecules or by-products for industrial applications [1, 3]. Such advantages are given by the culture conditions, nutrimental requirements, and isolation sources of yeast strains. Specifically, culture conditions of *C. tropicalis* are strain-dependent; however, its versatility to grow in variable conditions favors its uses and applications. In this case, the pH range covers between 4.0 and 6.0, with an incubation temperature from 25 to 35°C, under anaerobic or aerobic conditions with different carbon sources, including galactose, maltose, trehalose, sucrose, xylose, lactose, fructose, *n*-alkanes, lipids, and some lignocellulosic waste (i.e., sugarcane bagasse or wheat bran), although at laboratory level, the Sabouraud glucose or potato-dextrose are the conventional culture medium for yeast cultivation [1, 4].

Regarding its habitat, *C. tropicalis* has been identified in several ecological conditions such as arable soil, forest soil, freshwater, seawater, and even in hypersaline environments such as the Dead Sea. Likewise, it can be found as part of the human microbiota and skin, as well as in the genitourinary and respiratory tracts. So, it has been found that *C. tropicalis* can withstand and grow in environments with 10%–15% (w/v) of sodium chloride. This osmotolerance is not essential for the yeast survivability, however, highlights its biotechnological interest [5, 6].

Several microorganisms, such as nonconventional yeasts, are recurrently tested about their application in different biotechnological processes, taking into consideration that their use will economically benefit these processes due to its easy incubation and low maintenance and reactivation costs [7]. In addition to purity of the yeast strain, other parameters must be considered to ensure higher values of product efficiency and yield in the biotransformation process. Hence, parameters such as culture medium constitution, reactivation protocols and specific strain conservation must be involved as a part of the biotechnological applications of nonconventional yeasts such as *C. tropicalis* [8].

Cryopreservation in the ultra-freezer at −80°C or 196°C (in contact with liquid nitrogen) is the most used technique to keep microorganisms preserved, allowing the possibility to storing biological materials for an extended period of time before use it. Also, cryopreservation requires the use of cryoprotectants due to the low temperatures. These cryoprotectants are substances added to the culture medium in order to prevent severe damage at cellular level by freezing and thawing them [3, 9]. The selection of cryoprotectants and freezing conditions depends on the microorganism specie and strain, since maintaining cells growth capacity for a long storage period does not guarantee a successful application in industrial processes, in which yeasts such as *C. tropicalis* must present metabolic activity as fast as possible and in turn, produce the metabolites of interest [3]. With this in mind, there is a need to establish protocols that guarantee the correct storage and subsequent reactivation of microorganism cells with biotechnological applications, so that their use in bioprocesses contributes, in addition to the generation of biomolecules, to the development of more sustainable and profitable technologies [1].

Therefore, in the following lines, protocols for reactivation, isolation, and cryopreservation of *C. tropicalis* (Castellani) Berkhout 750™, a nonconventional yeast strain with applications in energy, food, chemical, and pharmaceutical industries, will be described, highlighting the necessary cultivation conditions as well as the handling of a KWIK-STIK™ unit as the source of the reference strain *C. tropicalis* (Castellani) Berkhout 750™.

2 Materials

2.1 Yeast Strain

Reference yeast strain of *Candida tropicalis* (Castellani) Berkhout derived from ATCC 750™ was purchased from Microbiologics (Saint Cloud, MN, USA). Lyophilized cells pellet was hydrated through a KWIK-STIK™ unit (Fig. 1), following the manufacturer's recommendations.

2.2 Surface Sterilization

Disinfectants solutions such as commercial liquid detergent, 70% (v/v) ethanol solution and 5% (v/v) sodium hypochlorite are used throughout the execution of the protocol.

2.3 Culture Medium and Microbiological Solutions

1. Microorganisms contained in KWIK-STIK™ require a nonselective, nutritive or enriched culture medium, and specific incubation conditions in order to optimize the growth and recover lyophilized cells of *C. tropicalis*. According to this, Sabouraud glucose agar and broth are used to cultivate the yeast cells (see Table 1).
2. Culture broth is also used as a cryopreservation medium once it is added with glycerol (used as cryoprotectant) in a 20:80 ratio (v/v, glycerol:culture broth) and sterilized by autoclaving for 15 min at 15 psi.

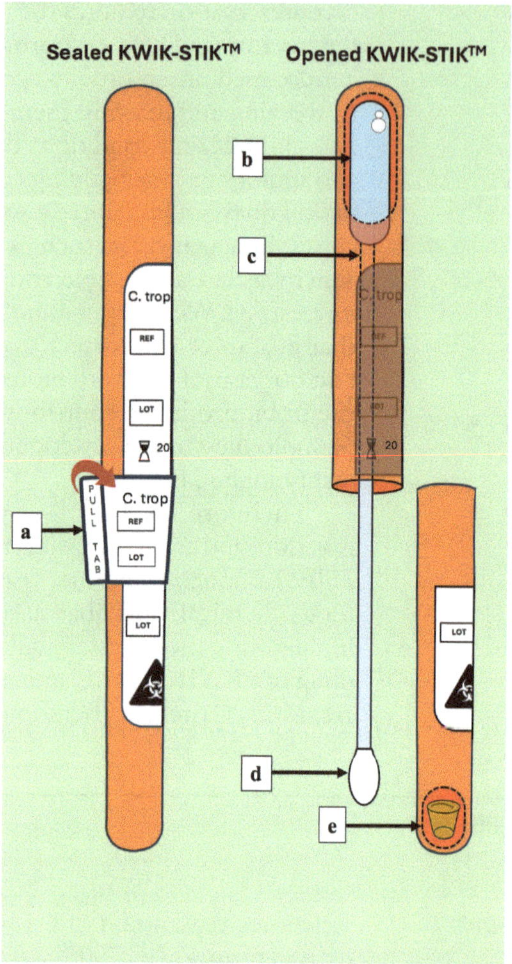

Fig. 1 Schematic diagrams of the sealed and opened KWIK-STIK™. Key components include pull-tab portion (**a**), inner glass ampoule (**b**) containing sterile hydration fluid, irrigation tube (**c**), sterile swab (**d**), and lyophilized pellet of yeast culture (**e**)

3. Phosphate buffer solution (PBS, 1×, pH 7.4) is prepared to be used in the washes of cell pellet obtained after yeast cells growth, prior to be mixed with the cryopreservation solution of glycerol:culture broth (*see* **Note 1**). Likewise, if the reagents for preparing the PBS are not available, an isotonic physiological solution of 0.85%–0.9% sodium chloride (NaCl), adjusted to the same pH value and sterilized under the same conditions, can also be used.
4. Methylene blue (MB) solution (0.002%) is used as a contrast dye in the staining process of yeast cells, facilitating the observation of the cellular structure [10]. This same staining medium is used to count viable cell number, since viable cell

Table 1
Composition of culture medium used for *Candida tropicalis* growth and recovery

Media constituents	Sabouraud dextrose broth	Sabouraud dextrose agar
Dextrose/glucose (g)	40.0	40.0
Enzymatic digest of casein (g)	5.0	5.0
Enzymatic digest of animal tissue (g)	5.0	5.0
Agar (g)	N/A	15.0
Distilled water (mL)	1000	1000
Final pH value	5.6 ± 0.2 at 25°C	

N/A not added.

exclude MB, reducing it to a colorless form, while dead cells absorb MB without reducing it [11].

2.4 Equipment

1. Laboratory equipment: analytical balance, electronic balance, magnetic stirrers, heating plates, potentiometers, refrigerator, autoclave, Bunsen burners, incubator, air oven, optical microscope, staining bridge, ultra-freezer (−80°C), laminar flow hood (for aseptic conditions).
2. Crystal material: volumetric flasks (100 and 1000 mL), Erlenmeyer flasks (250, 500 mL), beakers (250, 600, 1000 mL), graduated cylinder (100, 250, 1000 mL), graduate pipette (5 and 10 mL), borosilicate glass bottles (500 mL) with polypropylene screw cap and pouring ring, glass culture tubes with screw cap 15 × 150 mm.
3. Microbiological culture instruments: sterile and disposable plastic Petri dishes (90 × 15 mm), conical centrifuge tubes (15- and 50-mL capacity), cryogenic tubes with screw cap (2 mL capacity), cryogenic freezer storage box, nichrome inoculating loops, micropipettes (0.5–10, 20–200, and 100–1000 μL), and disposable micropipette's tips (0.1–20, 2–200, and 50–1000 μL).
4. Other materials: adhesive plastic wrap, paper towels, aluminum foil, adhesive tape, logbook, coverslip, slides, immersion oil, lighter, latex hoses, indelible marker, and trigger-spray bottle.

3 Methods

3.1 KWIK-STIK™ Microorganism Procedure

1. Before opening, let the KWIK-STIK™ pouch reach room temperature. Once temperature equilibrated, open the pouch at the notch and remove the KWIK-STIK™ unit (*see* **Note 2**). Crack the hydration capsule (just below the fluid meniscus)

located on the top of the KWIK-STIK™ unit, releasing the hydration fluid (*see* **Note 3**) (Fig. 2a, b).

2. Facilitate the flow of the hydration fluid holding vertically the KWIK-STIK™ to allow it to reach the bottom of the unit where the culture pellet is located. Avoid shaking the KWIK-STIK™ and do not cover the small vent hole on the top of the KWIK-STIK™ (Fig. 2c).
3. Once hydrated, dissolve the culture pellet in the hydration fluid using pinching movements (Fig. 2d).
4. Immediately after dissolving the cell pellet and having observed that the swab is generously impregnated with this cell solution, transfer the fluid to the Sabouraud dextrose agar by gently rolling the swab over one-third of the plate (Fig. 2e). Reassemble the KWIK-STIK™ unit and return it to the pouch for disposal, following the institutional protocols for the management and final disposal of biological-infectious waste.
5. Using a microbiological loop, streak the culture starting from the previous inoculation, to facilitate colony isolation and ensure culture purity through colonial and cellular differentiation under microscope observation (Fig. 2f).
6. Incubate the inverted inoculated primary plate at 30°C for 48 h under aerobic conditions (Fig. 2g).
7. Once grown on the plate, the purity of the culture is confirmed by colony and cellular characterization after staining with an MB (Fig. 2h, i). Stained cells are observed under optic microscope to ensure that they represent a single cell morphology with the desired characteristics (Fig. 2j).
8. Finally, with the purity of the culture confirmed, it is possible to continue either with the planned tests or instead with mass propagation prior to its use for biotechnological purposes (Fig. 2l).

3.2 Procedure for Cryopreservation of Yeast Culture

A proper maintenance of microbiological strains for quality control is essential to ensure the reproducibility and accuracy of test assays. So, a plan to preserve viability, purity, genotypic and phenotypic characteristics of microorganisms should be considered once a lyophilized cells have been reactivated. Mostly, nonselective agar is used to reactivate microorganisms from lyophilized pellets, since they offer the opportunity to observe the purity of the culture according to the characteristics of the individual cells or isolated colonies [1] (*see* **Note 4**). Also, the use of broth is not recommended because some undesirable microorganisms can easily be introduced without notice differences in the culture. However, broth is required when mass propagation of culture or quantification of viable cells is required, as

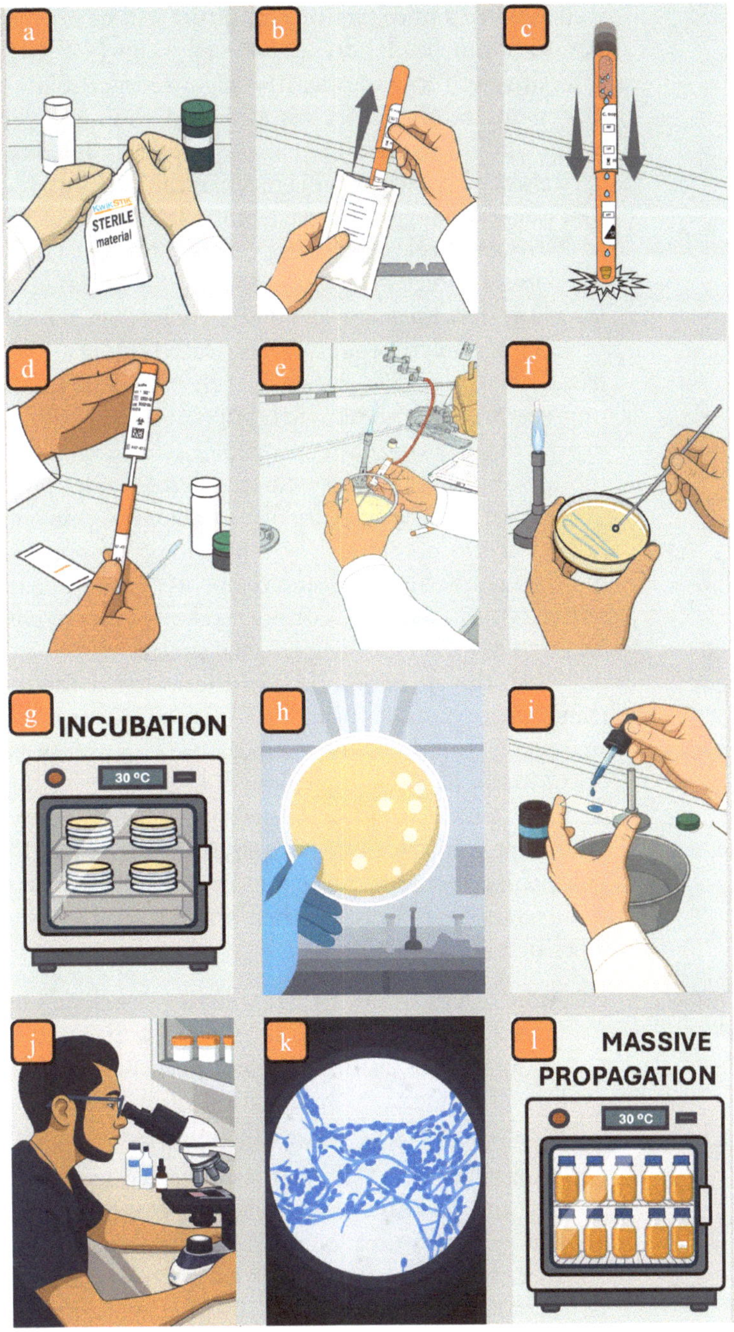

Fig. 2 Protocol for activation and propagation of *Candida tropicalis* (Castellani) Berkhout 750 obtained from KWIK-STIK™ for biotechnological purposes. Procedure starts with the acclimatization of KWIK-STIK™ pouch to room temperature, followed by the remotion of KWIK-STIK™ and cracking of the glass ampoule filled with the sterile hydration solution (**a,b**). Facilitate the flow of hydration fluid by keeping the stick vertical, avoiding covering the small vent hole on the top (**c**). Once at the bottom, the fluid will come into contact with the lyophilized culture and, once impregnated, the swab with culture is streaked on Sabouraud dextrose medium plates up to one third of the plate. The other parts of the plate are streaked with a microbiological loop (**d, e,** and **f**). Inoculated plates are incubated at 30°C under aerobic conditions (**g**). Colonial morphology of colonies forming units is studied and cellular structure evaluated by optical microscopy (**h–k**). Once purity is confirmed, the culture can be mass propagated and used for different biotechnological processes, or, if necessary, cryogenically preserved (**l**)

well as when a microbiological culture will be preserved for a longer period of time under cryogenic conditions [12, 13].

Although low temperatures pause microbial activity, adverse effects including severe osmotic and chemical stress are present along with the formation of intra and extracellular ice crystals. These crystals generate irreversible mechanical damage to cellular structure that decrease the postthaw viability. In this sense, cryopreservation of microbial species, including yeast strains, must be accompanied by the correct choice of cryoprotectants that maintain cellular viability for a long-term storage, avoiding genetic degenerations (commonly observed in microorganisms cultivated on agar plates with several passages). Chemical substances such as dimethyl sulfoxide (DMSO) or glycerol are the most used cryoprotectants; however, a prolonged exposure to DMSO affects cellular functions and interferes with the metabolism and enzymatic activity, while low concentrations of glycerol (<20%, v/v) are not sufficient to prevent crystallization [13]. Therefore, the choice of the correct cryoprotectant, as well as the cryogenic conditions and the protocol for preparing the culture to be cryopreserved are considered key points to ensure prolonged and stable storage of the microbial species.

With this in mind and considering that the reactivation and maintenance of strains from the KWIK-STIK™ is based on passages in agar plates stored at room temperature or in refrigeration, cryopreservation can be an aid in maintaining both the viability and genetic stability of microbiological reference cultures and with biotechnological applications such as *C. tropicalis* (Castellani) Berkhout 750™. The protocol proposed for the cryopreservation of pure microbial strains in order to preserve their phenotypic and genotypic purity is described below [14].

Cryopreservation Protocol for Yeast Cells Reactivated from KWIK-STIK™

1. After reactivation of *C. tropicalis* (Castellani) Berkhout 750™ from KWIK-STIK™, yeast cells are propagated in culture broth to obtain biomass for different purposes, either for biotechnological applications (generation of interest metabolites) or for preservation by cryogenic techniques.
2. Prior cryopreservation, it is necessary to evaluate the growth rate and to know the incubation conditions, specifically time and temperature. Starting from a single isolated colony, growth kinetics are performed by monitoring optical density (at 600 nm) or cell viability by plate counting, thus identifying the end of the exponential growth phase (*see* **Note 5**). Once this stage is identified, it is possible to obtain biomass with a greater capacity to tolerate the conditioning process prior to being exposed to the low temperatures of cryogenics (Fig. 3).
3. Once adjusted, the incubation time and temperature, cell pellet from second growth kinetic is obtained by centrifugation and the supernatant discarded.

4. Two washes with PBS are performed on the cell pellet, repeating the centrifugation to remove the remains of the used broth.
5. Washed cell pellet is resuspended in the cryogenic solution, constituted by sterile broth and the selected cryoprotectant (commonly glycerol) in a 20:80 (v/v) ratio.
6. Once homogenized, the cell solution is distributed in the cryogenic vials, being filled with variable volumes (e.g., 100, 200, 500, or 1000 μL), thus ensuring that the entire thawed volume is used in its entirety and that it is not necessary to return any

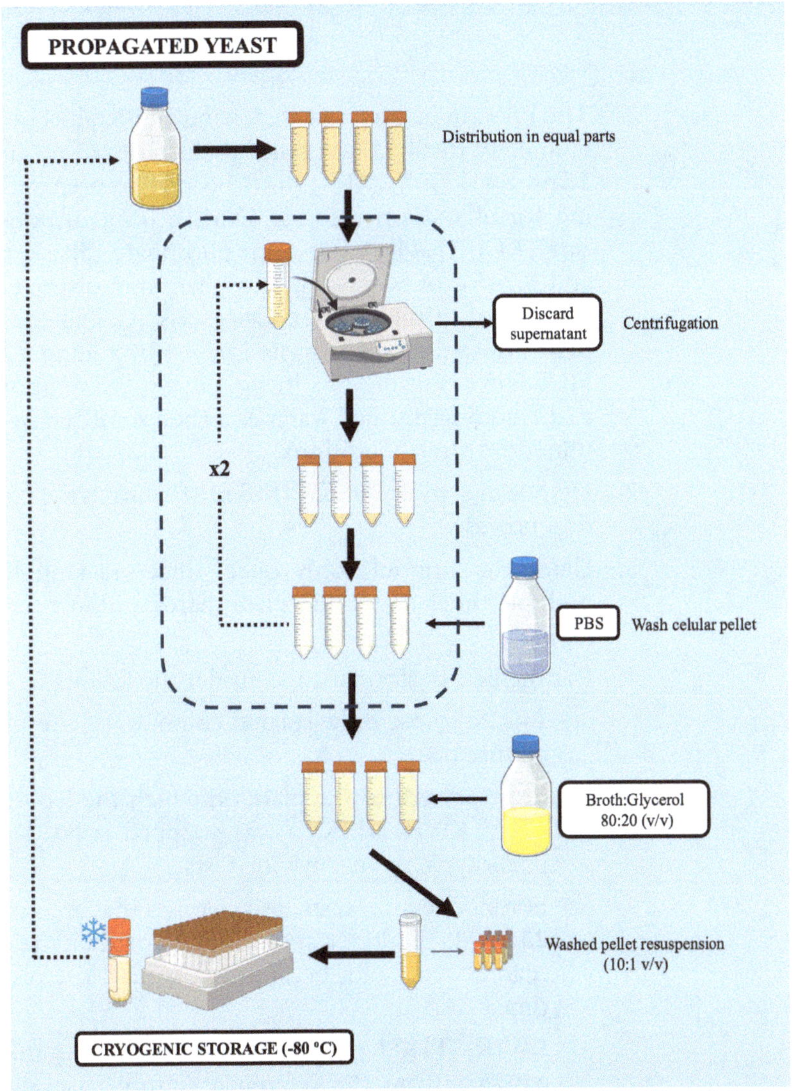

Fig. 3 Protocol for cryopreservation of *Candida tropicalis* Berkhout 750 reactivated from KWIK-STIK™.(created with BioRender.com)

surplus to cryopreservation, a fact that decreases the viability of the culture even when stored under the same conditions.

7. Immediately after being filled, vials are stored under cryogenic temperature of −80°C, being thawed when the culture needs to be reactivated for some analysis or to carry out its mass propagation, following the same protocols used to obtain the same vials.
8. The purity of the cell culture must be monitored throughout the cryopreservation process by plate streaking, performing colonial and cellular characterization under a microscope.

4 Notes

1. The PBS can be prepared dissolving a PBS tablet in the required volume of distilled water and then sterilized by autoclaving at 15 psi for 15 min. Also, this PBS can be prepared by dissolving 8 g of sodium chloride (NaCl), 0.2 g of potassium chloride (KCl), 1.44 g of sodium phosphate dibasic (Na_2HPO_4), and 0.245 g of potassium phosphate monobasic (KH_2PO_4) in 800 mL of distilled water. Once dissolved all the solutes, the pH of the solution is adjusted to 7.4 by adding 1 N NaOH or HCl under continuous stirring. Finally, the solution is brought to 1 L adding distilled water and then sterilized by autoclaving under the same conditions.
2. Do not dissemble the KWIK-STIK™ during the culture hydration process.
3. Crack the ampoule only once, since cracking the ampoule multiple times may cause glass shards and increase the risk of injury.
4. For the best performance, consider the following:
   (a) Ensure to use only isolated colonies for a test from a pure culture plate.
   (b) Yeast growth in the plate on which the lyophilized pellet (from KWIK-STIK™) was swapped is NOT suitable for phenotypic or antimycotic tests.
   (c) For biochemical tests, use colonies that are not more than 24 h old. While for antibiotic susceptibility tests, no more than 2 weeks old, because yeasts may lose resistance over time.
   (d) KWIK-STIK™ protocols recommend a minimal of two passages from the reference culture to evaluate growth promotion or for the performance of pharmacopeia tests.

5. In the first kinetic, the broth must be inoculated with a single colony isolated from a pure agar plate. Subsequently, the second kinetic is inoculated at 1% (v/v) from the cell culture from the first kinetic. For this second kinetic, a larger volume of medium is required depending on the quantity of vials to be prepared. Keep in mind that the cell pellet is dispersed in one-tenth of the initial volume of this kinetic using broth added with the cryoprotectant.

Conflict of Interest

The authors have no conflicts of interest to declare that are relevant to the content of this chapter.

References

1. Queiroz SS, Jofre FM, Bianchini IA, Boaes TS, Bordini FW, Chandel AK, Felipe MGA (2023) Current advances in *Candida tropicalis*: yeast overview and biotechnological applications. Biotechnol Appl Biochem 70:2069–2087. https://doi.org/10.1002/bab.2510
2. Tofalo R, Fusco V, Böhnlein C, Kabisch J, Logrieco AF, Habermann D, Cho GS, Benomar N, Abriouel H, Schmidt-Heydt M, Neve H, Bockelann W, Franz CMAP (2020) The life and times of yeasts in traditional food fermentations. Crit Rev Food Sci Nutr 60:3103–3132. https://doi.org/10.1080/10408398.2019.1677553
3. Casas-Godoy L, Arellano-Plaza M, Kirchmayr M, Barrera-Martínez I, Gschaedler-Mathis A (2021) Preservation of non-*Saccharomyces* yeasts: current technologies and challenges. Compr Rev Food Sci Foo Saf 20:3464–3503. https://doi.org/10.1111/1541-4337.12760
4. Shariq M, Sohail M (2019) Application of *Candida tropicalis* MK-160 for the production of xylanase and ethanol. J King Saud Univ Sci 31:1189–1194. https://doi.org/10.1016/j.ksus.2018.04.009
5. Zuza-Alves DL, Silva-Rocha WP, Chaves GM (2017) An update on *Candida tropicalis* based on basic and clinical approaches. Front Microbiol 8:1–25. https://doi.org/10.3389/fmicb.2017.01927
6. Zuza-Alves DL, Silva-Rocha WP, Francisco EC, de Araújo MCB, de Azevedo Melo AS, Chaves GM (2019) *Candida tropicalis* geographic population structure maintenance and dispersion in the coastal environment may be influenced by the climatic season and anthropogenic action. Microb Pathog 128:63–68. https://doi.org/10.1016/j.micpath.2018.12.018
7. Queiroz SS, Jofre FM, Mussatto SI, Felipe MGA (2022) Scaling up xylitol bioproduction: challenges to achieve a profitable bioprocess. Renew Sustain Energy Rev 154:111789. https://doi.org/10.1016/J.RSER.2021.111789
8. Arora R, Singh P, Sarangi PK, Kumar S, Chandel AK (2023) A critical assessment on scalable technologies using high solids loadings in lignocellulose biorefinery: challenges and solutions. Crit Rev Biotechnol 1–18. https://doi.org/10.1080/07388551.2022.2151409
9. Lima Costa ME, Pereira Rodrigues LA, dos Santos Silva M (2023) Comparison and validation of cryopreservation methods of *Candida* genus fungi: a brief review. J Bioeng Tech Health 6:174–176. https://doi.org/10.34178/jbth.v6i2.300
10. Yoda T (2022) Measurement and characterization of yeast cell size using a digital optical microscope. Processes 10:2396. https://doi.org/10.3390/pr10112396
11. Matsumoto A, Terashima I, Uesono Y (2022) A rapid and simple spectroscopic method for the determination of yeast cell viability using methylene blue. Yeast 39:607–616. https://doi.org/10.1002/yea.3819
12. Recommended culture methods for microorganisms. TIB-05617.P. www.microbiologics.com. Accessed August 2025
13. Ramasamy T, Tevatia R, Ali S, Muhle A, Knight-Connoni V, Chakraborty N (2025) Proteomic approach for evaluating cryoprotectant formulations for enhanced post-cryopreservation recovery of yeast. Sci Rep

15:15474. https://doi.org/10.1038/s41598-025-00534-0

14. Romero-Luna HE, Hernández-Sánchez H, Ribas-Aparicio RM, Cauich-Sánchez PI, Dávila-Ortíz G (2018) Evaluation of the probiotic potential of *Saccharomyces cerevisiae* strain (C41) isolated from tibicos by in vitro studies. Probiotics Antimicrob Prot 11:794–800. https://doi.org/10.1007/s12602-018-9471-2

Check for updates

Chapter 14

Preservation of *Fusarium oxysporum* Endophytes and Pathogenic of Vanilla

Mauricio Luna-Rodríguez and Dulce María Murrieta-Hernández

Abstract

The preservation of phytopathogenic fungi is essential for maintaining biological collections that support research and disease management strategies. Among the most important diseases affecting vanilla cultivation is root and stem rot caused by *Fusarium oxysporum* f. sp. *vanillae*. Since 2010, a collection of pathogenic and endophytic *F. oxysporum* strains associated with *Vanilla planifolia* and related *Vanilla* spp. has been maintained at the Universidad Veracruzana, Mexico. This chapter describes methodologies for strain isolation, morphological characterization, pathogenicity assessment, and long-term preservation. Protocols employing selective culture media, pathogenicity assays, and two preservation strategies—vegetative mycelium storage in minimal medium and chlamydospore conservation under low temperaturesș—are presented. These approaches ensure the maintenance of fungal viability, genetic stability, and pathogenic capacity, providing a reliable framework for phytopathological and molecular studies involving vanilla-associated *Fusarium* strains.

Keywords *Fusarium oxysporum*, *Vanilla planifolia*, Fungal preservation, Phytopathogenic fungi

1 Introduction

Biological collections represent essential resources for biodiversity conservation. Furthermore, in phytosanitary research, it is imperative to maintain collections of pathogenic species of interest while ensuring the preservation of their pathogenic character.

Among the primary challenges facing vanilla (*Vanilla planifolia* Jacks.) cultivation is root and stem rot caused by the fungus *Fusarium oxysporum* f. sp. *vanillae*, which has resulted in severe economic losses across various production regions worldwide. Since 2010, a collection of *F. oxysporum* strains has been established, isolated from the roots and stems of *V. planifolia* and *Vanilla* sp., exhibiting rot symptoms. The plant material was sourced from the Mexican states of Veracruz, Oaxaca, and Nayarit. This collection was developed for research and educational purposes; consequently, strain management

Marco A. Ramírez-Mosqueda et al. (eds.), *Methods for The Conservation of Microorganisms*, Methods in Molecular Biology, vol. 3090, https://doi.org/10.1007/978-1-0716-5623-5_14,

protocols ensure that specimens preserve their viability, genetic traits, and physiological characteristics, specifically their pathogenic capacity.

The collection is housed at the Faculty of Agricultural Sciences of the Universidad Veracruzana. It comprises strains that have demonstrated variations in pathogenicity, serving as a fundamental component for diverse studies concerning histological, genetic, transcriptomic, and metabolomic aspects of the interaction with vanilla plants. Additionally, physiological studies have been conducted to determine how factors such as temperature and carbon and nitrogen sources affect fungal development. The findings of these studies have been published in scientific journals and undergraduate and graduate theses, serving as a baseline for the development of disease control strategies.

2 Materials

2.1 Culture Media

1. Potato Dextrose Agar (PDA): A widely available commercial culture medium. Colony morphology and pigmentation on PDA frequently serve as useful secondary criteria for the identification of *Fusarium* species.
2. Spezieller Nährstoffarmer Agar (SNA) [1]: A nutrient-poor medium utilized for the identification and maintenance of *Fusarium* strains.
3. Czapek Dox Basal Medium: (KH_2PO_4 1 g, $MgSO_4 \cdot 7H_2O$ 0.5 g, KCl 0.5 g, $FeSO_4$. $7H_20$ 10 mg, Sucrose 30 g, Agar 20 g, trace elements solution 0.2 mL, 1 L distilled H_2O). Trace elements solution: Citric acid 5 g, $ZnSO_4{\cdot}7H_2O$ 5 g, $Fe(NH_4)_2(SO_4)_2{\cdot}6H_2O$ 1 g, $CuSO_4{\cdot}5H_2O$ 0.25 g, $MnSO_4.H_2O$ 50 mg, H_3BO_3 50 mg, $Na_2MoO_4{\cdot}2H_2O$ 50 mg, distilled H_2O 95 mL. The trace elements provide essential nutrients for various enzymatic activities.
4. Minimal medium (MM): This medium follows the standard formulation of Czapek–Dox basal medium supplemented with 2 g of $NaNO_3$ per liter, according to the modification by Correll et al. [1].
5. Carnation Leaf Agar (CLA) [2]: A natural culture medium. Most *Fusarium* species sporulate within 6–10 days on CLA, producing macroconidia that are highly uniform in size and shape.
6. Phosphate buffer solution (0.1 M).
7. Tartaric acid: Utilized for medium acidification (1.2–1.4 mL of 10% tartaric acid per 100 mL of medium).

3 Methods

3.1 Preparation of Solutions

1. Potato Dextrose Agar (PDA): Prepared according to the manufacturer's instructions.
2. Spezieller Nährstoffarmer Agar (SNA): The reagents are weighed, mixed, and the required volume of water is added. Finally, agar is incorporated, and the medium is heated until the agar is completely dissolved.
3. Czapek–Dox: A basal medium composed of salts and trace elements. The salts are weighed, the required amount of trace elements is added, and the preparation follows the same procedure as the SNA medium.
4. Carnation Leaf Agar (CLA): Prepared by aseptically placing sterile 3–5 mm^2 carnation leaf pieces into a Petri dish and adding 2% water agar [3].
5. Minimal Medium (MM): Prepared following the same procedure as described above.
6. Sterilization and Pouring: Once prepared, all culture media are sterilized by autoclaving at 121°C for 20 min. Subsequently, under a laminar flow hood, 20 - 25 mL of the medium is poured into presterilized Petri dishes. After the media have solidified, inoculation is performed.

3.2 Culture Collection Management

Isolation and Morphological Characterization

1. Isolation of strains can be performed on Potato Dextrose Agar (PDA) from root and stem samples, both symptomatic and asymptomatic, under aseptic conditions. To inhibit bacterial growth, PDA may be acidified with tartaric acid (1.2–1.4 mL of 10% tartaric acid per 100 mL of medium) or supplemented with antibiotic agents such as chloramphenicol or streptomycin, 100–250 mg/L [3].
2. Furthermore, if contaminating fungi develop in the cultures, samples of colonies exhibiting white, pink, or violet mycelia are collected using a dissecting needle or a microbiological loop and transferred to fresh culture media. Alternatively, cell dilutions (1:9) can be prepared in 0.1 M phosphate buffer, from which PDA cultures are established using the pour-plate method (*see* **Notes 1 and 2**).
3. Cultures are incubated between 26°C and 28°C under dark conditions with frequent monitoring to identify colonies with the desired characteristics, preferably from the higher dilutions. Subsequently, mycelial transfers to fresh PDA plates are

performed as necessary to ensure the procurement of a pure colony. Each strain must be identified with a registration code in a logbook, including origin data and isolation date (*see* **Note 3**).

3.3 Strain Preservation

Alternative I

1. For long-term preservation, strains are maintained in 15 or 50 mL plastic conical tubes containing minimal medium prepared from Czapek–Dox basal medium (*see* **Note 4**) [3].
2. The strains are stored in the laboratory at room temperature (*see* **Note 5**).
3. For reactivation, a portion of the agar containing mycelium is extracted and placed on PDA, then incubated at 28°C until mycelial development occurs.
4. Alternatively, a portion of the mycelium can be placed in a test tube, dissolved in sterile distilled water, and inoculated onto a vanilla stem fragment to induce disease (*see* **Note 6**).

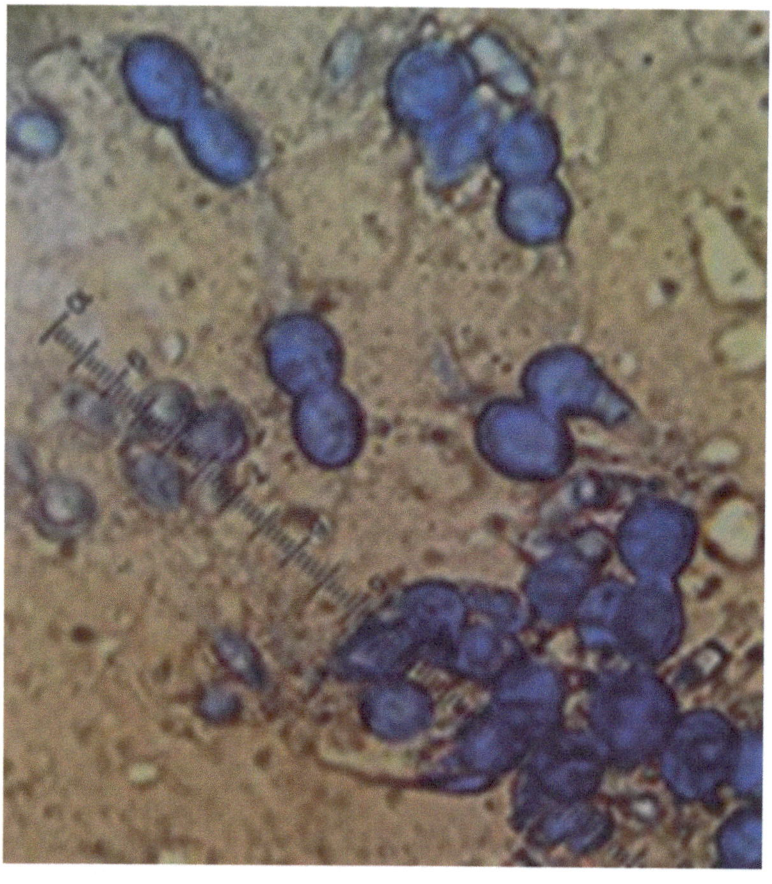

Fig. 1 Appearance of *Fusarium oxysporum* f. sp. *vanillae* chlamydospores stained with lactophenol cotton blue

Alternative II

1. Chlamydospores are resistance structures that enable the survival of several fungal species (*see* **Note 7**).
2. To utilize this as a preservation method, chlamydospore production is first induced on carnation agar during incubation at 28°C under dark conditions.
3. Subsequently, cell suspensions are prepared in sterile distilled water or sterilized soil-like substrates and stored between 4°C and 6°C.
4. Although spores may remain viable for several years, it is necessary to confirm viability at least every 2 or 3 years (Fig. 1).

4 Notes

1. In the preservation of microorganisms, ensuring the certainty of morphological characteristics is of paramount importance. For *F. oxysporum* f. sp. *vanillae*, the observation of microscopic structures developed on the various culture media recommended for the species by Leslie and Summerell [3] is suggested; these include Carnation Leaf Agar (CLA), Spezieller Nährstoffarmer Agar (SNA), and Potato Dextrose Agar (PDA).
2. It is common to isolate more than one *Fusarium* species from diseased plant material; among these, some isolates may be pathogenic, while others are saprophytes or endophytes that may act as secondary pathogens or bear no relation to the disease. In such cases, Koch's postulates must be applied to establish the ecological role of each strain concerning vanilla plants. The appropriate selection of isolation techniques can maximize the recovery of the primary pathogen [4].
3. In research involving *F. oxysporum* f. sp. *vanillae*, pathogenicity tests are conventionally performed on rooted vanilla cuttings following isolation. Alternatively, rapid assays on leaves and stem sections have shown consistent results and are thus proposed as alternative methods offering technical and temporal advantages.
4. As a medium where most *Fusarium* species grow vegetatively, it effectively reduces fungal metabolic activity until the strains are required for planned assays or subculturing.
5. Periodic supervision is recommended to ensure proper tube sealing and to prevent agar dehydration.
6. The stem must be previously disinfected by immersion in a 0.5% household bleach solution for 2–3 min, followed by rinses in sterile distilled water. It is recommended to reactivate the strain at intervals not exceeding 3 years and to validate pathogenic capacity before use in new research.

7. Their generation has been confirmed in strains of *F. oxysporum* f. sp. *vanillae* cultured on carnation agar, making this a viable alternative for this fungus.

Conflict of Interest

The authors have no conflicts of interest to declare that are relevant to the content of this chapter.

References

1. Correll JC, Klittich CJR, Leslie JF (1987) Nitrate non-utilizing mutants of *Fusarium oxysporum* and their use in vegetative compatibility tests. Phytopathology 77:1640–1646
2. Fisher NL, Burgess LW, Toussoun TA, Nelson PE (1982) Carnation leaves as a substrate and for preserving cultures of *Fusarium* species. Phytopathology 72:151–153
3. Leslie, JF, Summerell, BA, 2006. The *Fusarium* laboratory manual. Blackwell Publishing, Ioawa, USA
4. Lim G (1972) *Fusarium* populations in intensively cultivated soils. Trop Agric 49(2):159–164

Chapter 15

Preservation of Thermophilic Fungi: Laboratory Protocols for Short- and Long-Term Storage

Derya Berikten

Abstract

Fungal preservation is critical for safeguarding genetic stability and viability for research and industrial applications. Methods are generally categorized into short- and long-term storage. Short-term maintenance relies on constant sub-culturing, which is technologically simple but labor-intensive and carries risks of phenotypic changes, mutations, or contamination. Alternative intermediate methods include storage under mineral oil, distilled water stasis (Castellani method), and silica gel. For permanent preservation, lyophilization (freeze-drying) and cryopreservation at ultra-low temperatures (liquid nitrogen or −80°C freezers) are preferred as they effectively halt cellular metabolism. Thermophilic fungi present a unique challenge due to their extreme sensitivity to subminimum temperatures below 20°C. Recent research indicates that the freezing block method at −20°C and lyophilization are the most effective strategies for these isolates, maintaining viability for up to 5 years. The use of cryoprotectants, such as glycerol, skim milk, or trehalose, is essential in these protocols to minimize cellular damage caused by ice crystal formation. Ultimately, since no universal protocol exists, the choice of method must be tailored to the specific fungal species and available laboratory resources.

Keywords Thermophilic fungi, Fungal culture preservation, Cryopreservation, Lyophilization, Glycerol, Freezing block

1 Introduction

Thermophilic fungi are a unique group of microorganisms characterized by their ability to thrive at high temperatures, typically exhibiting a growth temperature minimum at or above 20°C and a maximum at or above 50°C [1–3]. These "heat-loving" organisms are found in various extreme environments and play a vital role in ecosystems by mediating nutrient cycling through the degradation of complex organic matter like cellulose and lignin [2]. Beyond their ecological roles, they have significant biotechnological and industrial applications, including the production of thermostable enzymes, secondary metabolites, antibiotics, and their use in processes such as tobacco fermentation and biodegradation [1, 4].

Marco A. Ramírez-Mosqueda et al. (eds.), *Methods for The Conservation of Microorganisms*, Methods in Molecular Biology, vol. 3090, https://doi.org/10.1007/978-1-0716-5623-5_15,

Despite their industrial value, the preservation of thermophilic fungi remains a substantial challenge in microbiology. Unlike mesophilic fungi, thermophiles are highly sensitive to subminimum temperatures and generally cannot survive at temperatures below 20°C [1]. Because they are not stable at low temperatures, standard refrigeration (e.g., storage at 4°C) often leads to a rapid loss of viability, making traditional stocking methods difficult and unreliable. Furthermore, there is currently no universal protocol for their preservation, as different species exhibit varying levels of sensitivity to cold and dehydration [1].

Preservation techniques are broadly divided into short- and long-term storage methods. Short-term maintenance usually involves constant subculturing (periodic transfer) to fresh agar media, a process that is simple and inexpensive but labor-intensive [1, 4]. However, repeated subculturing carries a high risk of contamination, media desiccation, and most critically the loss of phenotypic and genotypic stability, such as reduced pathogenicity, virulence, or sporulation ability [1, 5].

For long-term maintenance, more complex and permanent methods are required to halt or significantly slow cellular metabolism. These include lyophilization (freeze-drying) and cryopreservation at ultra-low temperatures, such as in liquid nitrogen (−196°C) or electric deep freezers (−80°C) [1, 6]. Recent research has demonstrated that specialized protocols, such as the freezing block method at −20°C and lyophilization, are particularly effective for thermophilic isolates, ensuring their viability and morphological stability for up to 5 years [1]. This manual aims to detail these laboratory protocols, providing a framework for the successful short- and long-term storage of these industrially significant fungal resources.

2 Materials

Prepare all culture media using distilled water. Sterilize by autoclaving at 121°C for 15 min unless otherwise stated. Prepare all solutions using ultrapure water and analytical grade reagents. Prepare and store all reagents at room temperature unless indicated otherwise. Diligently follow all waste disposal regulations when disposing waste materials.

2.1 Culture Media

1. Emerson's Yeast Extract Soluble Starch Agar (YpSsA): Soluble starch 15 g; Yeast extract 4 g; K_2HPO_4 1 g; $MgSO_4{\cdot}7H_2O$ 0.5 g; Agar 15 g; Distilled water 1 L; Adjust pH to 7.0 before sterilization (*see* **Note 1**). The sterilized medium was aseptically dispensed into Petri dishes inside a laminar airflow cabinet.
2. Sabouraud Dextrose Agar (SDA): Dextrose 40 g; Peptone 10 g; Agar 15 g; Distilled water 1 L; Final pH 5.6 ± 0.2. After

sterilization, the medium was aseptically poured into Petri dishes in a laminar airflow cabinet.

3. Liquid Glucose Medium: Glucose 1.5 g; Yeast extract 0.1 g; KH_2PO_4 0.2 g; K_2HPO_4 0.23 g; $MgSO_4 \cdot 7H_2O$ 0.1 g; Distilled water 100 mL; Adjust pH to 7.0 prior to sterilization (*see* **Note 2**).
4. Agar slants in screw-cap tubes for constant subculturing: SDA culture medium, weighed according to the above formula and with added distilled water, is boiled in a heated magnetic stirrer until the culture medium composition is completely dissolved. Adjust to pH 5.6 using hydrochloric acid or sodium hydroxide. Pour the hot, liquid medium (~4 mL) into screw-cap glass test tubes (16 × 100 mm). Close the screw caps (*see* **Note 3**). Autoclave the tubes at 121°C for 15 min under 15 p.s.i. pressure. After sterilization, tighten the screw caps before the medium cools down significantly. Immediately place the tubes on a slanted rack or a flat surface at an angle of 5° to 20° (Fig. 1). Allow the agar to cool and solidify (*see* **Notes 4 and 5**).

2.2 Preservation Solutions

1. Glycerol Cryoprotectant (15%, v/v): 15 mL glycerol + 85 mL sterile distilled water. Mix thoroughly and sterilize by autoclaving.
2. Skimmed Milk Solution (15%, w/v): 15 g skimmed milk powder + 100 mL distilled water. Sterilize by autoclaving and cool before use.

2.3 Equipment

1. Sterile screw-cap culture tubes (*see* **Note 6**)
2. Sterile cryovials (1–2 mL) (*see* **Note 7**)
3. Sterile cork borer (5 mm) (*see* **Note 8**)
4. Sterile bamboo toothpicks (*see* **Note 9**)
5. Inoculation needle

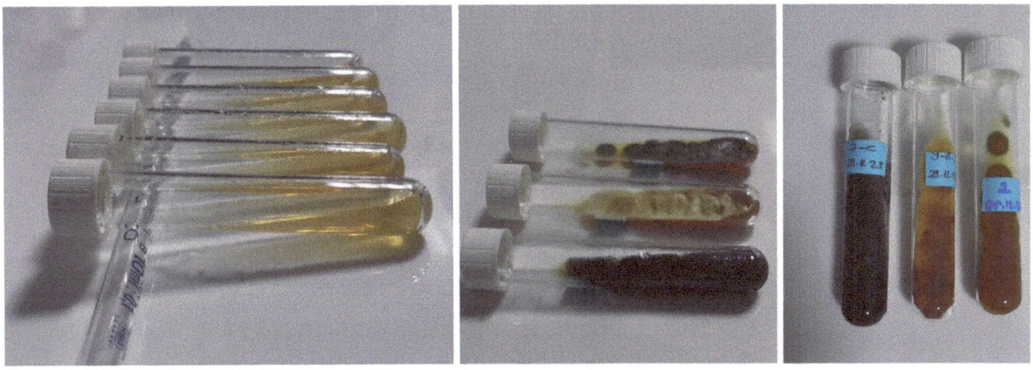

Fig. 1 Agar slants in screw-cap tubes prepared for constant subculturing, inoculated and grown fungal cultures on agar slants, and labeled slant agar tubes (from left to right)

6. Freezer (−20°C) (*see* **Note 10**)
7. Ultra-freezer (−80°C to −86°C) (*see* **Note 11**)
8. Lyophilizer (*see* **Note 12**)
9. Incubator (45°C) (*see* **Note 13**)

3 Methods

3.1 Preparation of Actively Growing Cultures

1. Inoculate fungal isolates onto YpSsA or SDA plates. Inoculum taken from an actively growing culture is transferred to a new culture medium under sterile conditions (*see* **Note 14**).
2. Incubate at 45°C until the colony margin shows active growth (*see* **Note 15**).
3. Use cultures not older than 7–10 days for preservation (*see* **Note 16**).

3.2 Preservation by Constant Subculturing

1. Under aseptic conditions, transfer a small sample (mycelium or spore) taken from a pure culture in the active growth phase to agar slants in screw-cap tubes using a sterile inoculation needle.
2. Incubate at 45°C until visible growth occurs (Fig. 1).
3. Store tubes at room temperature, +20°C, or +4°C (*see* **Note 17**).
4. Sub-culture every 6–8 weeks onto fresh medium (*see* **Note 18**).

3.3 Freezing in Glycerol at −20°C

1. Under aseptic conditions, harvest conidia and mycelium from a pure culture using a sterile loop.
2. Suspend biomass in cryovials containing 1 mL 15% glycerol (*see* **Note 19**).
3. Seal vials tightly and place directly at −20°C (*see* **Note 20**).
4. Many fungal isolates can remain viable for 4 to 5 years using this method (see [2]). However, some sensitive species, such as thermophilic fungi, may have limited survival times at this temperature to 18–24 months (*see* **Note 21**).

3.4 Freezing Block Method

1. Approximately five blocks of 5 mm diameter are cut from the actively growing edge portions of the pure colony in an agar medium using a sterile cork borer (Fig. 2).
2. Transfer blocks into 2 mL sterile cryovials without additives by using sterile bamboo toothpicks (*see* **Note 22**).
3. The tubes are tightly sealed and covered with parafilm to prevent moisture loss and contamination.

Fig. 2 Freezing block method steps: Cutting discs from the developing fungal colony with a cork borer, removing them with a toothpick, placing them in a cryovial, stock vial, and stock box (from left to right)

4. The prepared tubes are stored in a freezer at −20°C
5. Many thermophilic fungi, which are normally sensitive to low temperatures, can be preserved using this method with a viability rate of up to 76% for up to 60 months (*see* **Note 23**).

3.5 Lyophilization

1. Fungal cultures are grown until they reach a resting phase where they produce maximum spores (see [5]).
2. The mycelial structures and mostly spores of the fungus are scraped off the agar surface using a sterile spatula.
3. These scraped fungal materials are placed into 2 mL sterile cryotubes containing 1 mL of 15% sterile nonfat milk (*see* **Note 24**).
4. The prepared samples are frozen overnight in an ultrafreezer at −80°C to −86°C before the drying process.
5. The frozen cryotubes are placed in a freeze-dryer and dried under vacuum for 24 h.
6. The caps of the completed tubes are tightly closed, and they are stored at +4°C (*see* **Note 25**).

3.6 Preservation in Liquid Medium

1. Fungal isolates are inoculated into 50 mL Duran bottles containing 25 mL of liquid medium.
2. The inoculated vials are incubated at 125 rpm and 45°C for 7 days until sufficient growth is achieved.
3. After growth is complete, the bottle caps are tightly closed and wrapped tightly with parafilm to ensure airtightness.
4. The prepared cultures are stored at room temperature in a dark environment (*see* **Note 26**).

4 Notes

1. Growing cultures in freshly prepared culture media in the laboratory ensures healthier development.
2. 50 mL heat-resistant glass Duran laboratory bottles with screw caps are the most suitable option for this process. They are both autoclavable, and their tight-fitting caps minimize the risk of contamination.
3. Close the screw caps, then unscrew them a quarter of the way to allow pressure equilibration during sterilization.
4. Allow the agar to cool and solidify in this position to create a large surface area for fungal growth.
5. Store the prepared slants in the dark, typically at 4°C, until they are needed for inoculation.
6. Screw-cap culture tubes provide an airtight seal that minimizes moisture loss and contamination during long-term storage. They are suitable for agar slants and repeated handling during subculturing procedures.
7. Cryovials made of polypropylene are resistant to low temperatures and chemical stress from cryoprotectants. They are designed to maintain sample integrity during freezing and thawing cycles.
8. A sterile cork borer is used to excise uniform agar blocks from the actively growing margins of fungal colonies. Standardized block size ensures reproducibility in freezing block preservation methods.
9. Bamboo toothpicks can be sterilized in an autoclave inside a heat-resistant glass jar.
10. A conventional laboratory freezer is used for short- to medium-term storage and freezing block preservation. Temperature stability is critical to prevent repeated freeze–thaw stress on thermophilic fungi.
11. An ultra-freezer enables rapid freezing prior to lyophilization and improves survival by reducing ice crystal formation. It is particularly important for preserving fungal material before long-term storage.
12. The lyophilizer removes water under vacuum conditions, allowing long-term storage of fungal cultures in a stable, dried state. Proper sealing after drying is essential to prevent moisture uptake and loss of viability.
13. An incubator capable of maintaining stable high temperatures is essential for the growth and recovery of thermophilic fungi. Temperature fluctuations may adversely affect growth rate and morphological characteristics.

14. The culture to be stored must be a pure culture; this is usually achieved using actively growing tips from the edge of the colony or single spore isolates (see [2]).
15. Actively growing cultures are critical for successful preservation; aged cultures show significantly lower recovery rates.
16. Before proceeding with the storage process, the morphological characteristics, purity, and viability of the obtained culture must be checked and recorded.
17. After incubation, the caps of the tubes should be tightly closed, usually with parafilm, to prevent drying and the entry of mites.
18. Repeated subculturing increases the risk of contamination also, long-term serial subculturing (more than a year) may lead to physiological drift and should be avoided when alternative methods are available.
19. Glycerol concentrations above 15% may reduce viability in some thermophilic fungi.
20. Culture names/codes should be labeled on the tubes in a way that is resistant to −20°C conditions and freeze–thaw cycles.
21. Repeated freezing and thawing of isolates greatly reduces their viability. Therefore, only one test tube should be thawed for checking the stored lot, while the others should be left frozen.
22. Freezing block methods minimize osmotic stress by avoiding cryoprotectants and are particularly suitable for thermophilic fungi with low sporulation capacity.
23. Thermophilic fungi often exhibit reduced viability after exposure to temperatures below 0°C; therefore, freezing protocols should be validated individually for each species.
24. To ensure a homogeneous distribution after lyophilization, the fungal material must be well suspended in skim milk.
25. Lyophilization is most suitable for sporulating fungi and requires strict moisture control after drying. The tubes should be tightly sealed and wrapped securely with parafilm to protect them from moisture.
26. Liquid medium preservation should be considered a short-term method due to contamination risk.

Conflict of Interest

The authors have no conflicts of interest to declare that are relevant to the content of this chapter.

References

1. Berikten D (2021) Survival of thermophilic fungi in various preservation methods: a comparative study. Cryobiology 101:38–43
2. Smith D (2002) Culturing, preservation and maintenance of fungi. In: Waller JM, Lenné JM, Waller SJ (Eds.)Eds. Plant pathologist's pocketbook. CAB International, UK
3. Cooney DG, Emerson R (1964) Thermophilic fungi: An account of their biology, activities, and classification. W.H. Freeman, San Francisco
4. Bermeo Escobar LP, Castro-Ríos K (2021) Methods for the culture conservation of edible and medicinal fungi. J Microbiol Biotechnol Food Sci 10(4):620–625
5. Nakasone KK, Peterson SW, Jong SC (2004) Preservation and distribution of fungal cultures. In: Mueller GM, Bills GF, Foster MS (Eds.)Eds. Biodiversity of fungi: inventory and monitoring methods. Elsevier Academic Press, USA
6. Rohadi H, Ilyas M, Ekowati N (2020) Preservation technique of filamentous fungi based on inactive metabolism at Indonesian Culture Collection (InaCC). IOP Conf Ser: Earth Environ Sci 593(1):012001

Check for updates

Chapter 16

Strategies for the Use and Conservation of *Trichoderma* spp. for the Growth of Forest Trees in Nurseries

Rubén Barrera-Ramírez, Lorena Jacqueline Gómez-Godínez, Hipólito Jesús Muñoz-Flores, Magali Ruíz-Rivas and José Luis Aguirre-Noyola

Abstract

Deforestation and the degradation of forest ecosystems are critical problems affecting biodiversity and the sustainability of rural communities, especially in Latin America. Ecological restoration seeks to restore these ecosystems by using forest nurseries to produce native plants. However, diseases such as the damping-off complex, caused by fungi like *Rhizoctonia* sp., *Pythium* sp., and *Fusarium* sp., are common and affect seedlings in these nurseries. The use of biological agents, such as the fungus *Trichoderma*, has proven effective in controlling these pathogens and promoting plant growth. *Trichoderma* works through mechanisms such as nutrient competition, mycoparasitism, and the production of antimicrobial metabolites, thereby promoting plant health by activating their natural defenses. Strains such as *T. harzianum* and *T. viride* have been extensively studied and have shown good results in disease control. However, their use in forest nurseries has been limited by their high cost and limited knowledge of their efficacy. In Mexico, research has shown that inoculating seedlings with *Trichoderma* improves germination, growth, and disease resistance. Conserving *Trichoderma* through methods such as cryopreservation and freeze-drying is crucial to ensuring its long-term availability. Future research is expected to optimize commercial formulations and explore the use of native strains adapted to local conditions, contributing to the sustainability of forest ecosystems.

Keywords *Trichoderma* sp., Biological control, Forest nurseries, Damping-off, Ecological restoration

1 Introduction

According to the World Wildlife Fund (WWF, 2024) [1], we are reaching points of no return and irreversibly affecting the planet's life-support systems. This is because the effects of deforestation and the transformation of natural ecosystems, intensive land use, and climate change are already evident. The degradation of forest ecosystems is one of the major problems of the twenty-first century, especially in Latin America, where deforestation has significantly reduced the cover of native natural forests and other ecosystems.

Marco A. Ramírez-Mosqueda et al. (eds.), *Methods for The Conservation of Microorganisms*, Methods in Molecular Biology, vol. 3090, https://doi.org/10.1007/978-1-0716-5623-5_16,

Consequently, biodiversity, food security, and the sustainability of some rural communities have been directly affected [2].

In addition to increasing land use changes, forest ecosystems are currently facing latent problems [1]. In this sense, ecological restoration aims to help restore an ecosystem that has been degraded, damaged, or destroyed [3]. In forest ecosystems, a series of actions are carried out to restore their original functions and productivity and restore ecosystem services. This process begins with selecting the site to be restored, followed by designing and establishing soil and water conservation measures, acquiring native forest germplasm, and producing seedlings in forest nurseries. Subsequently, reforestation is carried out along with the protection and maintenance of the planted area. All these actions are constantly monitored and evaluated until the restoration objective is achieved [3].

However, some stages of this process, such as seedling production in nurseries, pose inherent problems, including mortality caused by damping-off diseases complex, in which several microorganisms interact, primarily fungi of the genera *Rhizoctonia*, *Pythium*, and *Fusarium*. This disease mainly affects seedlings. Damping-off can occur both in the field (direct seeding) and in seedbeds and generally occurs in relatively humid conditions [4], causing losses of up to 50%. Therefore, the most common treatment is the use of agrochemicals [5]. It has been determined that this disease has developed some resistance, leading to increased environmental pollution, higher production costs, and even harming human health. Consequently, nonpolluting biological control products have been adopted for their control [6].

As an alternative to chemical products, biological control agents, such as antagonists (e.g., bacteria and fungi), have been employed in agriculture, and their effectiveness in controlling various pathogens has been demonstrated; however, little research has been conducted on the control of forest pathogens [7].

Among biological control agents, the genus *Trichoderma* has been extensively researched for its ability to manage diseases caused primarily by pathogens such as *Fusarium*, *Pythium*, *Rhizoctonia*, *Sclerotinia*, and *Gaeumannomyces*, among others. However, the effectiveness of control can vary depending on the *Trichoderma* species used [8]. The genus *Trichoderma* is characterized by its ability to grow, develop, and colonize soils with diverse environmental characteristics. It also possesses different mechanisms of action against pathogens, such as competition for space and nutrients, mycoparasitism, antibiosis, and resistance induction, in addition to promoting plant growth [7, 8]. The most studied and widely used species in biological control are *Trichoderma harzianum* Rifai, *T. viride* Persoon, and *T. virens*, due to their rapid growth and substrate colonization, high antagonistic activity against various pathogens, and capacity to promote plant growth [7–9].

At the national level, only one commercial strain of *T. harzianum* has been used for *Fusarium* control in forest nurseries; however, its adoption in the forestry sector has been limited, mainly due to its high cost and limited knowledge of its potential benefits [10]. Likewise, the development of commercial bioproducts based on *Trichoderma* has shown reduced efficacy and poor adaptability in application areas, since the antagonistic fungi often come from regions with different climatic and soil conditions [11]. In this context, authors such as Moreno [7] suggest using native strains, better adapted to local environmental conditions, to enhance the effectiveness of biological control. However, little is known about the effects and efficacy of this antagonist when incorporated into different forestry activities; therefore, it is necessary to develop strategies for the use and conservation of *Trichoderma* sp. to improve the growth, development, and health of forest species used in the restoration of degraded ecosystems.

2 Biological and Ecological Characteristics of *Trichoderma* sp. in the Forest Context

2.1 General Description of Trichoderma *sp.*

The genus *Trichoderma*, belonging to the family Hypocreaceae (order Hypocreales, class Sordariomycetes, phylum Ascomycota), comprises a widely studied group of filamentous fungi known for their numerous beneficial effects. It was first postulated in 1794 by Persoon [12]. These characteristics have favored its use, particularly in agriculture, as one of the main components of microbial inoculants [13]. According to the most recent taxonomic classification, the genus comprises more than 400 described species, organized into eight main clades [14, 15].

The genus consists of predominantly soil-dwelling microorganisms, widely distributed globally and present in virtually all soil types, where they function as symbionts, saprotrophs, and mycoparasites associated with plants [16]. *Trichoderma* species exhibit remarkable endosymbiotic capabilities, allowing them to establish favorable relationships with plant hosts, compete effectively in complex microbial communities, and colonize a wide variety of soil ecosystems [15].

These characteristics facilitate their efficient establishment in the rhizosphere and their direct interaction with plant tissues. Furthermore, they exhibit high adaptive plasticity in response to dynamic environments, manifested through structural remodeling of chitin components on the cell surface and the synthesis of various metabolites that enhance plant defense responses and improve vigor by activating beneficial biochemical pathways [17].

The main mechanisms of action of *Trichoderma* in the forest context involve natural associations with plants, which begin with colonization of the root surface and, in some cases, the internal tissues of tree and seedling roots [18]. This process facilitates fungus-plant communication and the transfer of beneficial metabolites through bidirectional chemical signals. Subsequently, fungal adhesion and limited root penetration occur, mediated by the secretion of proteins that modify the plant cell wall, thereby enabling the establishment of a mutualistic interaction [19].

Trichoderma produces three types of propagules: hyphae, chlamydospores, and conidia, which are active against phytopathogens in different phases of the life cycle, from spore germination to sporulation. In addition, it exhibits a series of characteristics in its direct interaction with the pathogen that, according to Harman (2000) and Howell (2006), make this organism a good antagonistic agent against disease-causing fungi [20, 21].

Morphology of Trichoderma

The rapid growth rate and adaptability of *Trichoderma* sp. to diverse substrates have facilitated its isolation in soils from different regions of the world [22], including soils associated with bananas [23]. Several studies describe consistent morphological characteristics among species, such as yellowish to green mycelium, 1.0–3.0 μm conidia, and branched conidiophores in *T. harzianum* [24]; dark green coloration and ovoid conidia of approximately 3.6 μm [25]; and concentric rings and cottony to granular textures in *T. asperellum* [26].

The isolates exhibit accelerated growth and circular colonies, initially white and later green or bluish, with septate hyaline hyphae, compact conidiophores in a pyramidal arrangement, elongated phialides, and oval conidia with slightly rough walls [18]. These characteristics correspond primarily to *T. viride*, according to the taxonomic keys of Rifai (1969) and Gams and Bissett (1998) [27, 28].

These observations are consistent with previous descriptions of septate hyaline mycelium, branched conidiophores, and small, ovoid, unicellular conidia, grouped in terminal clusters that turn green upon maturation [29]. On artificial media, the colonies show white mycelium that darkens as sporulation increases; the spores are unicellular, hyaline, and 3–5 μm in size, produced by hyaline, branched conidiophores [30].

In addition to its morphological characteristics, *Trichoderma* maintains mutualistic relationships with plants, colonizing the initial epidermal layers [31]. This process involves adhesion, penetration, and tolerance to defensive metabolites such as phytoalexins and phenolic compounds [32]. Based on taxonomic keys, species such as *T. harzianum* exhibit light to dark green colonies with a powdery or cottony texture and pyramidal branching, while *T. asperellum* and *T. viride* lack pigmentation on the reverse side of the Petri dish [30–33].

According to authors such as Moreno [7], Martínez et al. [18], and Ahedo-Quero et al. [15], the morphology of the genus *Trichoderma* is as follows:

Macroscopic Morphology (Culture):

Growth: Very rapid, often covering the Petri dish in a few days.
Color: Initially a cottony white mycelium, changing to shades of green (light, bright, or dark green) as it sporulates, frequently with concentric rings. Appearance: It can be compact, velvety, or cottony.
Odor: Some species have a characteristic coconut or almond odor.

Microscopic Morphology
Hyphae: Hyaline, septate (walled), and thin.
Conidiophores: Branched, often pyramidal structures with hyaline branches, terminating in phialides.
Phialides: Bottle-shaped (lageniform), subulate or ampulliform conidiogenous cells, arranged in groups or singly.
Conidia (Phialospores): Unicellular, variable shape (ellipsoidal, ovoid, subglobose), smooth or verrucose walls, green or hyaline in color.
Chlamydospores: Thick-walled, generally globose, resistant spores that can be terminal or intercalary within the hypha.

Morphology of reproductive structures
Asexual: This is the primary form, through conidia produced on conidiophores.
Sexual: The sexual state (teleomorph), when present, corresponds to the genus *Hypocrea*.

Physiology

Trichoderma is a strictly aerobic fungus with broad thermal tolerance [34]. Isolates capable of growing at low temperatures have been documented, such as a strain recovered from Alaskan soil that grows at 4°C and tolerates temperatures up to 33°C [35]. The genus's response to temperature varies among species and isolates. *T. pseudokoningii* and *T. saturnisporum* have been reported to tolerate up to 40°C, while *T. viride* and T. polysporum reach 31°C and *T. harzianum* up to 38°C [36]. However, antagonistic activity can decrease considerably at 30°C, indicating that the optimal growth temperature does not necessarily coincide with the temperature of maximum biocontrol efficacy and demonstrating a close interaction between isolate, temperature, and antagonism.

Regarding soil conditions, *Trichoderma* species have low pH requirements, growing in soils with pH values between 5.5 and 8.5, with a preference for slightly acidic conditions (5.5–6.5). Soil moisture also influences their development, with an optimum near 60% of the water retention capacity; higher saturation levels reduce colonization and survival due to reduced oxygen availability [37]. Furthermore, these fungi contribute to the degradation of organic matter and the breakdown of other microorganisms in the soil [7], so they are often found in environments rich in organic residues and are classified as hypogeous, lignolytic fungi with predatory behavior [19].

2.2 Mechanisms of Action of *Trichoderma* in Association with Forest Trees

During the nursery production phase, forest seedlings are often exposed to various abiotic stressors, such as climatic variations, temperature, and water availability, as well as biotic stressors caused by phytopathogenic agents [19]. These conditions can significantly restrict the production of vigorous plants with high morphological and physiological quality [38]. In this context, inoculation with fungi of the genus *Trichoderma* improves plant acclimatization and increases their resistance or tolerance to stress by activating physiological response mechanisms, thereby promoting greater seedling growth and development [37].

Regarding photosynthetic processes in forest species, the macronutrients and micronutrients involved depend on a series of chemical and physiological processes that support an adequate nutritional supply. The interaction between *Trichoderma* sp. and the plant enhances resource use efficiency, reducing energy loss during transpiration [39]. Furthermore, an increase in photosynthetic capacity has been reported, attributed to genetic regulation induced by *Trichoderma* sp., which allows plants to develop greater resistance to adverse conditions and achieve higher quality indices [19].

Consequently, the incorporation of beneficial microorganisms, particularly fungi of the genus *Trichoderma*, represents a viable strategy for the sustainable production of forest plants and for the implementation of restoration programs in degraded ecosystems, thereby reducing the use of agrochemicals and industrial fertilizers. Although most studies on plant-microorganism interactions have focused on annual species (e.g., crops), there is a growing need to expand research in the forestry sector to highlight the importance and potential of using microorganisms, such as *Trichoderma* sp., within silvicultural systems.

3 Strategies for the Use of *Trichoderma* sp. in Forest Nurseries

In Mexico, the forestry industry contributes approximately 1.5% to the national GDP and is a key activity for economic development and employment in rural areas. Within this sector, the production of forest seedlings for the establishment of commercial forest plantations (PFC) and plantations for reforestation or restoration (PFR) is also part of this production chain [3].

However, to obtain high-quality seedlings, it is necessary to meet the growth parameters derived from morphological and physiological attributes that reflect the capacity for development and survival in the field [40]. Morphological attributes are used to evaluate the plant's developmental capacity, measured by height, crown diameter, and root development [41]. In contrast, physiological attributes are obtained through the plant's capacity to absorb water and nutrients, providing important information about the species' performance and seedling production efficiency [19], as vigorous seedlings with high field yields achieve greater survival.

To maximize seedling survival, growth, and adaptability, *Trichoderma* sp. is inoculated. When inoculated into seedlings and after colonization of the root system, changes in plant metabolism increase root development, growth, and nutrition [42]. The main mechanism of interaction between *Trichoderma* sp. and seedlings involves chemical signaling, driven by the production of compounds that modify the plant's transcriptome, proteome, and metabolome [43].

3.1 Methods of Inoculation and Application of Trichoderma sp.

For this interaction to occur, the seed, substrate, or post-production inoculation must be applied directly to the root.

The seed is coated with the fungus *Trichoderma* sp. in equal parts (1:1, *Trichoderma* sp.:water) so that it is completely coated before sowing. This aims to promote early colonization [34], as it can help promote germination and root development in forest seedlings [13].

Substrate inoculation is generally carried out before planting in forest nurseries. The materials used are usually perlite, vermiculite, peat moss, ground pine bark, and pine sawdust, all in different proportions and combinations, so that the combination allows the plant to perform its physiological functions. Additionally, commercial products that already contain *Trichoderma* sp. inoculum have been added [34].

For species of the genus *Pinus*, the commercial strain T-22 KRL-AG2 (*T. harzianum*) has been one of the most used to inoculate the substrate through injections in a proportion of $3.4\,g\,L^{-1}$ of water, with a dilution of 107 spores g^{-1} of dry weight [10], the same inoculum has also been used in drench (2.5 g inoculum L^{-1} of water at regular intervals) with the substrate forest soil, pine bark and volcanic sand in proportion 60–35–5, respectively, for tropical species such as *Enterolobium cyclocarpum* (Jacq.) Griseb, where high-quality plants and good aerial/root part ratios were obtained [41].

3.2 Practical Examples of Successful Applications in Mexico

Various forestry and agricultural studies have also demonstrated that pre-planting applications of *Trichoderma* sp. to seedling substrates promote the establishment of young roots, contributing to plant growth, development, and vigor [44]. The species *T. harzianum* has been identified as the most ecologically and economically important in Mexico, across agricultural and forest soils and undisturbed ecosystems. It also has significant biotechnological importance, as this strain is the most widely used across various sectors of the Mexican economy, including agriculture and forestry [15].

In a study conducted by Don Juan [45] at the forest nursery of the Autonomous University of Chapingo, Texcoco, Mexico, the biological efficacy of *T. harzianum* and *T. erinaceum* against the complex of fungi that cause damping-off (root and seedling rot and death) in *Pinus ayacahuite* var. veitchii Shaw. was evaluated. The results indicated that the application of these organisms significantly improved germination, height growth, and dry biomass of pine seedlings, compared to the control treatment without any inoculum application.

Roldán [46], under nursery conditions, evaluated the effect of *T. erinaceum* as an antagonistic fungus against *Fusarium oxysporum* in two pine species, *Pinus devoniana* Lindl. Moreover, *P. pseudostrobus* Lindl. was noted, and it was found that, for both pine species, root development improved by up to 36% compared to the control, in which no inoculation was performed.

In another study conducted by Romero-Arenas et al. [47], in the municipality of San Juan Tzicatlacoyan, Puebla, Mexico, the importance of adding native strains of *Trichoderma* to the rhizosphere of *Juniperus deppeana* Steud. Moreover, *Agave cupreata* Trel. & A. Berger, a forest species used for the restoration of arid soils, was evaluated; the inoculant, based on wheat grain with a native strain of *Trichoderma* sp., was demonstrated to be effective. (TS1P1) showed better rhizosphere adherence and development than a commercial inoculant.

In agroforestry systems (AFS) that include cacao (*Theobroma cacao* L.) established in Tabasco, Mexico, Torres-De la Cruz et al. [48] characterized the diversity of *Hypocrea/Trichoderma* present in the cacao rhizosphere, obtaining and identifying 96 *Hypocrea/Trichoderma* isolates using morphology and ITS sequences. The species found were *T. asperellum*, *T. brevicompactum*, *T. harzianum/H. lixii*, *T. koningiopsis/H. Koningiopsis*, *T. longibrachiatum/H. sagamiensis*, *T. pleuroticola*, *T. reesei/H. jecorina*, *T. espirale*, and *T. virens/H. virens*. Finally, they conclude that *Trichoderma harzianum/H. Lixii* was identified as the most abundant species in the agroforestry system (AFS); however, all the species found represent the first record for cacao AFS in Tabasco.

Similarly, Moreno [7] collected forest soil samples from different environments and geographic regions of Mexico to select *Trichoderma* sp. strains and evaluate their in vitro antagonism against *F. circinatum*. Of the 101 strains obtained, only 15 were selected with a high percentage of inhibition and degree of antagonism against *F. circinatum*. Moreno also notes that the collected *Trichoderma* sp. strains have the potential to control pathogens at the nursery level if inoculation is carried out in a climatic region like that of the sites where they were isolated.

For their part, Cerón et al. In 2024 [49], at a forest nursery in Texcoco, Mexico, researchers evaluated *Trichoderma harzianum* as a preventive control for *Fusarium circinatum* through direct substrate application. The results indicated that the pathogenicity of *F. circinatum* decreased in treatments where *T. harzianum* was added to the substrate, and it showed a positive response in the sawdust-based mixture, thus ensuring better quality *Pinus devoniana* Lindl. seedlings.

Therefore, the inoculation of *Trichoderma* sp. in seeds and substrates in forest nurseries is a fundamental tool for improving the health, growth, and quality of forest seedlings. Its use contributes to integrated disease management, strengthens root systems from early stages, and increases the likelihood of success in forest restoration, conservation, and reforestation projects for both forest-based and forest-based plantations.

4 Conservation and Challenges in the Use of *Trichoderma* sp. in Forest Nurseries

The proper preservation of filamentous fungi is an essential process in both mycology and biotechnology, as it ensures long-term viability, genetic stability, and functionality [50]. Among the most used techniques for this purpose are lyophilization, cryopreservation, and storage in inert media such as sterile distilled water or agar slants. Each of these methodologies has characteristics that vary in complexity, cost, and impact on the viability and functional capacity of the fungi [51]. Conservation processes directly influence fungal characteristics, such as germination rate, gene expression related to secondary metabolite production, ability to colonize roots, and effectiveness in inhibiting pathogens [52].

Several studies have demonstrated that freeze-drying can effectively preserve the viability of strains in the short term, provided that cryoprotectants are incorporated appropriately. For example, in *Trichoderma* atroviride, 100% survival was observed immediately after vacuum drying. After 3 months of cold storage, some strains maintained viability of up to 95.2% [53]. Although viability may decrease over time, the use of protectants, such as maltodextrin, has been shown to preserve long-term viability [54]. On the other hand, cryopreservation has proven to be a reliable method for maintaining high viability rates, ranging from 85% to 100%, even after 1 or 2 years of storage, using glycerol as a cryoprotectant [54].

Storage in sterile distilled water has proven to be equally effective. Strains of *Trichoderma* and other molds showed near 100% survival after 2 years at 4°C [55]. Viability in sterile water of 90%–95% has also been reported for more than a decade [56]. Regarding storage on agar slants at 4°C, the results were also positive in the short term, reaching 96% viability in *Trichoderma* cultures after 4 months [57]. These results suggest that methods that limit metabolic activity, such as cryopreservation or storage in cold liquid media, are more effective for long-term preservation compared to traditional subculture-based methods, which tend to show a more rapid loss of viability [54–57]. Conventional methods, such as storage in sterile water, saline solution, or agar, typically achieve a high survival rate (often exceeding

90%). However, advanced freezing techniques, such as cryopreservation in liquid nitrogen and freeze-drying with cryoprotectants, are recognized as the most effective for ensuring stable and prolonged viability [58, 59]. A recent study concluded that cryopreservation at −196°C, lyophilization, and ultra-freezing at −70°C are the best methods to ensure the viability and stability of entomopathogenic fungi for up to 8 years [58].

5 Future Perspectives and New Research on *Trichoderma* sp. in Forest Nurseries

The use of *Trichoderma* sp. in forest nurseries has become established as an effective strategy for both the biological control of soilborne pathogens and the stimulation of seedling growth in controlled environments. This fungus establishes various beneficial interactions with plants, such as root colonization, the synthesis of beneficial metabolites, and nutrient mobilization—processes that improve plant quality and increase their resistance to biotic and abiotic stress factors [16–19].

The genus *Trichoderma* is effective against various soilborne pathogens, such as *Fusarium oxysporum*, *Rhizoctonia solani*, *Pythium aphanidermatum*, *Sclerotinia* sp., and *Botrytis cinerea*, and achieves control levels comparable to those of chemical products [60, 61]. It also displays nematicidal activity against *Meloidogyne incognita* and *M. javanica*, inhibiting hatching and causing death in the juvenile stage [62]. Furthermore, it has been documented that it can contribute to the management of foliar pathogens and reduce the incidence of leaf-cutter ants through antagonism [63].

It also exerts entomopathogenic effects through direct parasitism on hemipterans such as *Aphis gossypii* and orthopterans such as *Locusta migratoria* [64, 65], and some isolates produce secondary metabolites with insecticidal or repellent activity, such as *Trichoderma* atroviride against *Drosophila melanogaster* [66].

Therefore, an emerging line of research is the optimization of viable and stable microbial formulations for use in forest nurseries. In this regard, biotechnology applied to the production of *Trichoderma*-based biocontrol agents must seek mechanisms to improve the viability and efficacy of commercial products, thereby extending their shelf life and ensuring performance under diverse environmental conditions [18]. This process is crucial for reliably and predictably integrating biological solutions into forest production systems.

Likewise, field studies and nursery-scale trials are needed to validate effects observed under laboratory or greenhouse conditions. Although the literature has documented growth promotion and pathogen reduction in crops, few detailed trials have been conducted on forest plants under commercial nursery practices.

Future research should address factors such as inoculant dosage, substrate compatibility, and interaction with native soil microbiota to maximize agronomic benefits [19].

Another promising perspective involves identifying and selecting endophytic *Trichoderma* strains from local forest ecosystems [7], as they may be better adapted to the specific conditions of each region and the target forest species.

Furthermore, advances in sequencing and genomic analysis techniques also offer opportunities to gain a deeper understanding of the molecular mechanisms of *Trichoderma*–plant interactions, as well as the genetic basis of traits of interest such as the production of antifungal compounds or stress tolerance [16]. Therefore, the use of "omics" approaches could facilitate the identification of genes associated with biocontrol efficacy and growth promotion, thereby enabling the development of more effective bioinoculants for forest species.

Due to global demand to reduce the use of chemical fungicides and fertilizers in sustainable forestry production, future research on the genus *Trichoderma* must increasingly integrate ecological management strategies that also consider the soil microbiota as a central component of nursery health and the long-term resilience of this forestry production system.

Conflict of Interest

The authors have no conflicts of interest to declare that are relevant to the content of this chapter.

References

1. WWF., 2024. Informe Planeta Vivo 2024. Un sistema en peligro. WWF, Gland, Suiza, p. 96. https://wwflac.awsassets.panda.org/downloads/informe_planeta_vivo_2024_completo.pdfp
2. Pacheco J (2019) La deforestación en América Latina: causas, consecuencias y desafíos para la restauración. Rev Estud Ambient 10(3):22–38
3. Sistema Nacional de Información Forestal (SNIF) (2025) Restauración Forestal. Secretaria de Medio Ambiente y Recursos Naturales-Comisión Nacional Forestal, Mexico. https://snif.cnf.gob.mx/restauracion/
4. Aprea AM, Gergoff Grozeff GE, Ruscitti MF, Giménez DO (2023) Enfermedades de las plantas leñosas en vivero. *Libros de Cátedra*, pp 307–316
5. Loa AJ, Pérez ME, Romero MDL, Aparicio UJ, Rosales BI, De la Mora, RT (2021) Aplicación del tratamiento para el control del Damping off en plantas forestales. In: Primer congreso Nacional de Cuerpos Académicos, Fortaleciendo redes de investigación. Instituto Tecnológico Superior de los Ríos, Balancán, Tabasco, Méx, pp 633–637
6. Miranda Benites CM (2023) Sensibilidad in vitro de seis cepas de *Fusarium circinatum* a funguicidas y a bioproductos de diferente modo de acción, 121 p
7. Moreno DF (2023) Potencial antagonista del género *Trichoderma* contra *Fusarium circinatum*. Tesis de maestría. Colegio de Postgraduados, Montecillo, Texcoco. Edo Méx, 68 p
8. Martín-García J, Paraschiv M, Flores-Pacheco JA, Chira D, Diez JJ, Fernández M (2017) Susceptibility of several Northeastern conifers to *Fusarium* circinatum and strategies for biocontrol. Forests 8(9):318. https://doi.org/10.3390/f8090318

9. Samaniego-Fernández LM, Harouna M, Corbea O, Rondón-Castillo AJ, Placeres-Espinosa I (2018) Aislamiento, identificación y evaluación de cepas autóctonas de *Trichoderma* spp., antagonistas de patógenos del suelo. Rev Protección Veg 33:1–11
10. García-Díaz SE, Aldrete A, Alvarado-Rosales D, Cibrián-Tovar D, Méndez Montiel JT (2019) *Trichoderma harzianum* Rifai as a biocontrol of *Fusarium circinatum* Nirenberg & O´Donnell in seedlings of *Pinus greggii* Engelm. ex Parl. in three substrates. Rev Chapingo Ser Cienc For Amb 25:353–367. https://doi.org/10.5154/r.rchscfa.2018.12.088
11. Savín-Molina J, Hernández-Montiel LG, Ceiro-Catasú W, Ávila-Quezada GD, Palacios-Espinosa A, Ruiz-Espinoza FH, Romero-Bastidas M (2021) Morphological characterization and biocontrol potential of *Trichoderma* species isolated from semi-arid soils. Rev Mex Fitopatol 39:435–451. https://doi.org/10.18781/r.mex.fit.2106-7
12. López-Ferrer DCU, Brito-Vega H, López-Morales D, Salaya-Domínguez JM, Gómez-Méndez E (2017) Papel de *Trichoderma* en los sistemas agroforestales-cacaotal como un agente antagónico. Trop Subtrop Agroecosyst 20(1):91–100
13. Kredics L, Büchner R, Balázs D, Allaga H, Kedves O, Racić G (2024) Recent advances in the use of *Trichoderma*-containing multicomponent microbial inoculants for pathogen control and plant growth promotion. World J Microbiol Biotechnol 40(5):162. https://doi.org/10.1007/s11274-024-03965-5
14. Cai F, Druzhinina IS (2021) In honor of John Bissett: authoritative guidelines on molecular identification of *Trichoderma*. Fungal Divers 107:1–69. https://doi.org/10.1007/s13225-020-00464-4
15. Ahedo-Quero HO, Aquino-Bolanos T, Ortiz-Hernandez YD, García-Sánchez E (2024) *Trichoderma* diversity in Mexico: a systematic review and meta-analysis. Diversity 16(1):68. https://doi.org/10.3390/d16010068
16. Kubiak A, Wolna-Maruwka A, Pilarska AA, Niewiadomska A, Piotrowska-Cyplik A (2023) Fungi of the *Trichoderma* genus: future perspectives of benefits in sustainable agriculture. Appl Sci 13(11):6434. https://doi.org/10.3390/app13116434
17. Kappel L, Münsterkötter M, Sipos G, Rodriguez C, Gruber S (2020) Chitin and chitosan remodeling defines vegetative development and *Trichoderma* biocontrol. PLoS Pathog 16(2):e1008320. https://doi.org/10.1371/journal.ppat.1008320
18. Martínez DY, Martínez-Coca B, Peteira-Delgado B, Reyes-Duque Y, Gil K, Simpson J, Herrera-Estrella A (2023) Caracterización morfo–cultural y variabilidad genética y molecular de aislamientos de *Trichoderma*: Variabilidad de aislamientos de *Trichoderma*. Biotecnia 25(2):194–203
19. Ferreira NCDF, Ramos MLG, Gatto A (2024) Use of *Trichoderma* in the production of forest seedlings. Microorganisms 12(2):237. https://doi.org/10.3390/microorganisms12020237
20. Harman G (2000) Myths and dogmas of biocontrol changes in perceptions derived from research on *Trichoderma harzinum* T-22. Plant Dis 84(4):377–393. https://doi.org/10.1094/pdis.2000.84.4.377
21. Howell C (2006) Understanding the mechanisms employed by *Trichoderma virens* to effect biological control of cotton diseases. Phytopathology 96(2):178–180. https://doi.org/10.1094/phyto-96-0178
22. Bononi L, Chiaramonte JB, Pansa CC, Moitinho MA, Melo IS (2020) Phosphorus-solubilizing *Trichoderma* sp. from Amazon soils improves soybean plant growth. Sci Rep 10(1):2858. https://doi.org/10.1038/s41598-020-59793-8
23. Correa-Delgado R, Brito-López P, Jaizme-Vega MC, Laich F (2024) Biodiversity of *Trichoderma* species of healthy and *Fusarium* wilt-infected banana rhizosphere soils in Tenerife (Canary Islands, Spain). Front Microbiol 15:1376602. https://doi.org/10.3389/fmicb.2024.1376602
24. Asis A, Shahriar SA, Naher L, Saallah S, Fatihah HNN, Kumar V, Siddiquee S (2021) Identification patterns of *Trichoderma* strains using morphological characteristics, phylogenetic analyses and lignocellulolytic activities. Mol Biol Rep 48:m3285–m3301. https://doi.org/10.1007/s11033-021-06321-0
25. Silva LG, Camargo RC, Mascarin GM, Nunes PSDO, Dunlap C, Bettiol W (2022) Dual functionality of *Trichoderma*: biocontrol of *Sclerotinia sclerotiorum* and biostimulant of cotton plants. Front Plant Sci 13:983127. https://doi.org/10.3389/fpls.2022.983127
26. Sebumpan R, Guiritan KR, Suan M, Abapo CJ, Bhat AH, Machado RA, Sumaya NH (2022) Morphological and molecular identification of *Trichoderma asperellum* isolated from a dragon fruit farm in the southern Philippines and its pathogenicity against the larvae of the super worm, *Zophobas morio* (Fabricius, 1776) (Coleoptera: Tenebrionidae). Egypt J Biol Pest Control 32(1):1–7. https://doi.org/10.1186/s41938-022-00548-0
27. Rifai A (1969) A revision of the genus *Trichoderma*. Mycological 116:1–56
28. Gams W, Bissett J (1998) Morphology and identification of *Trichoderma*. In: Kubicek

C, Harman G (Eds.)Eds. *Trichoderma* and *Gliocladium* volume 1: basic biology, taxonomy and genetics. Taylor & Francis, London, UK, 3–34
29. El-Sobky MA, Fahmi AI, Ragaa AE, El-Zanaty AM (2019) Genetic characterization of *Trichoderma* sp. Isolated from different locations of Menoufia, Egypt and assessment of their antagonistic ability. J Microb Biochem Technol 11(1):9–23
30. Barnett, H, Hunter, B, 1972. Illustrated genera of imperfect fungi. EE. UU. Burgess Publ. Co., p. 241
31. Yedidia I, Srivastva AK, Kapulnik Y, Chet I (2001) Effect of *Trichoderma harzianum* on microelement concentrations and increased growth of cucumber plants. Plant Soil 235:235–242. https://doi.org/10.1023/A:1011990013955
32. Ortuño NC, Miranda MC (2013) Selección de cepas de *Trichoderma* sp. generadoras de metabolitos secundarios de interés para su uso como promotor de crecimiento en plantas cultivadas. J Selva Andina Biosph 1:16–32
33. Chaverri P, Branco-Rocha F, Jaklitsch W, Gazis R, Degenkolb T, Samuels GJ (2015) Systematics of the *Trichoderma harzianum* species complex and the re-identification of commercial biocontrol strains. Mycologia 107(3):558–590. https://doi.org/10.3852/14-147
34. Carranza-Patiño MS, Castro-Mena YA, García-Conza JN, Ramos-Remache RA, Herrera- Feijoo RJ (2025) Sustratos ecológicos y *Trichoderma* especies en la germinación y desarrollo inicial Delaware *Theobroma cacao* L. Multidiscip Colab Rev 3(4):178–204. https://doi.org/10.70881/mcj/v3/n4/98
35. McBeath J, Adelman M (1981) Taxonomy of a new *Trichoderma* found in Alaska. Phytopathology 10:1151
36. Danielson R, Davey C (1973) Non nutritional factors affecting the growth of *Trichoderma* in culture. Soil Biol Biochem 5(5):495–504
37. Qin X, Xu J, An X, Yang J, Wang Y, Dou M, Wang M, Huang F, Fu Y (2023) Insight of endophytic fungi promoting the growth and development of woody plants. Crit Rev Biotechnol 1:1–22. https://doi.org/10.1080/07388551.2022.2129579
38. Puglielli G, Laanisto L, Gori A, Cardoso AA (2023) Woody plant adaptations to multiple abiotic stressors: where are we?. Flora 299:152221. https://doi.org/10.1016/j.flora.2023.152221
39. Oszako T, Voitka D, Stocki M, Stocka N, Nowakowska JA, Linkiewicz A, Malewski T (2021) *Trichoderma asperellum* protege eficazmente las hojas de *Quercus robur* contra *Erysiphe alphitoides*. Eur J Plant Pathol 159:295–308. https://link.springer.com/article/10.1007/s10658-020-02162-y
40. Peccatti A, Rovedder APM, Steffen GPK, Maldaner J, Missio EL, Witt CS, Dalcul LP (2019) Effect of *Trichoderma* sp. on the propagation of *Maytenus ilicifolia* Mart. ex Reissek. J Agric Sci 11:435–442
41. Talavera-Silva L, Godina-Rodríguez JE, Barrera-Ramírez R, Muñoz-Flores HJ (2025) Calidad de planta de parota (Enterolobium cyclocarpum) con diferentes tratamientos pregerminativos, tipos de sustratos e inoculación con Trichoderma spp. In: Ortega Reyes L (Ed.) Memoria de 60a Reunión científica. Instituto Nacional de Investigaciones Forestales Agrícolas y Pecuarias, San Luis Potosi, México, 1811–1813
42. Khan MR, Mohiddin FA (2018) *Trichoderma* : its multifarious utility in crop improvement. Crop Improvement through microbial biotechnology. Elsevier, Amsterdam, The Netherlands, 263–291 https://doi.org/10.1016/B978-0-444-63987-5.00013-X
43. Kashyap PL, Rai P, Srivastava AK, Kumar S (2017) *Trichoderma* for climate resilient agriculture. World J Microbiol Biotechnol 33:155. https://doi.org/10.1007/s11274-017-2319-1
44. Campos BF, Araújo AJC, Felsemburgh CA, Vieira TA, Lustosa DC (2020) *Trichoderma* contributes to the germination and seedling development of Açaí Palm. Agriculture 10(10):456. https://doi.org/10.3390/agriculture10100456
45. Don Juan MB (2006) Efectividad Biológica de *Trichoderma spp.* contra hongos causantes de Damping Off en viveros forestales. Tesis de Maestría, Universidad Autónoma de Chapingo, Texcoco Edo, de Méx, p 79
46. Roldán CMA (2008) Control biológico de *Fusarium oxysporum* en dos especies de pino. Tesis de Maestría, Universidad Autónoma de Chapingo, Texcoco Edo. de Méx. 86 p
47. Romero-Arenas O, Huato MAD, Rivera Tapia JA, Aldana F, Parraguirre Lezama C (2015) Evaluación de *Trichoderma* sp., en asociación con *Agave cupreata* y *Juniperus deppeana* en suelos del municipio San Juan Tzicatlacoyan, Puebla-México. In: V Congreso Latinoamericano de Agroecología-SOCLA, p 9
48. Torres-De la Cruz M, Ortiz-García CF, Bautista-Muñoz C, Ramírez-Pool JA, Ávalos-Contreras N, Cappello-García S, De la Cruz-Pérez A (2015) Diversidad de *Trichoderma* en el agroecosistema cacao del estado de Tabasco, México. Rev Mex Biodivers 86(4):947–961. https://doi.org/10.1016/j.rmb.2015.07.012
49. Cerón MAO, Acosta EH, Edith S, Díaz G, Morales AV (2024) Control biológico de *Fusarium circinatum* Nirenberg & ODonnell

1998 con *Trichoderma harzianum* Rifai 1969 en *Pinus devoniana* Lindl. Rev Mex Cienc For 15:83. https://doi.org/10.29298/rmcf.v15i83.1446
50. Lakshman DK, Singh V, Camacho ME (2018) Long-term cryopreservation of non-spore-forming fungi in Microbank™ beads for plant pathological investigations. J Microbiol Meth 148:120–126
51. Day JG, Stacey GN (2008) Biobanking. Mol Biotechnol 40(2):202–213
52. Broughton R, Buddie AG, Smith D, Ryan MJ (2012) The effect of cryopreservation on genomic stability in strains of the fungus *Trichoderma*. CryoLetters 33(4):299–306
53. Grzegorczyk M, Kancelista A, Łaba W, Piegza M, Witkowska D (2018) The effect of lyophilization and storage time on the survival rate and hydrolytic activity of *Trichoderma* strains. Folia Microbiol 63(4):433–441
54. Paul JS, Tiwari KL, Jadhav SK (2015) Long term preservation of commercial important fungi in glycerol at 4 C. Int J Biol Chem 9(2):79–85
55. McGinnis LK, Duplantis JrSC, Youngs CR (1993) Cryopreservation of sheep embryos using ethylene glycol. Anim Reprod Sci 30(4):273–280
56. Jones DM, Sutcliffe EM, Curry A (1991) Recovery of viable but non-culturable *Campylobacter jejuni*. Microbiology 137(10):2477–2482
57. Sarmiento Rangel Y, Hazel Vergel A, Cárdenas Caro DM (2013) Evaluación de la estabilidad de *Trichoderma* sp. y *Azotobacter* sp. conservados por diferentes métodos. Rev Colomb Biotecnol 15(1):150–158
58. Ayala-Zermeño MA, Berlanga-Padilla AM, Rodriguez-Velez JM, Sánchez-Rangel JC, Rodriguez-Velez B (2024) Viability of Hirsutella spp. strains preserved in mineral oil. Sci Fungorum 55, e1473. https://doi.org/10.33885/sf.2024.55.1473
59. Ryan MJ, Smith D (2007) Cryopreservation and freeze-drying of fungi employing centrifugal and shelf freeze-drying. In Day, J.G., Stacey, G.N. (eds). Cryopreservation and freeze-drying protocols, p.p. 127–140
60. Nofal AM, El-Rahman MA, Abdelghany TM, El-Mongy A (2021) Mycoparasitic nature of Egyptian *Trichoderma* isolates and their impact on suppression *Fusarium* wilt of tomato. Egypt J Biol Pest Control 31(1):1–8. https://doi.org/10.1186/s41938-021-00450-1
61. Shenouda ML, Cox RJ (2021) Molecular methods unravel the biosynthetic potential of *Trichoderma* species. Royal Soc Chem 11(6):3622–3635. https://doi.org/10.1039/DORA09627J
62. Ibrahim DS, Elderiny MM, Ansari RA, Rizvi R, Sumbul A, Mahmood I (2020) Role of *Trichoderma* sp. in the management of plant-parasitic nematodes infesting important crops. Management of phytonematodes: recent advances and future challenges. 259–278. https://link.springer.com/chapter/10.1007/978-981-15-4087-5_11
63. Do Nascimento MO, De Almeida Sarmento R, Dos Santos GR, de Oliveira CA, de Souza DJ (2017) Antagonism of *Trichoderma* isolates against *Leucoagaricus gongylophorus* (singer) Möller. J Basic Microbiol 57(8):699–704. https://doi.org/10.1002/jobm.201600755
64. Nawaz A, Gogi MD, Naveed M, Arshad M, Sufyan M, Binyameen M, Ali H (2020) In vivo and in vitro assessment of *Trichoderma* species and *Bacillus thuringiensis* integration to mitigate insect pests of brinjal (*Solanum melongena* L.). Egypt J Biol Pest Control 30(1):60. https://doi.org/10.1186/s41938-020-00258-5
65. Laib DE, Benzara A, Akkal S, Bensouici C (2020) The anti-acetylcholinesterase, insecticidal and antifungal activities of the entophytic fungus Trichoderma sp. isolated from Ricinus communis L. against Locusta migratoria L. and Botrytis cinerea Pers.: Fr. Acta Scientifica Naturalis 7(1):112–125. https://doi.org/10.2478/asn-2020-0011
66. Atriztán-Hernández K, Moreno-Pedraza A, Winkler R, Markow T, Herrera-Estrella A (2019) *Trichoderma atroviride* from predator to prey: role of the mitogen-activated protein kinase Tmk3 in fungal chemical defense against fungivory by *Drosophila melanogaster* larvae. Appl Environ Microbiol 85(2):e01825–18. https://doi.org/10.1128/AEM.01825-18

Chapter 17

Preservation of Fungi by Suspension in Liquid Medium: Fundamentals, Applications, and Challenges

Emma O. Fuentes-Ramírez, Rafael Uzárraga-Salazar, Carlos Alberto Cruz-Cruz, Tania García-Herrera and Gabriel R. Hernández-Martínez

Abstract

Preserving fungi in liquid media is an essential strategy for maintaining long-term viability, genetic stability, and phenotypic characteristics across diverse fungal species. Among the most widely used methods are storage in sterile distilled water, storage in mineral oil, and cryopreservation. Each technique is based on biological and biochemical principles, such as metabolic suppression, osmotic balance, protection against oxidative stress, and induction of quiescence. These principles minimize cell damage, maintain structural integrity, and ensure the recovery of highly viable cultures. The selection of a method depends on factors such as resource availability, storage period, characteristics of the preserved species, and the conservation objective, ranging from research to industrial or medical applications. The relevance of these techniques lies in their role in conserving fungal biodiversity over the long term and ensuring the availability of strains for scientific studies, industrial production, biotechnological applications, and biological control. Their correct implementation helps safeguard genetic resources against threats such as climate change, ensuring the reproducibility of research and continuity of production processes. Thus, preservation in liquid media represents not only a technical procedure but also a strategic tool for the bioeconomy, food security, and the development of sustainable solutions in health, agriculture, and industry sectors.

Keywords Fungal preservation, Cryopreservation, Distilled water, Mineral oil, Viability, Biodiversity

1 Introduction

Preservation is a set of protocols and techniques for storing and maintaining various microorganisms and cell cultures in a viable state for extended periods. Among the objectives of preservation are the genotypic and phenotypic stability of cultures. This ensures the maintenance of genetic, morphological, and physiological characteristics, enabling reproducible results in research and/or for industrial or medical applications of the species. Therefore, species preservation

Marco A. Ramírez-Mosqueda et al. (eds.), *Methods for The Conservation of Microorganisms*, Methods in Molecular Biology, vol. 3090, https://doi.org/10.1007/978-1-0716-5623-5_17,

is a fundamental practice for maintaining the richness of organisms and accessing them when desired [1].

Figure 1 shows the scientific productivity related to the conservation of fungi during the period 2000–2025. The limited number of studies on this topic may be due to mycological research being usually oriented towards biotechnological and biological control applications, leaving approaches focused exclusively on preservation techniques in the background.

In parallel, the network of co-occurrences shows an interdisciplinary and methodological panorama, and it is observed that the conservation of fungi impacts the biotechnological and environmental areas (Fig. 2).

The nodes and links show the frequency and strength of the association between terms, while the colors distinguish three well-defined thematic clusters. The red cluster group concepts related to the microbiology workflow, highlighting terms such as preservation, fungal isolation, temperature, water, and storage.

Similarly, the blue cluster concentrated on terms related to general microbiology, including mineral oil, virulence, fungal strain, and unclassified drugs. Meanwhile, the green cluster brings together concepts associated with biotechnological and applied uses. Taken together, the network structure shows that fungal conservation is not

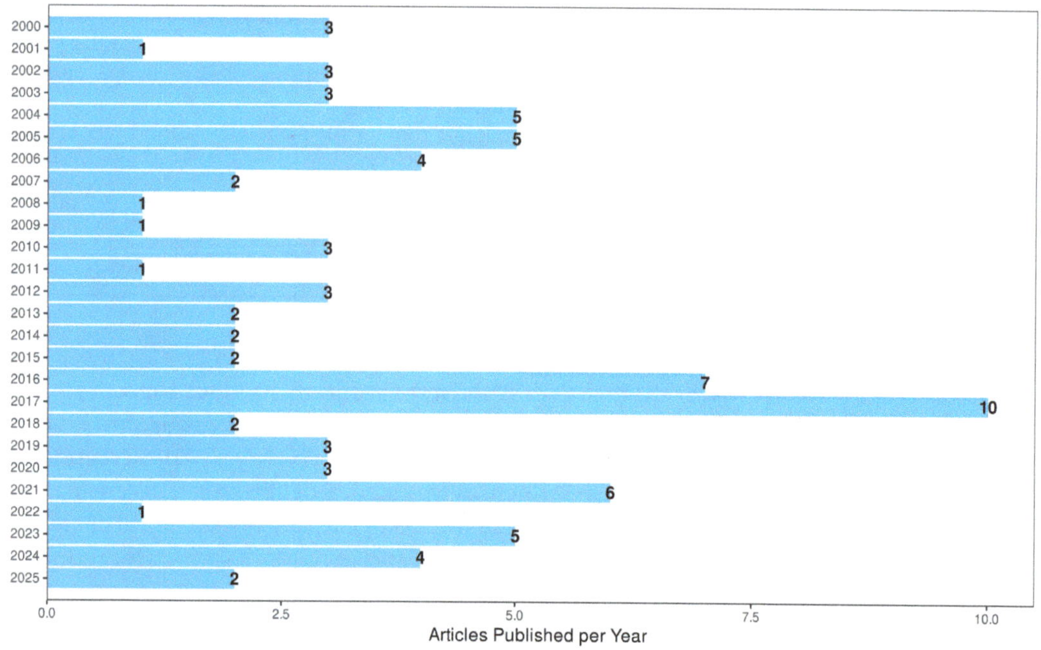

Fig. 1 Total number of publications indexed in Scopus® from 2000 to 2025 related to fungi and their preservation. The search included studies on fungal preservation, storage, cryopreservation, long-term storage, genetic stability, or quiescence, as well as publications that mentioned storage in sterile distilled water or under mineral oil. Only articles, reviews, and book chapters were considered, without language restrictions

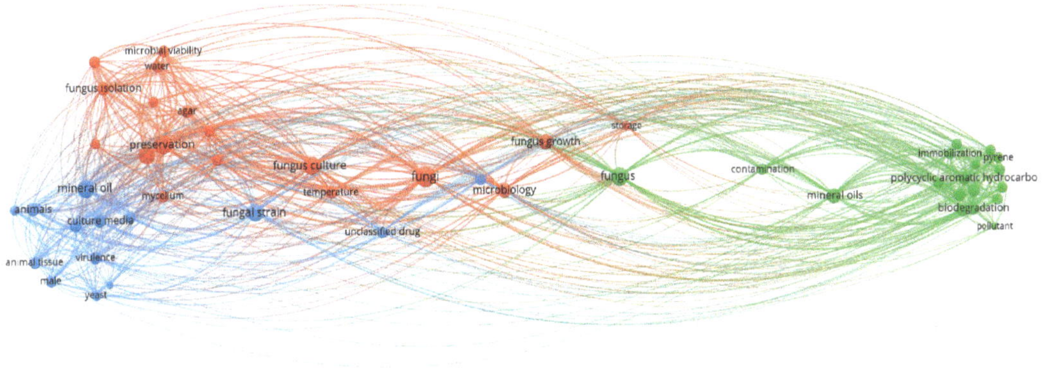

Fig. 2 Network visualization map of terms corresponding to publications indexed in Scopus® from 2000 to 2025 related to fungi and their preservation. The search included studies on fungal preservation, storage, cryopreservation, long-term storage, genetic stability, and quiescence, as well as publications that mentioned storage in sterile distilled water or under a mineral oil. Only articles, reviews, and book chapters were considered without language restrictions. Visualization generated using VOSviewer

an isolated field but rather a central node for applied research, including environmental, microbiological, and biotechnological studies.

The preservation of fungi has been practiced since the 19th century, given their diverse applications and the interest in accessing them at the desired time [2]. Since then, various methods have been proposed for the preservation of fungi and yeasts, emphasizing that each time a species needs to be preserved, different techniques must be evaluated, as it is not possible to apply a single process to all fungi. This is due to their metabolic, morphological, and genetic diversity. Additionally, it is necessary to consider the available infrastructure, time required, and purpose of preserving each species.

In this chapter, we discuss fungal preservation techniques in liquid media, their biochemical foundations, methodological aspects, and the applications of preservation and the preserved species.

2 Preservation Methods

Subculturing is one of the earliest methods of preserving fungi and was first reported in 1895, predominantly on solid media [3]. After that, alternative techniques were developed utilizing liquid media, including storage in sterile distilled water, storage in mineral oil, and cryopreservation, which are reviewed herein.

2.1 Storage in Sterile Distilled Water

This method was developed by Castellani in 1939 [4] and is still used by some groups because of its practicality and cost-effectiveness, coupled with the possibility of long-term preservation of some fungal species [5–8]. Since this storage method was proposed, several authors have reported the viability of numerous fungal species after decades of preservation. De Capriles et al. [5] reported a 90% viability of 594 fungal strains preserved for a period of 1–20 years by this method. Qiangqiang et al. [9] reported a viability of 89% for 78 fungal isolates after 12 years of storage in water. Borman et al. [10] reported that 90% of 179 pathogenic mold isolates remained viable after prolonged storage in sterile distilled water. Karabiçak et al. [11] evaluated the viability of pathogenic molds and yeasts after 1–12 years of preservation using this method, obtaining a viability of 94.7%. Owing to the simplicity of the technique and the low cost of inputs, storage in sterile distilled water is a successful and simple method for preserving many fungi, including those of medicinal and agricultural importance. Therefore, several research groups prefer this technique over ultra-freezing methods [12, 13].

2.2 Storage in Mineral Oil

This method involves placing approximately 1 cm agar squares containing the fungus to be preserved inside a test tube and covering them with a layer of mineral oil for storage at room temperature. This method maintains a certain level of humidity, preventing the fungus from dehydrating and slowing its metabolism and growth [14].

The first attempts at long-term preservation of fungi in mineral oil date back to the 1940s, when Henry [15] reported the viability of many fungi and yeasts after storage in mineral oil for 6 years. Subsequently, other groups attempted to preserve various species of fungi and yeasts for months or even years, highlighting the absence of morphological changes in recovered cultures [14–16]. But it was the study by Smith and Onions [17] that reported the longest preservation time in mineral oil, with 47 fungi remaining viable for 32 years without morphological changes. Therefore, according to various authors, storage in mineral oil is a successful and feasible method for the long-term preservation of fungi and yeasts without resorting to other, more expensive, and sophisticated techniques, such as ultra-freezing.

2.3 Cryopreservation

One objective of preservation is to slow cellular metabolism for extended periods; this is why cryopreservation is among the most widely evaluated techniques for numerous species of fungi and yeasts [1]. This technique involves suspending and storing cells at ultra-low temperatures, ranging from −50°C to −80°C, then −150°C, and finally −196°C, when cultures are preserved in liquid nitrogen. Cryopreservation requires more infrastructure, but it is an alternative for preserving fungal species that cannot be freeze-dried or whose morphology changes after storage in water or oil for years [18].

Pioneering experiments in fungal cryopreservation were conducted in 1960, where filamentous fungi were successfully preserved [19]. Subsequently, Elliot [20] developed a variant of fungal cryopreservation using polypropylene straws, and 10 years later Pasarell and McGinnis [21] reported on the viability of fungal cultures maintained at −70°C. Furthermore, other groups developed new variants of the method, such as Homolka's group, which used perlite as support for mycelium [22], and Kitamoto's group, which grew mycelium on sawdust [23].

Using this method and its variants, several groups have evaluated the preservation of hundreds of fungal strains after prolonged storage periods. Homolka et al. [24] cryopreserved 442 basidiomycete strains, achieving 100% viability, and subsequently tested the method on 30 basidiomycete strains after 10 years of storage [25]. Eichlerová et al. [26] cryopreserved *Pleurotus ostreatus* and *Trametes versicolor* for 12 years in liquid nitrogen. In general, cryopreservation in liquid nitrogen (−196°C) or in ultralow-temperature freezers (−70°C to −80°C) with cryoprotectants such as glycerol or DMSO is the preferred method for long-term storage, as it almost completely halts metabolism and maintains genetic and phenotypic stability. However, storage in sterile water and mineral oil remains an economical and effective method for species.

3 Principles of Fungi Preservation in Liquid Media

Fungal preservation methods in liquid media are based on various biochemical and biological principles that maintain the morphological and genetic integrity of the preserved spores, hyphae, or fungal cells. The primary biochemical and biological principles include metabolic suppression, osmotic balance, oxidative stress, and cellular quiescence. The principles are described below.

3.1 Metabolic Suppression

All methods of preserving fungal species in liquid media are based on the suppression of metabolic activity, that is, the reduction of cellular metabolism. This is crucial to prevent the accumulation of toxic byproducts, DNA damage, protein denaturation, and nutrient depletion, which could compromise fungal viability over time [27, 28]. For example, preserving fungal species in distilled water and sterile saline solution results in a metabolically inactive state due to nutrient deprivation [29]. Preserving fungal species by storing them in mineral oil decreases metabolic activity due to reduced oxygen supply, resulting in a microaerobic state in the medium [30].

On the other hand, in cryopreservation of fungi, metabolism is suppressed at ultralow temperatures, usually −80°C, or by storage in liquid nitrogen. Additionally, cryopreservation uses cryoprotective agents to mitigate freezing-induced cellular damage [31]. Cryoprotectants are molecules that can reduce ice crystal formation by binding to water molecules during freezing and thawing, thereby reducing mechanical damage to membranes and various cellular structures [32, 33]. Some of the most used cryoprotectants for fungal preservation are glycerol, dimethyl sulfoxide (DMSO), sucrose, polyethylene glycol (PEG), and trehalose.

3.2 Osmotic Balance

Fungal cells are susceptible to changes in the external osmotic environment, especially during low-temperature storage or freeze–thaw cycles, where water fluxes across the membrane can cause cell lysis or plasmolysis. Therefore, preservation media must be formulated to avoid drastic osmotic changes that could compromise cell wall integrity, membrane function, or intracellular structures [34, 35].

Fungal cell walls, although relatively robust, differ in composition depending on the species, developmental stage, and environmental adaptation. Most fungal cell walls are composed of glucans, chitin, mannans, and proteins, which confer resistance to mechanical and osmotic stresses [36]. However, these walls remain vulnerable to osmotic shocks. For example, hyphae of filamentous fungi are more prone to rupture under hypotonic conditions during thawing. In contrast, spores or chlamydospores, owing to their compact morphology and lower water content, are more resilient [37].

To counteract these effects, osmoprotective agents, such as sucrose, sorbitol, mannitol, or salts (e.g., NaCl and KCl), are added to the liquid medium. These solutes modulate the water activity (aw) of the medium, creating an isotonic environment that reduces the movement of water into or out of fungal cells during storage [38]. For example, sucrose concentrations of 10%–15% have been shown to prevent membrane disruption and improve the viability of Agaricus subrufescens mycelia after cryopreservation [39]. Similarly, 5% trehalose with 5% DMSO has been used to protect *Saccharomyces cerevisiae* during cryopreservation, reducing metabolic changes [40]. Ultimately, osmotic conditioning is essential for species that do not form resistant spores and for preservation strategies targeting metabolically active cells.

3.3 Oxidative Stress

Reactive oxygen species (ROS), such as superoxide anions (O_2-), hydrogen peroxide (H_2O_2), and hydroxyl radicals (OH•), are byproducts of normal aerobic metabolism; however, they can accumulate under stressful conditions, such as freezing, thawing, or dehydration. These ROS can cause lipid peroxidation, protein oxidation, and nucleic acid damage, ultimately leading to loss of membrane integrity and cell death [41].

Liquid media can be supplemented with exogenous antioxidants to mitigate oxidative stress during fungal preservation. For example, ascorbic acid is widely used as an ROS scavenger; it donates electrons to neutralize free radicals and stabilizes other antioxidant systems [42]. Glutathione, a natural intracellular tripeptide, is occasionally added to cryopreservation media to maintain the redox potential and protect thiol-containing enzymes [43]. In filamentous fungi, the addition of antioxidants has been shown to increase recovery rates and reduce morphological abnormalities following freeze–thaw cycles [44]. Furthermore, naturally ROS-resistant structures, such as chlamydospores or conidia, tend to accumulate endogenous antioxidants, suggesting that oxidative protection is key to their long-term survival.

3.4 Induction of Cellular Quiescence

A fundamental biological principle in the liquid preservation of fungi is the intentional limitation of nutrients in the medium to induce a physiological state of quiescence or dormancy. Quiescence is a reversible state of metabolic downregulation in which fungal cells reduce energy expenditure, halt cell division, and enter a survival mode that enhances their resistance to stressors, such as desiccation, freezing, or oxidative stress [45].

Cellular quiescence has been observed in several species. For example, when *Saccharomyces cerevisiae*, *Aspergillus fumigatus*, and *Neurospora sitophila* are placed in liquid media with minimal or low nitrogen content, they often cease mitotic activity and undergo morphological adaptations, such as spore formation or cell wall thickening [46, 47]. These structural changes protect cells and enhance preservation efficiency. For example, *Candida albicans* enters a dormant state in media with limited phosphate or glucose, thereby improving its survival during prolonged periods [48]. Storing Trichoderma harzianum in sterile distilled water at 4°C maintains viability and dormancy for several months, with high recovery rates [49]. In another study, *Pleurotus ostreatus* showed improved long-term viability when stored in saline solution, distilled water, and without carbon or nitrogen sources, maintaining its productivity rate and biological efficiency [34].

4 Selection of the Liquid Medium

In preservation using a liquid medium, the selection of the medium is determined by viability, genetic stability, the practicality of the method, the preservation period, and the preservation objective. All these aspects must be evaluated and, in turn, depend on the strain to be preserved and on whether it has been previously preserved or whether a preservation protocol needs to be developed or modified [10]. The most important criteria are mentioned as follows:

4.1 Characteristics of the Species to be Preserved

The diversity of fungi and yeasts is one of the richest on the planet; therefore, there is a wide variety of fungal species that tolerate certain storage conditions, such as ultra-freezing temperatures, while others do not [13, 31, 50, 51].

For species of greatest interest in preservation, given their applications, some storage conditions are already known. For example, in filamentous fungi and some yeasts, such as *Malassezia* spp., cryoprotectants, such as DMSO and glycerol, have shown positive results [1, 4, 33, 50, 52]. In contrast, thermophilic fungal species do not tolerate ultra-freezing temperatures and can only be stored in agar blocks at −20°C [51]. In the case of entomopathogenic fungi, they have been reported to tolerate −20°C in glycerol [53], whereas mycorrhizal species can be stored at 5°C in sterile distilled water for up to 20 years [28, 31, 54]. In contrast, phytopathogenic fungi, such as some oomycetes, are usually successfully stored in sterile distilled water [13, 33]. Therefore, it is necessary to evaluate the preservation conditions for each species.

4.2 Objective and Preservation Period

Although the preservation objective is generally long-term storage (years or decades), it is sometimes necessary to evaluate shorter-term periods (days or months), depending on the working group's needs. The latter is determined by the infrastructure available.

For short-term storage, storage in sterile distilled water is generally considered viable for many species [11, 28, 33, 51, 55]. However, this method has been successfully evaluated over more extended periods [5, 8, 54]. Therefore, it is necessary to review the assessed conditions for each strain and determine which are favorable for the species' viability and genotypic stability.

Regarding long-term storage, the liquid media most recommended for cryopreservation include glycerol, DMSO, and sugars such as sucrose and glucose [28, 31, 50, 56, 57]. On the other hand, for the cryopreservation of species that can tolerate ultra-freezing (−70°C, −80°C and −196°C), liquid nitrogen has been evaluated, as it interrupts metabolic activity for prolonged periods and, therefore, reduces the possibilities of genotypic and phenotypic variations [1, 31, 56, 58]. On the other hand, mineral oil is an economical and simple medium that has been evaluated in cultures for up to 32 years [27, 28, 33].

5 Preservation Procedure

Fungal preservation procedures in liquid media have been continuously evaluated for almost 100 years by various authors, who have sometimes modified the general preservation techniques [8, 10–13, 27, 28, 32, 33, 39, 51, 54, 55, 58–61]. However, the steps for the three primary liquid preservation methods—storage in sterile distilled water, storage in mineral oil, and cryopreservation—can be listed. For each case, it is necessary to review the literature to assess the required modifications and their outcomes.

5.1 Storage in Sterile Distilled Water

The general steps of this storage protocol are as follows (Fig. 3):

1. Inoculum selection: This step is crucial because cultures must be young and vigorous to be preserved. If necessary, cultures in Petri dishes can be used as an inoculum.

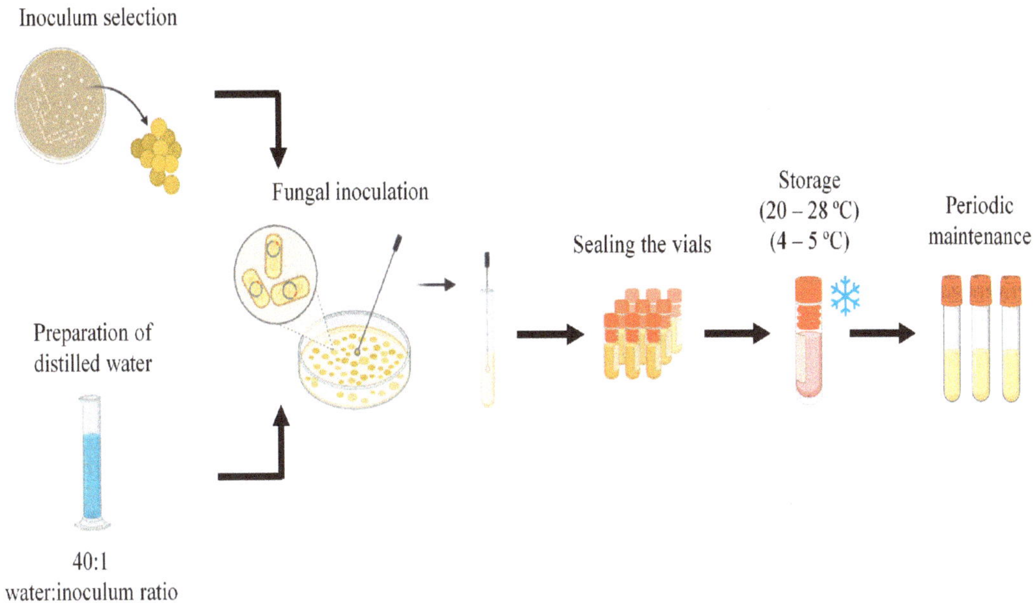

Fig. 3 Protocol for preserving fungi in sterile water. Created with Biorender.com

2. Preparation of distilled water: The water must be sterilized. In general, the water/inoculum ratio should be 40:1. Exceeding this inoculum ratio compromises the ability to withstand prolonged storage.
3. Fungal inoculation: Inoculation can be performed in three ways: (i) covering the surface of the agar containing the fungus with water; (ii) cutting small agar blocks of approximately 1–25 mm³ and transferring them to sterile water; or (iii) for sporulating fungi, suspending them directly in sterile water without agar.
4. Sealing the vials: Screw caps or rubber stoppers are recommended to prevent water evaporation. A paraffin film can be added to prevent contamination of the sample.
5. Storage conditions: Tubes are stored upright at room temperature (20°C–28°C) or, if possible, refrigerated at 4°C–5°C for long-term viability.
6. Periodic maintenance: Water levels are monitored periodically, and sterile water is added if evaporation is observed.

5.2 Storage in Mineral Oil

Preservation of fungi in mineral oil is a simple, feasible technique for long-term storage of fungal cultures. This method is useful for fungi that do not tolerate freeze-drying or for which cryopreservation is prohibitively expensive [16].

The general steps for preserving cultures in mineral oil are as follows (Fig. 4):

1. Culture selection: Young, vigorous cultures should be used for agar slants.
2. Mineral oil preparation: The mineral oil should be medical grade and sterilized twice on consecutive days to ensure the elimination of bacterial spores.
3. Inoculation: The culture is aseptically covered with sterile oil to a depth of approximately 1 cm above the medium surface.
4. Sealing the vials: The tubes are tightly capped, and several layers of paraffin film are applied as an additional barrier against evaporation, mites, or other contaminants.
5. Storage: Tubes are stored upright on racks, ideally refrigerated at 4°C to maintain longer viability, although they can also be stored at room temperature.

5.3 Cryopreservation

The general steps involved in cryopreservation are as follows (Fig. 5):

1. Inoculum selection: Young, vigorous cultures should be used, as with the previous two methods. A concentration of at least 107 colony-forming units (CFU)/mL is recommended.
2. Sample format: Filamentous fungi can be cut into small agar blocks (approximately 1–3 mm^3) from an active culture. For sporulating fungi, spores are collected and suspended in a cryoprotectant. For mycelia, the culture can be harvested from

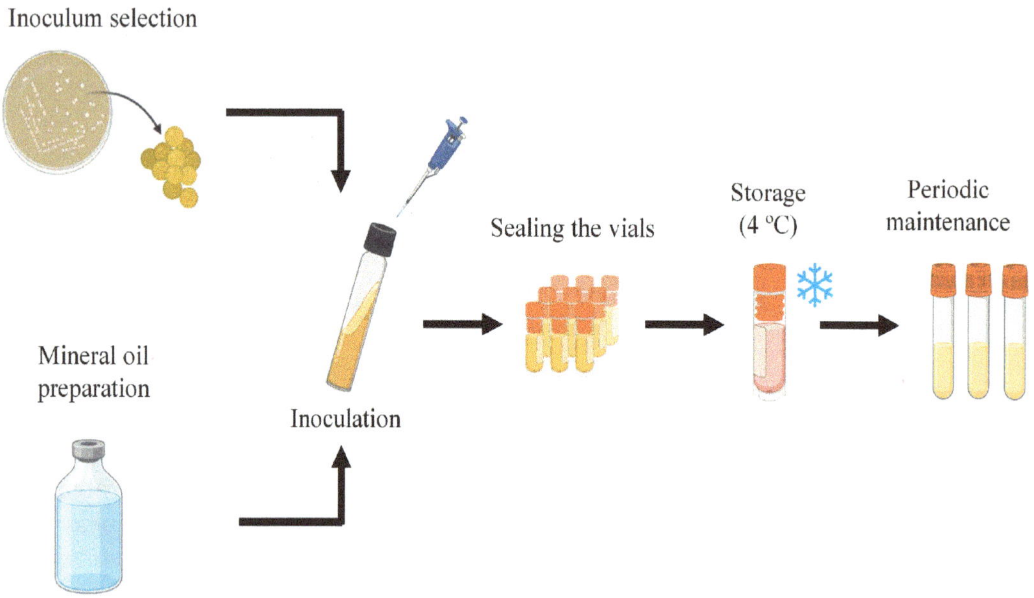

Fig. 4 Protocol for preserving fungi in mineral oil.Created with Biorender.com

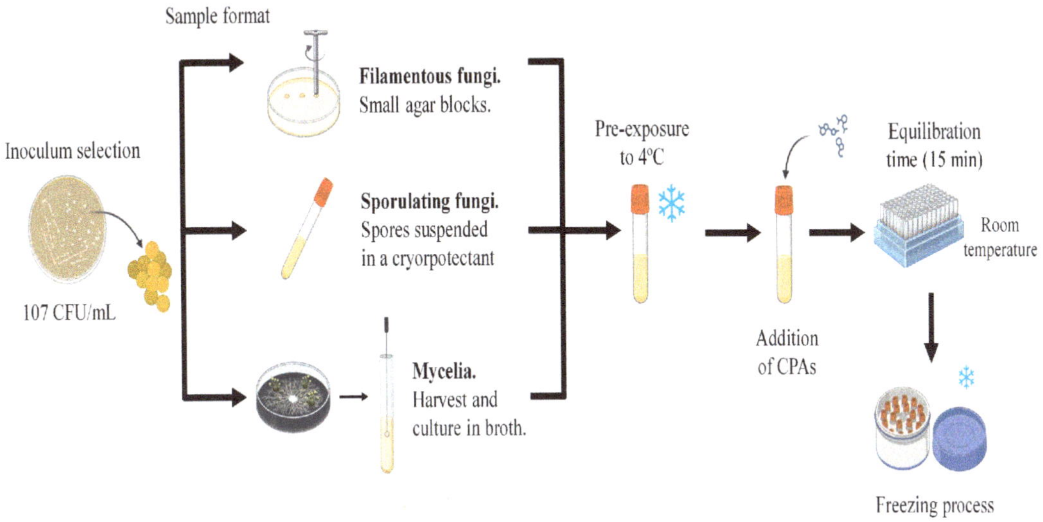

Fig. 5 Cryopreservation protocol for fungi. Created with Biorender.com

agar or grown in broth and mixed before freezing. In some protocols, fungi are grown directly in cryovials to avoid excessive handling that could damage the hyphae.

3. Prerequisites: Pre-exposure to low temperatures (4°C) or heat shock is recommended, as it can increase the freezing resistance of some fungi.
4. Addition of cryoprotectants (CPAs): These are added to cell suspensions. For cultures on solid media, they are dispensed directly into cryovials containing culture blocks, ensuring a significantly larger CPA volume than the inoculum. For liquid cultures, undiluted CPA is added and mixed rapidly to minimize osmotic damage.
5. Equilibration time: Incubate for 15 min at room temperature to allow the CPAs to penetrate the cells before freezing, as prolonged exposure can be toxic.
6. Freezing process: There are two methods, rapid and slow freezing. Slow freezing involves a gradual decrease in temperature, typically at 1°C/min. Rate-controlled freezers or devices such as the "Mr. Frosty" (with isopropyl alcohol) are used to maintain the desired cooling rate.

Using the described methods, the preservation of numerous fungal species has been evaluated, including entomopathogens, endophytic fungi, molds, yeasts, several basidiomycetes with a wide range of applications, mycorrhizal arbuscular fungi, ectomycorrhizal fungi, pathogens, and thermophilic fungi, as well as those of commercial importance and medicinal use [8, 10–13, 27, 28, 32, 33, 39, 51, 54, 55, 58–61]. It is essential to review the literature before preserving a particular species and to evaluate the methods according to the preservation needs.

6 Factors Affecting the Viability and Longevity of Fungi Preserved in Liquid Medium

The preservation of fungi in liquid media is a widely used method in research, industrial applications, and biodiversity conservation to maintain viable fungal cultures for extended periods. However, several factors influence the viability and longevity of preserved strains, including strain-specific characteristics, preservation method conditions, chemical additives, and reactivation protocols. A thorough understanding of these factors is essential for the successful conservation of fungal species. The main factors are discussed below.

6.1 Selection of the Preservation Method

The choice of the liquid preservation method is crucial for achieving the viability and stability of fungal species. Below, we discuss the main advantages and disadvantages of the most used methods, listed in Table 1.

Preservation in sterile distilled water has been widely used because it is an effective and economical method for preserving fungal cultures [27]. This technique is valued for its simplicity, cost-effectiveness, and ability to maintain fungal viability and its stability. However, its effectiveness can be influenced by factors such as temperature,

Table 1
Advantages and disadvantages of fungal preservation methods in liquid media

Method	Advantages	Disadvantages	References
Sterile distilled water	Easy to implement and cost-effective. Do not require specialized equipment.	The viability depends on fungi species, which can be from months to years.	[11, 62]
Mineral oil	Simple and cheap. Limit the exposition to oxygen and dehydration, which reduces the oxidative stress, improving the viability.	It is not for all fungi species. May delay the recovery of certain fungi. Requires periodic subculturing for viability.	[12, 63]
Cryopreservation	For long-term periods. Maintain viability and genetic stability.	Expensive and requires special equipment and manipulation. Requires cryopreservatives.	[31]
Liquid nitrogen.	Reliable for long-term periods. Maintain the genotype and phenotype.	Expensive and requires cryopreservatives. Requires highly technique knowledge and management.	[64, 65]

storage medium, and the specific fungal species being preserved. For example, pathogenic moulds and yeasts have been reported to be effectively preserved and maintain their viability at 4°C [11, 62]. In comparison, mycorrhizal fungi are best stored and recovered at 18°C, especially for Basidiomycotina species [66].

Storage in mineral oil is an inexpensive, easy-to-implement method; it is widely used for the long-term preservation of fungi; however, the viability of fungi depends on the species being preserved [63]. This method has given positive results for nonsporulating fungi or mycelia, making it ideal for species that cannot be freeze-dried or cryopreserved [27]. The fungi preserved with mineral oil exhibit reduced metabolism, thereby increasing their viability for years. However, this method does not entirely stop fungal growth, which can alter their morphological and physiological characteristics in the long term [33]. Examples of fungal species that can be successfully preserved include wood-rotting basidiomycetes, except for Zygomycotina [27].

Cryopreservation methods involve temperatures below −20°C, usually between −80°C and −150°C. This technique is preferred for long-term storage of fungal species, as it maintains viability and genetic stability [31]. The use of additives or cryoprotectants is essential to minimize cell damage during freezing, thawing, and reactivating cultures. This method has been used to preserve several basidiomycetes [31], filamentous fungi [10], yeasts [67], and pathogenic and entomopathogenic fungi [55, 68].

On the other hand, liquid nitrogen preservation is a widely used method for long-term preservation. It has been evaluated for several species, including phytopathogens [69], mycorrhizal fungi [28], anaerobes [64], saprophytes [27], and edible and medicinal mushrooms [70]. This method involves preserving fungal species at ultra-low temperatures ranging from −150°C to −196°C, thereby significantly extending their viability and stability [65].

6.2 Storage Conditions

Storage temperature and conditions are critical for improving the viability of fungal strains in conservation protocols. The temperature intervals reported are the following:

Room temperature (20°C–25°C). Suitable for short-term storage of the samples. However, it increases metabolic activity and the risk of contamination [11]. At these temperatures, phytopathogenic fungi and yeasts can be preserved in both distilled water and mineral oil [10, 33, 69]. Refrigeration temperature (4°C). At this temperature, fungal metabolism decreases, which increases viability from months to years. Some mycorrhizal and saprotrophic basidiomycetes have been reported to survive for 20 years at this temperature [28, 31]. Ultra-low temperatures (−20°C to −196°C). These temperatures give the best results for long-term viability and, at the same time, minimize metabolic degradation, pathogenicity, and sporulation of some fungal species [31, 64]. For example, thermophilic fungi are viable after 5 years of storage at −20°C [51], and entomopathogenic fungi have been preserved at −196°C and −70°C with positive results [71]. Also, several ectomycorrhizal basidiomycetes have been preserved at −196°C using liquid nitrogen vapor [72].

The addition of different chemical compounds has been reported to improve the viability of fungi preserved in liquid media [27, 31]. For example, glycerol has been widely used to protect fungal cells during storage and to maintain viability at concentrations of 4%–50% for months to years [32, 73]. Additionally, glycerol concentrations of 10%–20% are suitable for preserving filamentous fungi at ultra-low temperatures (from −20°C to −70°C) [31, 74]. A mixture of glycerol (85%) and skimmed milk (10%) was effective in preserving the viability of several strains of phytopathogenic fungi in liquid nitrogen for up to 9 years [69]. Another compound widely used as a preservative is dimethyl sulfoxide (DMSO), which, unlike glycerol, is less toxic to fungal strains, especially basidiomycetes [31], and helps maintain the viability, sporulation, and pathogenicity of phytopathogenic and filamentous species [69]. Sato et al. [75] used a modified perlite protocol with 5% v/v DMSO to preserve cryosensitive ectomycorrhizal basidiomycetes, which could not be maintained with standard cryopreservation protocols, achieving good viability. In contrast, Sato et al. [72] used 10% DMSO in a mixture with 10% trehalose, following a vermiculite protocol in liquid nitrogen, and observed 100% viability of 12 ectomycorrhizal basidiomycete strains after 1 year of conservation. Additionally, other preservatives, such as trehalose, glucose, sucrose, and proline, have been reported at concentrations ranging from 5% to 40% [31].

6.3 Avoiding the Risk of Contamination

An essential factor in preservation is controlling contamination during the preparation and storage of preserved cultures [30, 76]. Contamination by rapidly growing microorganisms, such as bacteria, mites, or fungi, can compromise the viability of preserved strains. One of the most widely used strategies to control this risk is the use of antibiotics or antifungals; however, it must be ensured that the concentrations used do not harm the strains of interest to be preserved [30]. The conservation methods most prone to contamination are periodic subculturing, conservation in distilled water, and mineral oil, as these methods require periodic verification that the fungal strains are viable [9, 77]. In contrast, cryopreservation methods minimize the risk of contamination because fungal cultures can remain preserved for years without the need for subculturing, although there is a minimal risk when cryopreserving and transporting the cryobank [61]. Some of the recommendations that can minimize contamination in preserved fungi are maintaining good laboratory practices, using a biological safety cabinet, periodically checking stored cultures, especially when stored in distilled water and mineral oil, properly preparing the inoculum, and using appropriate containers and a secure seal [12, 30, 66].

In conclusion, preserving fungi in liquid media depends on factors such as storage conditions, the method used, and the composition of the culture medium. All these factors influence viability. On the other hand, simple techniques such as distilled water and mineral oil are inexpensive and easy to implement; they can also be suitable for the long-term conservation of many fungal species. Cryopreservation methods, on the other hand, are the most appropriate for maintaining the long-term viability of fungi; however, they require equipment, infrastructure, and trained personnel for their implementation. In many cases, they require cryopreservatives, such as glycerol and DMSO. The objectives of preservation should determine the choice of method and conditions.

7 Recovery and Reactivation of Preserved Fungi

Recovery and reactivation protocols for preserved fungi are essential to ensure the viability of these strains. In the case of preservation in water or oil, strain reactivation is performed directly in the fresh medium and at the temperature conditions appropriate for the strain in question [27, 66]. Some authors have reported that, for preservation in water, a concentration step of spores or hyphae is required, typically performed by centrifugation. On the other hand, in the case of cryopreservation, it has been reported that rapid thawing at a temperature of 35°C–40°C for 1–2 min can minimize strain damage and restore their viability [28, 31]. Furthermore, in cryopreservation, the use of cryopreservatives, such as trehalose and DMSO, has been reported to improve the recovery of strains post-thawing [28, 31]. The reactivation media used in these preservation methods are nutrient-rich media, such as Potato dextrose agar (PDA), sabouraud dextrose agar (SDA), malt extract agar (MEA), and yeast extract peptone dextrose (YPD). The choice of medium depends on the fungal strain to be reactivated [31, 33, 77].

7.1 Postreactivation Viability Assessment

Postpreservation viability is assessed by observing fungal growth in the recovery medium [27, 51]. For sporulating fungi, spore germination is evaluated, whereas for non-sporulating or mycelial fungi, the radial growth rate (mm/d) is measured [78]. For fungi preserved in suspensions, colony-forming units (CFU) can be measured after inoculation onto agar plates [67]. In all cases, the contamination of recovered fungal strains must be monitored [51, 67]. Another test to verify the viability of preserved fungi is morphological examination, which consists of the observation of the macroscopic characteristics (color, texture, diameter, and topology of the fungal strains) and microscopic characteristics (morphology of spores, conidia, or hyphae, and color pigments) of the strains [51, 60]. Physiological and metabolic properties are indicators of the viability of preserved cultures. It is common to measure the main enzymatic activities in preserved cultures, such as laccases, peroxidases, proteases, lipases, and amylases, using the Analytical Profile Index ZYM [27, 28, 78]. Similarly, some metabolites are indicators of the viability of fungal strains, such as gallic acid, tannic acid, tyrosine, guaiacol, pigments, and secondary metabolites [11, 78]. Virulence and pathogenicity tests can be performed on phytopathogenic and entomopathogenic fungi to evaluate their ability to cause diseases in target insects or host plants [68, 79]. Techniques have also been used to verify the genetic stability of conserved strains, such as amplified fragment length polymorphism (AFLP) analysis or PCR fingerprinting [28, 56, 57, 80].

8 Applications of Preservation

Various efforts have been made to estimate the total number of fungal species on Earth. Some authors report the number of species ranges from 1.5 million to 3.3 million, while others estimate it at 6.3 million [81–84]. These estimates are primarily based on molecular analysis tools such as high-throughput DNA sequencing [84, 85].

However, only approximately 150,000 species have been described, representing <10% of the potential species worldwide [86]. Some authors have reported that over the last 10 years, an average of 2000 new fungal species have been described annually [87], while in 2011, the World Data Centre for Microorganisms (WDCM) recorded just over 500,000 fungal species in its collections worldwide [88].

Given the number of fungal species, it is crucial to identify and preserve them in germplasm banks that can be accessed whenever the vast wealth of existing fungi is desired [82, 89, 90]. Therefore, long-term preservation methods for fungi and yeasts are of great importance, as they enable the conservation of genetic resources and their access whenever necessary. Among the studies that can be conducted with preserved fungi and yeasts are: (1) Effects of climate change on the metabolism and physiology of each species. (2) Proteomic, transcriptomic, and metabolomic studies under different environmental conditions. (3) Evolutionary and phylogenetic analyses.

Furthermore, by accessing fungal diversity and ensuring the morphological, physiological, and genetic integrity of preserved species, the reproducibility of research results can be guaranteed with greater confidence [13, 59, 60]. In addition to the importance of conserving genetic diversity, the preservation of fungi and yeasts has various applications, mainly in the industrial and biotechnological fields, biomedicine and health, and agriculture.

- Industrial and biotechnological development: In the industry, it is essential to standardize production processes, as well as raw materials and inputs, so that the characteristics of the obtained product and established quality standards can be maintained. By preserving fungal and yeast species, it is ensured that the production processes in which they are used maintain quality standards [4, 8, 25, 56, 70, 91, 92].
- Biomedicine and health: The preservation of medically necessary fungi and yeasts is essential, as they can be used for epidemiological and virulence studies and for the development of antifungal treatments [13, 93].
- Agriculture: In this area, preserved fungi are used to study phytopathogenic species and develop strategies to control root diseases. They are also used to develop fungicides and implement biological control strategies [31, 59, 94, 95]. On the other hand, entomopathogenic fungi are also of interest. Due to their sporulation capacity and virulence, they are crucial for the management of agricultural pests [53, 55, 59, 95, 96]. Preserving and studying them enables comprehensive biological control management in the face of an uncertain future due to climate change. Mycorrhizal fungi are also of agricultural interest, as they are primarily used for ecological restoration, as inoculum for economically significant forest tree seedlings, and to produce edible mushrooms [28, 59, 97].

Table 2 lists the species of fungi and yeasts that have been most frequently preserved, their main applications, and the preservation methods used for each species.

9 Challenges and Future Perspectives

More than 100 years after the initiation of fungal preservation techniques in liquid media, valuable information has been obtained about the conditions and different variations of each protocol to achieve the viability and genetic and phenotypic stability of preserved cultures. However, the current challenges stem primarily from the identification and description of more fungal species each year [27, 82].

Among the challenges and prospects are the continuous optimization of techniques, thereby reducing the infrastructure costs required to maintain fungal germplasm banks for years, even decades, to ensure the preservation of this genetic wealth for the future [31, 64].

Aspects such as the selection and development of cryoprotectants, culture conditioning, and evaluation and monitoring of viability and genetic stability are key points for optimizing the different techniques [24, 28].

Table 2
Applications for preserved fungi and yeasts

Application	Species name	Preservation method	References
Pathogens of importance in biomedicine and health.	*Microsporum canis* *Blastomyces dermatitidis* *Histoplasma capsulatum var. capsulatum* *Trichophyton tonsurans* *Aspergillus fumigatus* *Cladophialophora bantiana* *Trichophyton interdigitale* *Madurella mycetomatis* *Cladophialophora carrionii* *Paracoccidioides brasiliensis*	Sterile distilled water, cryopreservation	[10, 69]
Phytopathogens of importance in agriculture.	*Fusarium oxysporum* *Phytophthora infestans* *Rhizoctonia solani* *Botrytis cinerea* *Alternaria alternata* *Alternaria solani* *Colletotrichum gloeosporioides* *Curvularia lunata* *Lasiodiplodia theobromae* *Rosellinia necatrix*	Sterile distilled water, cryopreservation	[67, 98]
Entomopathogens of importance in agriculture.	*Beauveria bassiana* *Metarhizium anisopliae* *Paecilomyces fumosoroseus* *Drechmeria coniospora* *Metarhizium acridum* *Metarhizium rileyi* *Isaria javanica* *Hirsutella thompsonii* *Hirsutella citriformis* *Lecanicillium lecanii* *Zoophthora radicans*	Sterile distilled water, cryopreservation	[55, 66, 98]

Table 2 (Continued)

Application	Species name	Preservation method	References
Mycorrhizal organisms of agricultural interest.	*Laccaria*	Cryopreservation	[28]
	Suillus		
	Boletus		
	Lactarius		
	Paxillus		
	Thelephora		
	Rhizophagus		
	Gigaspora		
	Entrophospora		
	Glomus		
	Acaulospora		
	Scutellospora		
	Cantharellus cibarius		
Development of functional and pharmaceutical products.	*Lentinula edodes*	Sterile distilled water, cryopreservation	[54, 91, 99]
	Agaricus bisporus		
	Agaricus subrufescens		
	Ganoderma lucidum		
	Pleurotus ostreatus		
Ligninolytic basidiomycetes. Wood biodegradation and enzyme production.	*Trametes versicolor*	Sterile distilled water, cryopreservation	[8, 26]
	Pleurotus ostreatus		
	Fomitopsis lilacinogilva		
	Serpula lacrymans		
Production of secondary metabolites and extracellular enzymes, organic acids, and antibiotics.	*Aspergillus*	Sterile distilled water, freeze-drying, mineral oil, and storage at 20°C and −70°C	[9, 60]
	Penicillium		
Edible mushrooms.	*Pleurotus* spp.	Sterile distilled water	[8]
	Pleurotus djamor		
	Pleurotus pulmonarius		
	Agaricus spp.		

In the face of climate change, prioritizing long-term preservation efforts worldwide is necessary. Currently, the largest WDCM collections are in East Asia, North America, and Europe [90]. Still, space is needed to preserve the enormous fungal wealth of Latin America and Africa, two central regions noted for their biodiversity. Therefore, the challenge is to combine efforts to identify and preserve the majority of species to ensure a bioeconomy based on the exploitation of fungal and yeast genetic resources in the coming decades.

10 Conclusion

Liquid media preservation methods are valuable and accessible alternatives for the long-term maintenance of many fungal species. Storage in sterile water and mineral oil is a relevant technique when infrastructure is limited and has been successfully evaluated for long-term periods. However, it is recommended that these techniques be complemented by more sophisticated methods, such as cryopreservation, when fungal species tolerate storage at ultra-low freezing temperatures.

Continuous evaluation and adaptation of protocols are essential whenever a new species needs to be preserved. Because fungal metabolism is so diverse, their responses to storage conditions differ across techniques. Therefore, it is not possible to preserve all fungi using a single standardized technique.

The current challenge is to optimize techniques to preserve fungi long-term with fewer resources while maintaining their viability and genetic and phenotypic stability.

Conflict of Interest

The authors have no conflicts of interest to declare that are relevant to the content of this chapter.

References

1. Wolkers, WF, Oldenhof, H, 2015. Cryopreservation and freeze-drying protocols. Humana Press. pp. 3–25
2. Onions AH (1971) Chapter IV Preservation of fungi. In: Booth C (Ed.) Methods in microbiology. Academic Press, vol 4:113–151
3. Kaur L, Dhanda S, Sodhi HS, Kapoor S, Khanna PK (2011) Storage and preservation of temperate mushroom cultures, Agaricus bisporus and Pleurotus florida. Indian J Microbiol 51(2):234–238
4. Harirchi S, Rousta N, Varjani S, Taherzadeh MJ (2023) Sampling, preservation, and growth monitoring of filamentous fungi. In: Mohammad J, Taherzadeh JA, Ferreira, Ashok P (Eds.) Current developments in biotechnology and bioengineering. Elsevier, 149–180
5. de Capriles CH, Mata S, Middelveen M (1989) Preservation of fungi in water (Castellani): 20 years. Mycopathologia 106(2):73–79
6. Burdsall HH Jr, Dorworth EB (1994) Preserving cultures of wood-decaying

Basidiomycotina using sterile distilled water in cryovials. Mycologia 86(2):275–280
7. Jones RJ, Sizmur KJ, Wildman HG (1991) A miniaturised system for storage of fungal cultures in water. Mycologist 5(4):184–186
8. Richter DL, Kangas LC, Smith JK, Laks PE (2010) Comparison of effectiveness of wood decay fungi maintained by annual subculture on agar and stored in sterile water for 18 years. Can J Microbiol 56(3):268–271
9. Qiangqiang Z, Jiajun W, Li L (1998) Storage of fungi using sterile distilled water or lyophilization: comparison after 12 years: Vergleich der Aufbewahrung von Pilzen in sterilem destilliertem Wasser oder Lyophilisierung nach 12 Jahren. Mycoses 41(5-6):255–257
10. Borman AM, Szekely A, Campbell CK, Johnson EM (2006) Evaluation of the viability of pathogenic filamentous fungi after prolonged storage in sterile water and review of recent published studies on storage methods. Mycopathologia 161(6):361–368
11. Karabıçak N, Karatuna O, Akyar I (2016) Evaluation of the viabilities and stabilities of pathogenic mold and yeast species using three different preservation methods over a 12-year period along with a review of published reports. Mycopathologia 181:415–424
12. Humber RA (2012) Preservation of entomopathogenic fungal cultures. Manual of techniques in invertebrate pathology, vol 2, Academic Press, pp 317–32
13. Castro-Rios K, Bermeo-Escobar LP (2021) Methods for the culture conservation of edible and medicinal fungi. J Microbiol Biotechnol Food Sci 10(4):620–625
14. Fennell DI (1960) Conservation of fungous cultures. Bot Rev 26(1):79–141
15. Henry BS (1947) The viability of yeast cultures preserved under mineral oil. J Bacteriol 54(2):264
16. Perrin PW (1979) Long-term storage of cultures of wood-inhabiting fungi under mineral oil. Mycologia 71(4):867–869
17. Smith D, Onions AH (1983) A comparison of some preservation techniques for fungi. Trans Br Mycol Soc 81(3):535–540
18. Nakasone KK, Peterson SW, Jong SC (2004) Preservation and distribution of fungal cultures. In: Muller GM, Bills GF, Foster MF (Eds.) Biodiversity of fungi. Elsevier, 3–11
19. Hwang SW (1960) Effects of ultra-low temperatures on the viability of selected fungus strains. Mycologia 52(3):527–529
20. Elliott TJ (1976) Alternative ampoule for storing fungal cultures in liquid nitrogen. Cryobiology 13(4):465–470
21. Pasarell L, McGinnis MR (1992) Viability of fungal cultures maintained at −70 degrees C. J Clin Microbiol 30(4):1000–1004
22. Homolka L, Lisá L, Eichlerová I, Nerud F (2001) Cryopreservation of basidiomycete strains using perlite. J Microbiol Methods 47(3):307–313
23. Kitamoto Y, Suzuki A, Shimada S, Yamanaka K (2002) A new method for the preservation of fungus stock cultures by deep-freezing. Mycoscience 43(2):143–149
24. Homolka L, Lisá L, Nerud F (2006) Basidiomycete cryopreservation on perlite: evaluation of a new method. Cryobiology 52(3):446–453
25. Homolka L, Lisá L, Eichlerová I, Valášková V, Baldrian P (2010) Effect of long-term preservation of basidiomycetes on perlite in liquid nitrogen on their growth, morphological, enzymatic and genetic characteristics. Fungal Biol 114(11-12):929–935
26. Eichlerová I, Homolka L, Tomšovský M, Lisá L (2015) Long term storage of Pleurotus ostreatus and Trametes versicolor isolates using different cryopreservation techniques and its impact on laccase activity. Fungal Biol 119(12):1345–1353
27. Homolka L (2014) Preservation of live cultures of basidiomycetes—recent methods. Fungal. Biol 118(2):107–125
28. Lalaymia I, Cranenbrouck S, Declerck S (2014) Maintenance and preservation of ectomycorrhizal and arbuscular mycorrhizal fungi. Mycorrhiza 24(5):323–337
29. Humber RA (1997) Fungi: preservation of cultures. In: Lacey LA (Ed.) Manual of techniques in insect pathology. Academic Press, 269–279
30. Jong SC, Birmingham JM (2001) Cultivation and preservation of fungi in culture. In: Systematics and evolution. Springer, Berlin, Heidelberg, pp 193–202
31. Eichlerová I (2025) The latest trends in cryopreservation of fungi. Mycol Prog 24(1):1–18
32. Paul JS, Tiwari KL, Jadhav SK (2015) Long term preservation of commercial important fungi in glycerol at 4°C.. Int J Biol Chem 9(2):79–85
33. Singh SK (2017) Ex situ conservation of fungi: a review. In: Satyanarayana T, Deshmukh S, Johri B (Eds.) Developments in fungal biology and applied mycology. Springer, Singapore, 543–562
34. Escobar LPB, González JPP, Orjuela M, Castro-Ríos K (2020) Effect of culture preservation methods in the stability and nutritional characteristics of Pleurotus ostreatus. Asian J Microbiol Biotech Environ Sci 22(2):359–368
35. Gostinčar C, Gunde-Cimerman N (2023) Understanding fungi in glacial and hypersaline environments. Annu Rev Microbiol 77(1):89–109
36. Ene IV, Walker LA, Schiavone M, Lee KK, Martin-Yken H, Dague E, Gow NAR, Munro

CA, Brown AJ (2015) Cell wall remodeling enzymes modulate fungal cell wall elasticity and osmotic stress resistance. mBio 6(4):e00986–15
37. Magan N (2007) Fungi in extreme environments. In: Kubicek C, Druzhinina I (Eds.) Environmental and Microbial Relationships. The Mycota, Springer, Berlin, Heidelberg, vol 4
38. Thomas J, Saha JK (2021) Evaluation of antagonistic property and osmotic stress tolerance of yeast based biocontrol formulation. Pharma Innov J 10(5):1569–1573. https://www.thepharmajournal.com/archives/?year=2021&vol=10&issue=5&ArticleId=6455. Accessed 15 Aug 2025
39. Zaghi Júnior LL, Lopes AD, Cordeiro FA, Colla IM, Bertéli MBD, Valle JS, Linde GA, Colauto NB (2018) Cryopreservation at −75° C of Agaricus subrufescens on wheat grains with sucrose. Braz J Microbiol 49(2):370–377
40. Ramasamy T, Tevatia R, Ali S, Muhle A, Knight-Connoni V, Chakraborty N (2025) Proteomic approach for evaluating cryoprotectant formulations for enhanced post-cryopreservation recovery of yeast. Sci Rep 15(1):15474
41. Kranner I, Birtic S (2005) A modulating role for antioxidants in desiccation tolerance. Integr Comp Biol 45(5):734–740
42. Brown AJ, Cowen LE, Di Pietro A, Quinn J (2017) Stress adaptation. Microbiol Spectr 5(4):FUNK-0027-2016
43. Zhang C, Zhang H, Wang L, Yao H (2007) Validation of antifreeze properties of glutathione based on its thermodynamic characteristics and protection of baker's yeast during cryopreservation. J Agric Food Chem 55(12):4698–4703
44. Abrashev R, Miteva-Staleva J, Gocheva Y, Stoyancheva G, Dishliyska V, Spasova B, Krumova E, Angelova M (2025) Cell response to oxidative stress in Antarctic filamentous fungi. Appl Sci 15(9):5149
45. Gray JV, Petsko GA, Johnston GC, Ringe D, Singer RA, Werner-Washburne M (2004) "Sleeping beauty": quiescence in Saccharomyces cerevisiae. Microbiol Mol Biol Rev 68(2):187–206
46. Dagenais TRT, Keller NP (2009) Pathogenesis of Aspergillus fumigatus in invasive aspergillosis. Clin Microbiol Rev 22(3):447–465
47. Niloy MAHM, Islam S, Ferdous T, Rahman S, Yesmin S, Bin Rasul S, Khandakar J (2025) Deciphering the role of substrate carbon to nitrogen ratio in preventing orange mold contamination caused by Neurospora sitophila in mushroom cultivation. Front Ind Microbiol 2:1508079
48. Rodaki A, Bohovych IM, Enjalbert B, Young T, Odds FC, Gow NA, Brown AJ (2009) Glucose promotes stress resistance in the fungal pathogen Candida albicans. Mol Biol Cell 20(22):4845–4855
49. Bueno L, Gallardo R (1998) Preservación de hongos filamentosos en agua destilada estéril. Rev Iberoam Micol 15(3):166–168
50. Ryan MJ, Smith D (2007) Cryopreservation and freeze-drying of fungi employing centrifugal and shelf freeze-drying. In: Day JG, Stacey GN (eds) Cryopreservation and freeze-drying protocols. Methods in Molecular Biology. Humana Press, pp 127–140
51. Berikten D (2021) Survival of thermophilic fungi in various preservation methods: a comparative study. Cryobiology 101:38–43
52. Tedeschi R, DePaoli P (2010) Collection and preservation of frozen microorganisms. In: Dillner J (ed) Methods in biobanking, methods in molecular biology, vol 675, Humana Press, Totowa, NJ, pp 313–326
53. Oliveira I, Pereira JA, Bento A, Baptista P (2011) Viability of Beauveria bassiana isolates after storage under several preservation methods. Ann Microbiol 61(2):339–344
54. Richter DL (2008) Revival of saprotrophic and mycorrhizal basidiomycete cultures after 20 years in cold storage in sterile water. Can J Microbiol 54(8):595–599
55. Ayala-Zermeño MA, Gallou A, Berlanga-Padilla AM, Andrade-Michel GY, Rodríguez-Rodríguez JC, Arredondo-Bernal HC, Montesinos-Matías R (2017) Viability, purity, and genetic stability of entomopathogenic fungi species using different preservation methods. Fungal Biol 121(11):920–928
56. Voyron S, Roussel S, Munaut F, Varese GC, Ginepro M, Declerck S, Marchisio VF (2009) Vitality and genetic fidelity of white-rot fungi mycelia following different methods of preservation. Mycol Res 113(10):1027–1038
57. Broughton R, Buddie AG, Smith D, Ryan MJ (2012) The effect of cryopreservation on genomic stability in strains of the fungus Trichoderma. CryoLetters 33(4):299–306
58. Senanayake IC, Rathnayaka AR, Marasinghe DS et al. (2020) Morphological approaches in studying fungi: collection, examination, isolation, sporulation and preservation. Mycosphere 11(1):2678–2754
59. Singh SK, Singh PN, Gaikwad SB, Maurya DK (2018) Conservation of fungi: a review on conventional approaches. In: Sharma S, Varma A (eds) Microbial resource conservation. Soil Biology, vol 54, Springer, Cham, pp 223–237
60. Freire AKL, dos Santos Bentes A, de Lima Sampaio I, de Lima AM, Botineli LF, da Rocha LC (2016) Availability and morphological characteristics of endophytic fungi held

in different methods of preservation. Sci Res Essays 11(6):76–79
61. Zalesky T, Bradshaw AJ, Bair ZJ, Meyer KW, Stamets P (2024) Fungal cryopreservation across 61 genera: practical application and method evaluation. Mycologia 116(5):865–876
62. Sakr N (2018) Evaluation of two storage methods for fungal isolates of Fusarium sp. and Cochliobolus sativus. Acta Phytopathol Entomol Hung 53(1):11–18
63. Sharma SK, Kumar R, Vaishnav A, Sharma PK, Singh UB, Sharma AK (2017) Microbial cultures: maintenance, preservation and registration. In: Varma A, Sharma A (Eds.) Modern tools and techniques to understand microbes. Springer, Singapore, 335–367
64. Vinzelj J, Joshi A, Young D et al. (2022) No time to die: comparative study on preservation protocols for anaerobic fungi. Front Microbiol 13:978028
65. Webb KM, Holman G, Duke S, Greene S, McCluskey K (2018) Frozen fungi: cryogenic storage is an effective method to store Fusarium cultures for the long-term. Ann Appl Biol 173(2):133–140
66. Smith D (1993) Long-term preservation of test strains (fungus). Int Biodeterior Biodegradation 31(3):227–230
67. Nascimento CRS, Souto ASS, Galvão RM, Lazera MS, Trilles L (2020) Genotypic and phenotypic stability of mixed primary isolates of Cryptococcus gattii and Cryptococcus neoformans: a comparative analysis of four preservation methods. Biopreserv Biobank 18(3):196–203
68. de Souza Rabello VB, Corrêa-Moreira D, Santos C et al. (2022) Preservation methods in isolates of Sporothrix characterized by polyphasic approach. J Fungi 9(1):34
69. Dahmen H, Staub T, Schwinn FJ (1983) Technique for long-term preservation of phytopathogenic fungi in liquid nitrogen. Phytopathology 73(2):241–246
70. Homolka L (2012) Methods of cryopreservation in fungi. In: Gupta V, Tuohy M, Ayyachamy M, Turner K, O'Donovan A (eds) Laboratory protocols in fungal biology. Fungal Biology, Springer, New York, NY, pp 9–16
71. Ayala-Zermeño MA, Berlanga-Padilla AM, Regla-Márquez CF, Lino-López GJ, Muñiz-Paredes F, Montesinos-Matías R, Sánchez-González JA (2023) Long-term preservation and genetic stability of entomopathogenic fungal species. J Microbiol Methods 208:106711
72. Sato M, Inaba S, Noguchi M, Nakagiri A (2020) Vermiculite as a culture substrate greatly improves the viability of frozen cultures of ectomycorrhizal basidiomycetes. Fungal Biol 124(8):742–751
73. Sriram S, Roopa KP, Savitha MJ (2011) Extended shelf-life of liquid fermentation derived talc formulations of Trichoderma harzianum with the addition of glycerol in the production medium. Crop Prot 30(10):1334–1339
74. Hwang SW (1968) Investigation of ultra-low temperature for fungal cultures. I. An evaluation of liquid-nitrogen storage for preservation of selected fungal cultures. Mycologia 60(3):613–621
75. Sato M, Inaba S, Sukenobe J, Sasaki T, Inoue R, Noguchi M, Nakagiri A (2019) A modified perlite protocol with a mixed dimethyl sulfoxide and trehalose cryoprotectant improves the viability of frozen cultures of ectomycorrhizal basidiomycetes. Mycologia 111(1):161–176
76. Gill JS, Sen S (2023) Evaluation of purity, viability, stability and biochemical characteristics of clinical fungal isolates preserved by three different methods. IP Int J Med Microbiol Trop Dis 4(4):203–205
77. Iqbal S, Ashfaq M, Malik AH, Khan KS, Mathew P (2017) Isolation, preservation and revival of Trichoderma viride in culture media. J Entomol Zool Stud 5(3):1640–1646
78. Croan SC, Burdsall HH Jr, Rentmeester RM (1999) Preservation of tropical wood-inhabiting basidiomycetes. Mycologia 91(5):908–916
79. El-Naggar AA, Yassin MA (2021) Viability, growth and virulence of Harpophora maydis isolates preserved under different storage methods. Egypt J Phytopathol 49(2):93–102
80. Crahay C, Munaut F, Colpaert JV, Huret S, Declerck S (2017) Genetic stability of Ectomycorrhizal fungi is not affected by cryopreservation at −130 C or cold storage with repeated sub-cultivations over a period of 2 years. Mycorrhiza 27(6):595–601
81. Hawksworth DL (2001) The magnitude of fungal diversity: the 1.5 million species estimate revisited. Mycol Res 105(12):1422–1432
82. Sette LD, Pagnocca FC, Rodrigues A (2013) Microbial culture collections as pillars for promoting fungal diversity, conservation and exploitation. Fungal Genet Biol 60:2–8
83. Hawksworth DL, Lücking R (2017) Fungal diversity revisited: 2.2 to 3.8 million species. Microbiol Spectr 5(4):FUNK-0052-2016
84. Baldrian P, Větrovský T, Lepinay C, Kohout P (2022) High-throughput sequencing view on the magnitude of global fungal diversity. Fungal Divers 114(1):539–547
85. Blackwell M (2011) The Fungi: 1, 2, 3… 5.1 million species?. Am J Bot 98(3):426–438
86. Phukhamsakda C, Nilsson RH, Bhunjun CS et al. (2022) The numbers of fungi: Contributions from traditional taxonomic studies and challenges of metabarcoding. Fungal Divers 114(1):327–386

87. Cheek M, Nic Lughadha E, Kirk P et al. (2020) New scientific discoveries: plants and fungi. Plants People Planet 2(5):371–388
88. Karaduman AB, Atli B, Yamac M (2012) An example for comparison of storage methods of macrofungus cultures: Schizophyllum commune. Turk J Bot 36(2):205–212
89. Lalaymia I, Cranenbrouck S, Draye X, Declerck S (2012) Preservation at ultra-low temperature of in vitro cultured arbuscular mycorrhizal fungi via encapsulation-drying. Fungal Biol 116(10):1032–1041
90. Ryan MJ, McCluskey K, Verkleij G, Robert V, Smith D (2019) Fungal biological resources to support international development: challenges and opportunities. World J Microbiol Biotechnol 35(9):139
91. Linde GA, Luciani A, Lopes AD, do Valle JS, Colauto NB (2018) Long-term cryopreservation of basidiomycetes. Braz J Microbiol 49(2):220–231
92. Nagpal R, Puniya AK, Sehgal JP, Singh K (2012) Survival of anaerobic fungus Caecomyces sp. in various preservation methods: a comparative study. Mycoscience 53(6):427–432
93. Colauto NB, Cordeiro FA, Geromini KVN et al. (2012) Viability of Agaricus blazei after long-term cryopreservation. Ann Microbiol 62(2):871–876
94. Houseknecht JL, Suh SO, Zhou JJ (2012) Viability of fastidious phytophthora following different cryopreservation treatments. Fungal Biol 116(10):1081–1089
95. Al-Bedak OA, Sayed RM, Hassan SH (2019) A new low-cost method for long-term preservation of filamentous fungi. Biocatal Agric Biotechnol 22:101417
96. López Lastra CC, Hajek AE, Humber RA (2002) Comparing methods of preservation for cultures of entomopathogenic fungi. Can J Bot 80(10):1126–1130
97. Danell E, Flygh G (2002) Cryopreservation of the ectomycorrhizal mushroom Cantharellus cibarius. Mycol Res 106(11):1340–1342
98. Ryan MJ (1999) The effect of preservation regime on the physiology and genetic stability of economically important fungi. Dissertation, University of Kent
99. Mata G, Gaitán-Hernández R (2005) Viability in spawn stocks of the white button mushroom, Agaricus bisporus, after freezing in liquid nitrogen without a cryoprotectant. Int J Agric Technol 1(1):153–162

Check for updates

Chapter 18

Techniques and Strategies for the Conservation of Arbuscular Mycorrhizal Fungi

Alejandra Ortiz-Gasca, Lorena Jacqueline Gómez-Godínez, José Luis Aguirre-Noyola, Magali Ruíz-Rivas, Dora Trejo-Aguilar and Wendy Sangabriel-Conde

Abstract

Arbuscular mycorrhizal symbiosis between fungi of the phylum Glomeromycota and the roots of terrestrial plants is fundamental for nutrient and water absorption, thereby improving plant nutritional efficiency. Arbuscular mycorrhizal fungi (AMF) influence key ecological processes, such as nutrient cycling and carbon stabilization in soil, making them crucial to agricultural sustainability. AMF conservation is carried out using various techniques. One of the most common is the use of trap crops, where host plants that favor efficient fungal colonization are selected. This method allows for maintaining the genetic and functional diversity of AMF, although it faces limitations such as the possibility of contamination or loss of diversity. Additionally, solid substrates such as soil, peat, or compost are used, which allow viable propagules to be preserved without losing their infective capacity. However, these methods can exhibit variability due to environmental factors such as pH and humidity. For long-term conservation, alternatives such as cryopreservation and combining encapsulation with controlled desiccation are being explored. This method has proven effective, though its success depends on the fungus's characteristics. Monoxenic culture with transformed roots has also improved the conservation of arbuscular mycorrhizal fungi (AMF), reducing contamination and increasing the reproducibility of propagules. Despite these advances, AMF conservation continues to face challenges. The future of their conservation will likely include hybrid strategies, combining cryopreservation, periodic maintenance, and functional monitoring.

Keywords Conservation, Arbuscular mycorrhizal fungi, Cryopreservation

1 Introduction

Arbuscular mycorrhizal symbiosis is one of the oldest, formed between obligate biotrophic fungi of the phylum Glomeromycota and the roots of most terrestrial plants [1]. It is estimated that this association is established with approximately 80% of vascular plants [2], including a very high proportion of cultivated species. It is characterized by intracortical colonization and the formation of arbuscules as a key structure for nutrient exchange between symbionts [3].

Marco A. Ramírez-Mosqueda et al. (eds.), *Methods for The Conservation of Microorganisms*, Methods in Molecular Biology, vol. 3090, https://doi.org/10.1007/978-1-0716-5623-5_18,

Arbuscular mycorrhizal fungi (AMF) depend on carbon from associated plants in the form of sugars and lipids [4]. In exchange for carbon, the fungi provide nutrients, micronutrients, and water [5]. They can modulate ecological processes such as nutrient cycling [6] and influence productivity [7].

It is estimated that they provide more than 50% of phosphorus uptake and up to 35% of the total water transpired daily by plants [8, 9]. AMF develop networks of hyphae in the soil, extra-radical mycelium (ERM), used to collect nutrients inaccessible to plants, such as nitrogen and phosphorus [10], as well as to explore the soil [11]. Nutrients are transferred through intra-radical mycelium (IRM) to specialized structures called arbuscules formed inside the cells of the host plant [12].

Furthermore, AMF influence the structure of the rhizosphere microbiome and contribute to ecosystem functions such as carbon cycling (via the transfer and stabilization of plant-derived carbon in the soil) [13]. In ecosystems and agroecosystems, they modulate nutrient cycles and microbial networks [14], and their presence has been associated with changes in carbon stabilization and storage, as well as the formation of organic fractions [15].

Currently recognized as biostimulant microorganisms due to their ability to improve nutritional efficiency, physiological performance, and plant resilience to adverse conditions [16], their application has reduced dependence on phosphate fertilizers and other chemical inputs, aligning with sustainability goals. Agrochemicals not only improve plant physiological performance but also provide systemic benefits that directly impact the profitability and sustainability of agronomic management.

2 Conventional Methods for Conserving AMF Under Field Conditions

In practice, conservation usually refers to preserving viable inoculum (propagules: spores, hyphae, and colonized root fragments) or maintaining active strain communities with host plants (traps, pots, and plots). This allows the fungus to operate in its natural biological mode and avoids the bottleneck of in vitro culture for recalcitrant isolates.

The choice of an optimal host plant for AMF multiplication or evaluation should be based on a set of physiological, ecological, and agronomic characteristics that favor efficient and stable colonization by the symbiont. According to the International Collection of Vesicular and Arbuscular Mycorrhizal Fungi (INVAM), selecting clearly mycotrophic species is recommended, as these tend to support greater fungal genotype diversity and promote intense and persistent colonization [17]. This trait is particularly relevant when seeking to conserve or multiply fungal consortia with high genetic variability.

From an agronomic perspective, the plant must exhibit high compatibility with controlled greenhouse conditions and broad tolerance to soils with varying nutrient compositions [18]. A root system that completely colonizes the substrate volume within 2–3 months promotes prolonged interaction with the fungus and enables complete colonization cycles [19]. Species with high photosynthetic efficiency and high phosphorus requirements, such as those with C4 metabolism, are often particularly suitable, as they allocate greater carbon fluxes to the root system [17].

Finally, the selected species must be relatively resistant to pests and pathogens, including soil-transmitted plant-parasitic nematodes. This characteristic reduces experimental interference, prevents root biomass loss, and ensures that observed variations in mycorrhizal colonization are not due to uncontrolled biotic damage. For this reason, INVAM recommends using Sudan grass (*Sorghum* × *drummondii*), sorghum, maize (which tends to limit the number of species in a fungal community in trap crops), and onion. However, the Bever–Schultz laboratory also uses perennial prairie plants that are highly sensitive to colonization [19]. Multiple studies warn that host-fungus selectivity exists, so the choice of trap crop can bias which species are conserved or recovered. In this regard, integrating more than one host (e.g., combining a grass with alternative hosts such as white clover (*Trifolium repens*) or species like *Allium porrum*/*Plantago lanceolata* can increase the recovery and stability of fungal diversity when the goal is to preserve taxonomic richness and not just spore production [21, 22]. Furthermore, it has been reported that the diversity of arbuscular mycorrhizal fungi (AMF) is reduced after long-term subculture with a single host plant species [23].

Among the limitations of propagation using trap crops are drift due to management (fertilization, stress, and changes in microbiota), the risk of contamination, and the cost of maintenance (space and time).

2.1 Preservation Methods on Solid Substrates

In agronomic applications, the most common preservation method is the storage of propagules on solid substrates to maintain viability without triggering microbial collapse or loss of infectivity [24].

The use of soil and sand mixtures has allowed the preservation of the local mycorrhizal community and the maintenance of soil conditions relatively similar to those at the site of origin, thereby favoring the ecological stability of the fungal consortium [25]. However, this system exhibits high intrinsic variability, driven by factors such as pH, moisture, nutrient content, and the accompanying microbiota. These fluctuations can induce a gradual reduction in the density of viable propagules and their infective potential.

Peat and compost, for their part, offer significant physical advantages, such as greater water retention capacity, improved aeration, and easier handling [26]. However, their high microbial activity can pose a risk if humidity and temperature are not properly controlled, as certain saprophytic or antagonistic microorganisms can compete for resources, degrade fungal structures, or even prey on mycorrhizal propagules [27]. The viability/infectivity of inocula on peat-based substrates has been tested under various temperature regimes (−20°C, 5°C, 20°C, and outdoor conditions), where it was observed that viability/infectivity decreases over time, but temperature and exposure drastically alter the rate of this decline, with the response varying between species [28].

In contrast, inert substrates, such as vermiculite or perlite, combined with precolonized roots, provide greater experimental reproducibility and control of variables [29]. These systems are commonly used as "inoculum carriers" because they facilitate standardization of propagule density, reduce microbiological interference, and allow for better preservation of infectivity under controlled conditions [30]. Furthermore, infectivity can change with wetting–drying cycles (not always in the same direction, depending on the species), which is important when the inoculum is stored in materials with variable moisture content [31].

2.2 Conservation in Monoxenic Cultures

Historically, conservation in axenic cultures has been an obstacle, as arbuscular mycorrhizal fungi (AMF) are obligate symbionts. The practical revolution was the use of transformed roots ("hairy roots" induced by *Agrobacterium rhizogenes*, Ri T-DNA) in monoxenic systems: the fungus grows in association with roots (Root organ culture, ROC) under controlled conditions, with high reproducibility and reduced contamination [32]. Monoxenic culture with transformed roots allows the fungus to be maintained in controlled plates or systems and to produce relatively standardized propagules [33, 34]. These systems are key for germplasm banks, strain standardization, and physiological experiments [35, 36]. The classic and robust host has been carrot roots (*Daucus carota*) due to their stability, sustained growth, and compatibility with various arbuscular mycorrhizal fungi (AMF) [37, 38]. Various sources have been used as inoculants: field spores (after isolation and sterilization), spores from previous cultures, and colonized root fragments/hyphae from a previous system (less common in strict monoxenia) [38, 39].

The methodological literature highlights that one of the critical steps in this culture method is decontamination without killing the propagule and ensuring spore germination on synthetic media. For this purpose, different disinfectants have been used, such as chloramine-T in combination with a surfactant, and washes with antibiotics such as streptomycin and gentamicin [34, 36].

Therefore, a high failure rate in the first generation is accepted in practice. Once the fungus produces spores in vitro, subculturing becomes much more efficient, as documented for *Gigaspora decipiens* using field spores [39].

A significant technical advance was the use of two-compartment systems, where a proximal compartment contains roots (enriched medium), and a distal compartment contains a minimal mineral medium so that the extraradical mycelium can develop without direct competition from the root [40]. In these arrangements, a notable increase in hyphal growth and spore production has been observed in the distal compartment, which is useful for both experimentation and isolation.

The base media used in ROC are diverse, and although there is no "universal" one, the most commonly used are MSR (modified Strullu-Romand) and related variants for routine root maintenance and AMF (ammonium foliar nutrient). MSR is used to support cultural routines in physiological studies [41, 42]. Although the response to culture is not linear, there is consensus on critical variables to consider: medium pH, which affects germination, mycelial expansion, and sporulation [43, 44], and is best maintained in slightly acidic ranges (depending on the taxon). Carbon availability (sucrose) for the root influences root vigor and exudation [44]. Phosphorus and nitrogen concentrations in the medium are also important; excess phosphorus can reduce mycorrhizal dependence, as extraradical mycelium modifies its microenvironment by consuming nitrate and altering pH [40].

Among the contributions of this cultivation method to agroecology is the study of tripartite interactions with certain pathogens [45–47]. For example, the interaction between *Glomus intraradices* and *Pratylenchus coffeae* has shown that the arbuscular mycorrhizal fungus (AMF) provides protective effects, without the nematode necessarily altering internal or external mycorrhizal development [48].

One aspect of monoxenic AMF cultivation that has received little attention is the potential for phenotypic drift in in vitro cultures. Studies have documented that certain isolates progressively increase their colonization efficiency after successive subcultures in ROC systems, suggesting adaptation to the axenic environment and the physicochemical conditions of the medium [49]. Although this response may be favorable for standardized production of propagules, from a conservation perspective, it raises the question: Does the maintained isolate faithfully represent the original ecological phenotype, or does it correspond to a "domesticated" variant optimized for artificial conditions? In this context, prolonged maintenance should be accompanied by periodic biological quality controls [50].

3 AMF Preservation Techniques Under Laboratory Conditions

The most critical phase for preserving the biological quality of the material occurs between collection and processing. Temperature and the substrate's moisture content during storage significantly influence the germination and subsequent colonization of propagules. Moderate cold storage (≈4°C) can preserve infectivity for weeks to months, compared with variable environmental conditions [51]. Likewise, variations in soil moisture content during storage directly affect spore viability and infectivity [52]. On the other hand, for roots intended for DNA analysis, silica gel is frequently used as a rapid desiccant and for temporary storage [53], especially in the field when ultra-freezing is unavailable, or there is a risk of contamination from ethanol preservation.

When evaluating mycorrhizal colonization by microscopy, structural preservation may take priority over viability. Ethanol (70%–95%) has traditionally been used to preserve roots until staining and quantification [54]; however, this practice eliminates propagule viability. Recently, controlled drying of fine roots has been proposed as a practical, field-based alternative for preserving mycorrhizal structures without the use of liquid preservatives. Giraldo-Kalil et al. [55] demonstrated that oven-dried roots retain identifiable structures (arbuscules, vesicles, and hyphae) for subsequent microscopic analysis, facilitating transport, and storage under limited conditions. For long-term preservation, encapsulation-drying is among the most promising approaches. This involves encapsulating propagules in vitro in sodium alginate beads, previously hardened with $CaCl_2$ and subjected to controlled dehydration, before finally storing them at −80°C or liquid nitrogen [56]. The goal is to reduce the free water content and prevent the lethal formation of ice crystals. Lalaymia et al. [56] demonstrated that survival after cryopreservation depends on the degree of dehydration and the mycorrhizal taxon, indicating that optimization should be performed for each isolate.

To maintain functional infectivity, live systems (in vivo or ROC) and refrigerated solid substrates remain the most robust methods. For long-term preservation, encapsulation-drying combined with cryopreservation offers promising prospects, although it requires taxon-specific adjustments. Conservation methods tend to favor species with greater capacity to grow under artificial conditions, leading to the loss of less competitive taxa and a reduction in functional diversity.

4 Challenges and Future Perspectives in AMF Conservation

Despite the methodological advances achieved in recent decades, AMF conservation continues to face structural, physiological, and ecological limitations that prevent the establishment of universal, fully stable protocols. Prolonged maintenance in monoxenic culture can induce evolutionary or adaptive changes, introducing uncertainty about the stability of the preserved material [50]. In parallel, research on cryopreservation has shown that, although techniques such as encapsulation-desiccation allow for the recovery of viability after storage at ultra-low temperatures, the response is highly taxon-dependent and requires specific optimization for each isolate [56]. This evidence underscores that successful conservation should not be evaluated solely by structural survival, but rather by the comprehensive preservation of symbiotic performance. Furthermore, the production and storage of commercial inoculants have underscored the need to establish quality standards for propagule density, infective stability, and functional reproducibility [57, 58]. However, even in applied contexts, challenges persist related to the loss of diversity and the dominance of species better adapted to artificial conditions, which could limit ecological effectiveness in the field.

On the other hand, the increasing use of molecular tools has expanded the capacity to monitor genetic stability and population structure during conservation [32, 59]. These approaches open the possibility of integrating genomic and transcriptomic assessments into quality control for mycorrhizal germplasm banks. Future challenges lie not only in perfecting storage methods but also in developing integrated conservation strategies that combine optimized cryopreservation, periodic maintenance validated by bioassays, and molecular monitoring of genetic stability. Furthermore, it is necessary to differentiate protocols according to the final objective, avoiding extrapolations from experiments.

5 Conclusions

The conservation of arbuscular mycorrhizal fungi (AMF) is fundamental to ensuring their continued availability in agricultural applications and their role in ecosystems. Although significant progress has been made in conservation methods, such as trap cultures, solid substrates, and cryopreservation, challenges related to propagule stability and the conservation of AMF functional diversity remain. There is no single, universal protocol for AMF conservation, as techniques must be tailored to each fungus and its specific conservation objectives. In this regard, the use of hybrid systems that combine approaches such as optimized cryopreservation, periodic maintenance of AMF, and continuous functional monitoring is the most promising strategy. These systems would ensure the preservation not only of the structural viability of the propagules but also of their symbiotic and functional capacity within ecosystems. The incorporation of emerging technologies, such as molecular monitoring and genomic assessments, could improve the quality and genetic stability of arbuscular mycorrhizal fungi (AMF) during their conservation. Furthermore, the development of quality standards for the production and storage of commercial inoculants will allow for more effective control over propagule density and infective stability. In conclusion, the future of AMF conservation will depend on combining traditional techniques with technological innovations, ensuring the preservation of their ecological identity and functionality in the field.

Conflict of Interest Statement

The authors have no conflicts of interest to declare that are relevant to the content of this chapter.

References

1. Brundrett MC, Tedersoo L (2018) Evolutionary history of mycorrhizal symbioses and global host plant diversity. New Phytol 220(4):1108–1115
2. Wipf D, Krajinski F, van Tuinen D, Recorbet G, Courty PE (2019) Trading on the arbuscular mycorrhiza market: from arbuscules to common mycorrhizal networks. New Phytol 223(3):1127–1142
3. Cargill RI, Shimizu TS, Kiers ET, Kokkoris V (2025) Cellular anatomy of arbuscular mycorrhizal fungi. Curr Biol 35(11):R545–R562
4. Smith SE, Smith FA (2012) Fresh perspectives on the roles of arbuscular mycorrhizal fungi in plant nutrition and growth. Mycologia 104(1):1–13
5. Begum N, Qin C, Ahanger MA et al (2019) Role of arbuscular mycorrhizal fungi in plant growth regulation: implications in abiotic stress tolerance. Front Plant Sci 10:1068
6. Chen B-D, Fu W, Wu S-L, Zhu Y-G (2024) Involvements of mycorrhizal fungi in terrestrial ecosystem carbon cycling. Chin J Plant Ecol 48(1):1–20
7. Camenzind T, Aguilar-Trigueros CA, Heuck MK et al (2024) Progressing beyond colonization strategies to understand arbuscular mycorrhizal fungal life history. New Phytol 244(3):752–759
8. Li H, Smith FA, Dickson S, Holloway RE, Smith SE (2006) Plant growth depressions in arbuscular mycorrhizal symbioses: not just caused by carbon drain? New Phytol 170:809–818
9. Kakouridis A, Hagen JA, Kan MP et al (2022) Routes to roots: direct evidence of water transport by arbuscular mycorrhizal fungi to host plants. New Phytol 236:210–221
10. González-Osorio H, Góngora Botero CE, Jaramillo Padilla SP, Osorio W (2022) Plant growth and phosphorus uptake of coffee seedlings through mycorrhizal inoculation. Agron Colomb 40(1):77–84
11. Kokkoris V (2026) Functional ecology of arbuscular mycorrhizal fungal networks. Soil Biol Biochem 210:109900
12. Bhupenchandra I, Sharma S, Devi YP et al (2025) Structural and functional diversity of arbuscules in plant–AMF symbiosis. Mycorrhiza 35:145–160
13. Kakouridis A, Powell JR et al (2023) Mycorrhizal fungi and soil carbon stabilization. Glob Change Biol 29:3456–3472
14. Vieira CK, Marascalchi MN, Rozmoš M et al (2025) Arbuscular mycorrhizal fungal highways—what, how and why? Soil Biol Biochem 202:109702
15. Ao D, Wang B, Wang Y, Chen Y, Liang C, An S (2025) Arbuscular mycorrhizal fungi communities and glomalin mediate particulate and mineral-associated organic carbon formation in grassland patches. Commun Earth Environ 6(1):553
16. Delaeter M, Magnin-Robert M, Randoux B, Lounès-Hadj Sahraoui A (2024) Arbuscular mycorrhizal fungi as biostimulant and biocontrol agents: a review. Microorganisms 12(7):1281
17. INVAM (Sitio Oficial) (2026). International collection of (vesicular) arbuscular mycorrhizal fungi (INVAM): host plant choices. University of Kansas
18. Morton JB, Bentivenga SP, Wheeler WW (1993) Germ plasm in the International Collection of Arbuscular and Vesicular-arbuscular Mycorrhizal Fungi (INVAM) and procedures for culture development, documentation, and storage. Mycotaxon 48:491–528
19. Walker C, Vestberg M (1994) A simple and inexpensive method for producing and maintaining closed pot cultures of arbuscular mycorrhizal fungi. Agric Sci Finland 3:233–240
20. Koziol L, Bever JD (2015) Mycorrhizal response trades off with plant growth rate and increases with plant successional status. Ecology 96(7):1768–1774
21. Yao Q, Gao JL, Zhu HH et al (2010) Evaluation of the potential of trap plants to detect arbuscular mycorrhizal fungi using PCR-DGGE analysis. Soil Sci Plant Nutr 56(2):205–211

22. Liu R, Wang F (2003) Selection of appropriate host plants used in trap culture of arbuscular mycorrhizal fungi. Mycorrhiza 13(3):123–127
23. Trejo-Aguilar D, Lara-Capistrán L, Maldonado-Mendoza IE et al (2013) Loss of arbuscular mycorrhizal fungal diversity in trap cultures during long-term subculturing. IMA Fungus 4(2):161–167
24. Bitterlich M, Rouphael Y, Graefe J, Franken P (2018) Arbuscular mycorrhizas: a promising component of plant production systems provided favorable conditions for their growth. Front Plant Sci 9:1329
25. Smith, SE, Read, DJ, 2010. Mycorrhizal Symbiosis. Academic Press. (Nota: en tu texto yo la cité como 2008; si usarás 2010/edición específica, ajusta el año también en el texto.)
26. Cavagnaro TR (2014) Impacts of compost application on the formation and functioning of arbuscular mycorrhizas. Soil Biol Biochem 78:38–44
27. Casbis GM, Torres ÓGV, Rodríguez MA, Nava HS (2022) Propiedades físicas y químicas de sustratos en función de su granulometría y componente orgánico-mineral. Acta Agrícola y Pecuaria. 8: e0081007. https://doi.org/10.30973/aap/2022.8.0081007
28. Püschel D, Kolaříková Z, Šmilauer P, Rydlova J (2019) Survival and long-term infectivity of arbuscular mycorrhizal fungi in peat-based substrates stored under different temperature regimes. Appl Soil Ecol 140:98–107
29. Douds JrDD, Nagahashi G, Pfeffer PE, Reider C, Kayser WM (2006) On-farm production of AM fungus inoculum in mixtures of compost and vermiculite. Bioresour Technol 97(6):809–818
30. Kumar, A., Singh, R., & Adholeya, A. (2017). Biotechnological advancements in industrial production of arbuscular mycorrhizal fungi: achievements, challenges, and future prospects. In Satyanarayana, T., Deshmukh, S.K, Johri B.N. (eds) Developments in fungal biology and applied mycology. Springer, Singapore. 413-431. https://doi.org/10.1007/978-981-10-4768-8_21
31. Braunberger PG, Abbott LK, Robson AD (1996) Infectivity of arbuscular mycorrhizal fungi after wetting and drying. New Phytol 134(4):673–684
32. Tisserant E, Malbreil M, Kuo A et al (2013) Genome of an arbuscular mycorrhizal fungus provides insight into the oldest plant symbiosis. Proc Natl Acad Sci USA 110(50):20117–20122
33. Romero-Ceciliano M, Andrade-Torres A, Artavia-Salazar E, Solís-Ramos LY (2025) Establishing monoxenic culture of arbuscular mycorrhizal fungus Glomus sp. through in vitro root organ culture and Swietenia macrophylla in vitro cultures. Agriculture 15(7):673
34. Gagou E, El Yeznasni H, Chafai W et al (2026) Monoxenic root organ culture enables high-yield production of viable indigenous Rhizophagus irregularis inoculum for arid oasis agroecosystems. Microbiol Res 17(1):28
35. Bécard G, Fortin JA (1988) Early events of vesicular–arbuscular mycorrhiza formation. Can J Bot 66:147–155
36. Bécard G, Piché Y (1992) Establishment of vesicular-arbuscular mycorrhiza in root organ culture: review and proposed methodology. Methods Microbiol 24:89–108
37. Declerck S, Strullu DG, Plenchette C (1996) In vitro mass-production of the arbuscular mycorrhizal fungus Glomus versiforme associated with Ri T-DNA transformed carrot roots. Mycol Res 100(10):1237–1242
38. Cranenbrouck, S., Voets, L., Bivort, C., Renard, L., Strullu, DG., Declerck, S. (2005). Methodologies for in Vitro Cultivation of Arbuscular Mycorrhizal Fungi with Root Organs. In: Declerck, S., Fortin, J.A., Strullu, DG. (eds) In Vitro Culture of Mycorrhizas. Soil Biology, vol 4. Springer, Berlin, Heidelberg. https://doi.org/10.1007/3-540-27331-X_18
39. Bidondo LF, Pergola M, Silvani V et al (2012) Continuous and long-term monoxenic culture of the arbuscular mycorrhizal fungus Gigaspora decipiens in root organ culture. Fungal Biol 116(6):729–735
40. Fortin JA, Bécard G, Declerck S et al (2002) Arbuscular mycorrhiza on root-organ cultures. Can J Bot 80(1):1–20
41. Rufyikiri G, Thiry Y, Declerck S (2003) Contribution of hyphae and roots to uranium uptake and translocation by arbuscular mycorrhizal carrot roots under root-organ culture conditions. New Phytol 158(2):391–399
42. Declerck S et al (2018) Advances in root organ culture systems for AMF. Mycorrhiza 28:1–15
43. Costa FA, Haddad LSA, Kasuya MCM et al (2013) In vitro culture of Gigaspora decipiens and Glomus clarum in transformed roots of carrot: influence of temperature and pH. Acta Sci Agron 35:315–323
44. Radha V, Marimuthu RP, Kumutha K (2015) Sporulation of arbuscular mycorrhizal fungus, Glomus intraradices, through root organ culture. Appl Biol Res 17(1):38–47
45. Benhamou N, Fortin JA, Hamel C et al (1994) Resistance responses of mycorrhizal Ri T-DNA-transformed carrot roots to infection by Fusarium oxysporum f. sp. chrysanthemi. Phytopathology 84(9):958–968

46. Nogales A, Camprubí A, Estaún V, Marfà V, Calvet C (2010) In vitro interaction studies between Glomus intraradices and Armillaria mellea in vines. Span J Agric Res 8:62–68
47. Giachero ML, Marquez N, Gallou A et al (2017) An in vitro method for studying the three-way interaction between soybean, Rhizophagus irregularis and Fusarium virguliforme. Front Plant Sci 8:1033
48. Elsen A, Declerck S, De Waele D (2003) Use of root organ cultures to investigate the interaction between Glomus intraradices and Pratylenchus coffeae. Appl Environ Microbiol 69(7):4308–4311
49. Goh D, Martin JG, Banchini C, MacLean AM, Stefani F (2022) RocTest: a standardized method to assess the performance of root organ cultures in the propagation of arbuscular mycorrhizal fungi. Front Microbiol 13:937912
50. Kokkoris V, Hart M (2019) In vitro propagation of arbuscular mycorrhizal fungi may drive fungal evolution. Front Microbiol 10:2420
51. Liu X, Ye G, Feng Z et al (2024) Cold storage promotes germination and colonization of arbuscular mycorrhizal fungal hyphae as propagules. Front Plant Sci 15:1450829
52. Ruiz-Lozano JM, Azcón R (1996) Viability and infectivity of mycorrhizal spores after long term storage in soils with different water potentials. Appl Soil Ecol 3(2):183–186
53. Guillen-Otero T, Lee SJ, Chen CW, Szoevenyi P, Kessler M (2023) A metabarcoding protocol targeting two DNA regions to analyze root-associated fungal communities in ferns and lycophytes. Appl Plant Sci 11(3):e11523
54. Freschet GT, Pagès L, Iversen CM et al (2021) A starting guide to root ecology: strengthening ecological concepts and standardising root classification, sampling, processing and trait measurements. New Phytol 232(3):973–1122
55. Giraldo-Kalil LJ, Carrillo-Saucedo SM, Ramírez GC, Paz H (2025) Dry root preservation: a field-friendly method for assessing tree arbuscular mycorrhizal colonization. Rhizosphere 36:101227.
56. Lalaymia I, Cranenbrouck S, Draye X, Declerck S (2012) Preservation at ultra-low temperature of in vitro cultured arbuscular mycorrhizal fungi via encapsulation–drying. Fungal Biol 116(10):1032–1041
57. Hart MM, Aleklett K, Chagnon PL, Egan C, Ghignone S, Helgason T, Waller L (2015) Navigating the labyrinth: a guide to sequence-based, community ecology of arbuscular mycorrhizal fungi. New Phytol 207(1):235–247
58. Berruti A, Lumini E, Balestrini R, Bianciotto V (2016) Arbuscular mycorrhizal fungi as natural biofertilizers: let's benefit from past successes. Front Microbiol 6:1559
59. Ropars J, Toro KS, Noel J et al (2016) Evidence for sexual reproduction in arbuscular mycorrhizal fungi. Nat Microbiol 1:16033

Chapter 19

Cryopreservation Method for *Aurantiochytrium limacinum*

Ayushi Dalmia and Ajay W. Tumaney

Abstract

Aurantiochytrium limacinum is a marine protist that produces high levels of omega-3 fatty acid DHA and is thus of biotechnological importance. Here, we describe a cryopreservation technique to preserve *A. limacinum* at two different temperatures (−80°C) and liquid nitrogen (−196°C). At −80°C, cells were preserved in 25% (v/v) glycerol; at −196°C (liquid nitrogen), in 10% (v/v) methanol, 15% (v/v) glycerol, or 15% (v/v) dimethyl sulfoxide. The cells were stable for up to 12 months of storage. The method preserves the genetic integrity and functionality of the cells for biotechnological applications.

Keywords *Aurantiochytrium limacinum*, DHA, Cryoprotectant, Cryopreservation, Liquid nitrogen

1 Introduction

Aurantiochytrium limacinum is a DHA-producing marine thraustochytrid and holds biotechnological importance [1–3]. *A. limacinum* is generally maintained by subculturing, which is resource intensive and also risks contamination and genetic drift [4, 5]. Having a method to preserve culture for a longer period emerges as a valuable technique. Cryopreservation is one of the most acceptable preservation methods [6, 7]. Cryopreservation suspends cellular metabolism at ultralow temperature [8]. Using suitable cryprotective agents that minimize ice crystal formation and osmotic damage. Long-term cryopreservation is generally performed at two different temperatures: −80°C using a mechanical freezer or −196°C using LN_2. Cryopreservation below the glassy stage, i.e., −135°C is preferred; however, having a method to preserve at −80°C is also of great value for organizations that do not have a liquid nitrogen storage facility [9]. For storage at −80°C, glycerol was the cryoprotectant of choice, where as in the case of storage in LN_2, methanol, glycerol, or DMSO were found to be effective. The procedure to preserve includes harvesting cells in log phase, equilibrating with a cryoprotectant, and directly storing at −80°C in a freezer for −80°C storage, or performing two-step cooling for LN_2 storage [10, 11].

Marco A. Ramírez-Mosqueda et al. (eds.), *Methods for The Conservation of Microorganisms*, Methods in Molecular Biology, vol. 3090, https://doi.org/10.1007/978-1-0716-5623-5_19,

The two-step cooling for LN_2 involved first keeping samples in a 1°C cooler with isopropanol in an −80°C freezer for rate-controlled cooling, and then plunging the samples into LN_2 for rapid freezing [12, 13]. Revival is equally important, and thawing cells correctly ensures better survival rates. For revival, the cells are thawed in a water bath, washed with fresh media, and inoculated into fresh medium, then incubated in an incubator shaker for growth. This method shows viability up to 12 months of storage. This method preserves the morphological and physiological characteristics of *A. limacinum*, as evidenced by microscopic observations and DHA content measurements after revival [14].

2 Materials

Prepare all solutions using ultrapure water (18 MΩ-cm at 25°C) and use analytical grade reagents.

2.1 Culture Preparation and Maintenance

1. *Aurantiochytrium limacinum*: *Aurantiochytrium limacinum* SR 21 culture is obtained from the American Type Culture Collection (ATCC).
2. Artificial Sea Water Medium: 5 $g L^{-1}$ glucose, 1 $g L^{-1}$ yeast extract, 1 $g L^{-1}$ peptone, and 32.31 $g L^{-1}$ of artificial seawater medium (III) from HiMedia Labs (India). Measure 600 mL of water in a 1 L graduated cylinder. Weigh 5 g of glucose, 1 g of yeast extract, and 1 g of peptone and add to the measured water. Weigh 32.31 g of Artificial seawater nutrient medium from HiMedia Labs (India) and add to the water and mix till the solids dissolve. The composition of 32.31 g of the artificial seawater nutrient medium (III) from Himedia Labs contains sodium chloride 25 g, magnesium sulfate 3.5 g, sodium nitrate 750 mg, potassium phosphate dibasic 26.30 mg, magnesium chloride·6H_2O 2 g, calcium chloride·2H_2O 500 mg, potassium chloride 500 mg, sodium carbonate 20 mg, citric acid 3 mg, ferric ammonium citrate 3 mg, magnesium EDTA 0.50 mg, cyanocobalamin (vitamin B12) 0.01 mg, boric acid 2.86 mg, manganese chloride·4H_2O 1.81 mg, sodium molybdate·2H_2O 0.252 mg, znc sulfate·7H_2O 0.222 mg, copper sulfate·7H_2O 0.079 mg. The pH of the solution is adjusted to 7.5 using 1 N NaOH (*see* **Note 1**). The volume of the medium is up to 1000 ml. Transfer the medium to a borosilicate bottle and autoclave at 15 lbs (121°C) for 20 min. Store prepared medium at 4°C (*see* **Note 2**).
3. Orbital shaker: Orbital shakers with a capacity to maintain 25°C and capable of running at 170 rpm can be selected.

2.2 Cryopreservation

Cryoprotectants

1. Glycerol solution (−80°C): 50% (v/v) in ASW medium. Glycerol is autoclaved and cooled. Aseptically, 50 ml of glycerol is measured in a measuring cylinder, then made up to 100 ml with ASW medium. The prepared solution is stored in a sterile Falcon tube.
2. Dimethyl sulfoxide (DMSO) (LN_2) solution: 30% (v/v) DMSO in ASW medium. Cell culture-grade DMSO is used to ensure sterility. As a precautionary measure, DMSO is filtered by passing it through 0.22 micron PTFE filters using a syringe (*see* **Notes 3 and 4**). In a sterile measuring cylinder inside a laminar flow hood, 30 ml of DMSO is measured, and the volume is made up to 100 ml with sterile ASW medium. The prepared solution is stored in a sterile airtight Falcon tube.
3. Glycerol solution (LN_2): 30% (v/v) in ASW medium. 30 ml of autoclaved glycerol is measured, and the volume is made up to 100 ml with an ASW medium. The prepared solution is stored in a sterile Falcon tube.
4. Methanol solution (LN_2): 20% (v/v) in ASW medium. Methanol is filtered by passing through 0.22-μm PTFE filters using a syringe (*see* **Notes 3 and 4**). In a sterile measuring cylinder inside the laminar flow, 20 ml of methanol is measured, and the volume is made up to 100 ml with sterile ASW medium. The prepared solution is stored in a sterile airtight Falcon tube.
5. Mechanical freezer −80°C.
6. Liquid Nitrogen (LN_2).
7. Liquid Nitrogen Container: Cryocan.
8. Cryovials (*see* **Note 5**).
9. 1 Degree Celsius cooler (*see* **Note 6**).

2.3 Revival

1. 70% ethanol: Add 70 ml of ethanol to 30 ml of water to obtain 70% (v/v) ethanol solution.
2. Water bath: A temperature-controlled water bath.

2.4 Post-Thaw Analysis

1. Phase contrast microscope.
2. Glass slide.
3. Cover slip.
4. Isopropanol.
5. Homogenizer.
6. Hexane.

7. 12% (w/v) sodium sulfate (*see* **Note 7**).
8. Nitrogen evaporator.
9. Boron trifluoride methanol.
10. Acetic acid.
11. Gas chromatograph connected to flame ionization detector (GC-FID).
12. Evans blue solution (0.1% in PBS, w/v) (*see* **Note 8**).
13. Hemocytometer.

3 Methods

All the steps are to be performed at room temperature unless otherwise specified (*see* **Note 9**).

3.1 Culture Preparation

1. Inoculate *A. limacinum* SR 21 into ASW medium from cultures in log phase. Incubate at 25 ± 2°C in an orbital shaker at 170 rpm and grow for 24 h. Pellet cells by centrifuging at 5000 × *g* for 6 min and resuspend in 1 ml of fresh ASW medium (*see* **Note 10**).

3.2 Cryopreservation at −80°C

1. Pipette 250 μl of culture into a cryovial and gently add 250 μl of glycerol solution (50%, v/v) to achieve a final concentration of 25% (v/v) glycerol, and gently mix by pipetting.
2. Incubate for 15 min to allow for equilibration (*see* **Note 11**). Transfer the tubes to a −80°C mechanical freezer for long-term storage.

3.3 Cryopreservation at −196°C (LN_2)

Any of the three cryoprotectants can be used to preserve the cells in liquid nitrogen:-

1. Methanol: add 250 μl of 20% (v/v) methanol solution to 250 μl of culture, then gently mix by pipetting. Equilibrate at room temperature for 10 min.
2. DMSO: add 250 μl of 30% (v/v) methanol solution to 250 μl of culture, then gently mix by pipetting. Equilibrate at room temperature for 30 min.
3. Glycerol: add 250 μl of 30% (v/v) methanol solution to 250 μl of culture, then gently mix by pipetting. Equilibrate at room temperature for 15 min.

After equilibration, transfer the cryovials into a 1°C cooler (*see* **Notes 12, 13, and 14**) and transfer the cooler to a mechanical freezer at −80°C for 90 min. Plunge the cryovials into LN_2 in a cryocan for rapid freezing and long-term storage (*see* **Notes 15, 16, and 17**) (Fig. 1a–d).

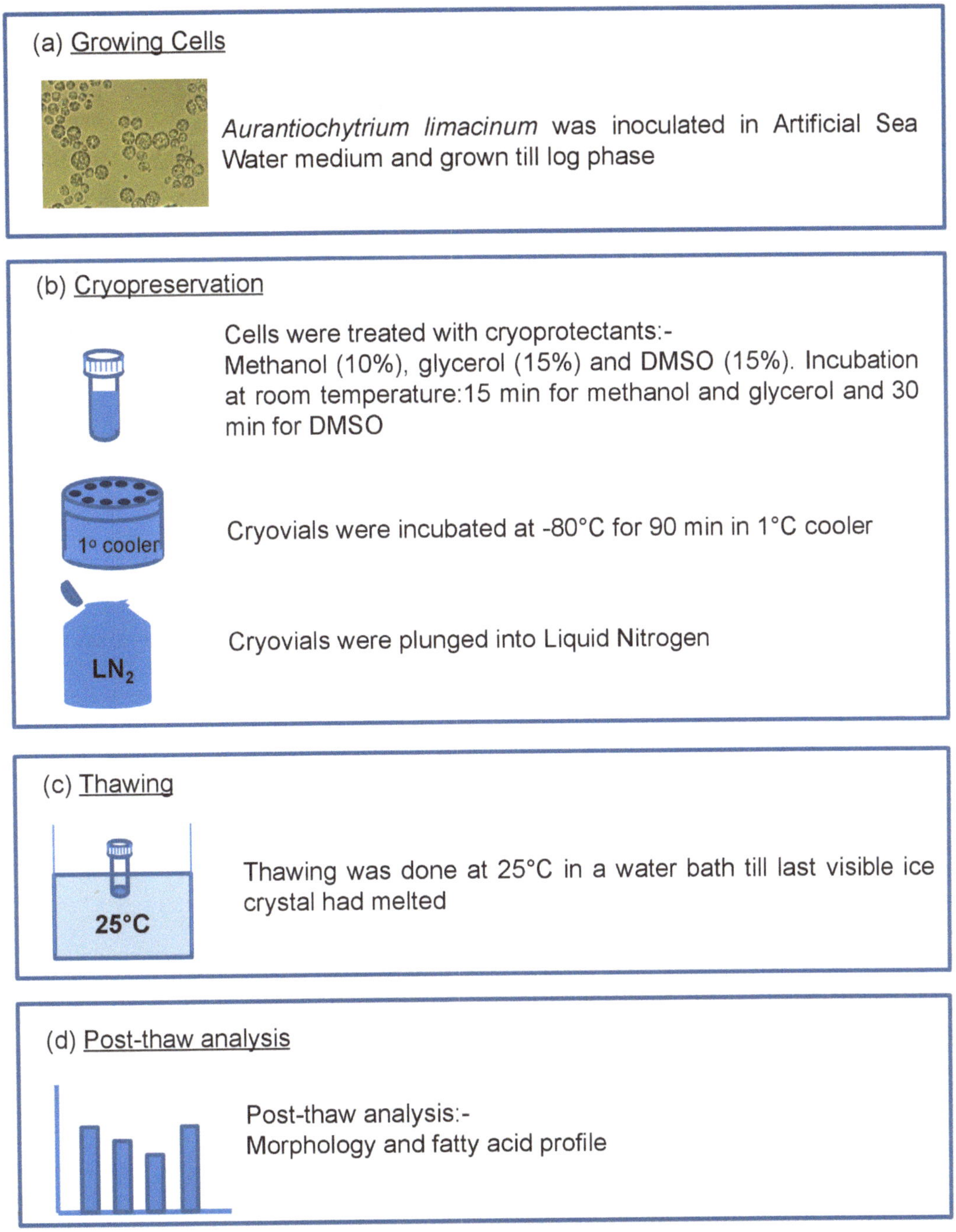

Fig. 1 Overview of the workflow adopted for cryopreservation, revival, and post-thaw analysis of *A. limacinum* cells stored in liquid nitrogen. (**a**) Growing of cells for cryopreservation, (**b**) Treatment of cells with various cryoprotectants: methanol (10%, v/v), glycerol (15%, v/v), and DMSO (15%, v/v). 15 min of incubation at room temperature was performed for methanol and glycerol, and 30 min for DMSO. Cryovials were incubated at −80°C in a 1°C cooler for 90 min, then plunged into liquid nitrogen. (**c**) Thawing was done in a 25°C water bath. (**d**) Post-thaw assessment was carried out, which included viability assessment, growth curve, morphological observation, and fatty acid profile. (Reproduced from Ref. [14] with permission from Springer Nature)

3.4 Thawing

1. Remove cryovial from storage canister and place immediately in an ice bath (*see* **Note 18**).
2. Thaw by placing the tubes in a 25°C water bath until no visible ice crystals remain (*see* **Note 19**).
3. Wipe tubes with 70% ethanol to reduce the risk of contamination. Pellet at 4000 × *g* for 5 min.
4. Discard the supernatant.
5. Resuspend the pellet in fresh ASW medium.
6. Inoculate 10 ml of ASW media with the resuspended cell pellet and keep growing to revive the cells.
7. Maintain at 25 ± 2°C and 170 rpm in an incubator shaker.

3.5 Post-Thaw Analysis (see Note 20)

Morphology Assessment

1. Grow the revived culture for 24 h or until the cell density at 660 nm is around 1.
2. Pellet cells (*see* **Note 21**) and wash with 500 μl PBS.
3. Resuspend the cell pellet in PBS, mount the cells on a glass slide, place a cover slip on the sample, and observe under a phase-contrast microscope at 400× magnification (Fig. 2a–e).

Fatty Acid Profile

1. After revival, on attaining considerable growth, the cells are harvested, and the lipid is extracted. To extract lipids, 2 ml of

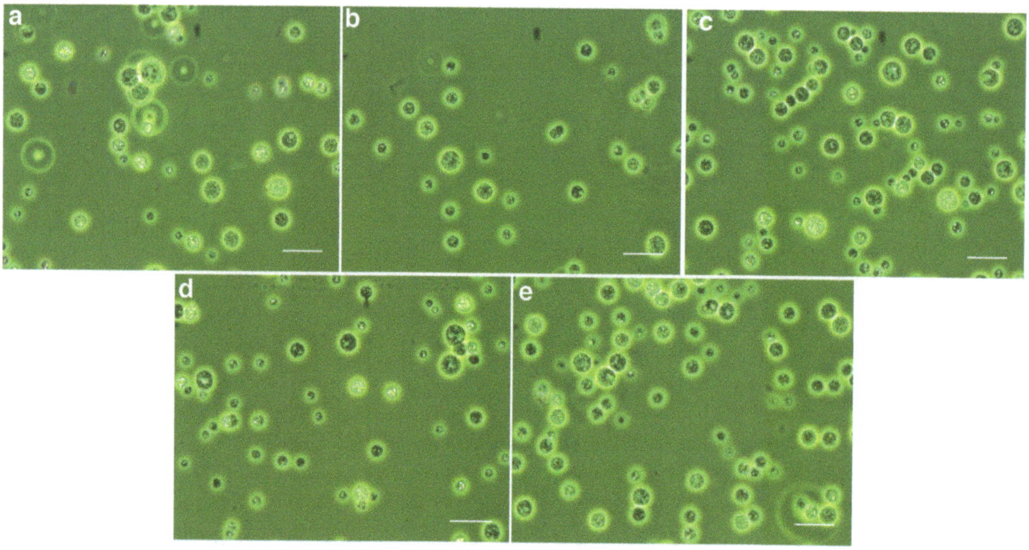

Fig. 2 Microscopic images of cells on revival. *A. limacinum* cells stored in liquid nitrogen were revived and grown in fresh medium. Upon sufficient growth, the cells were examined under a microscope for morphological changes. The phase-contrast images were taken at 400× magnification using a BX51 Olympus microscope. (**a**) glycerol at −80°C; and cells stored in liquid nitrogen; (**b**) control; (**c**) DMSO; (**d**) glycerol; and (**e**) methanol (scale bar = 25 μm. (Reproduced from Ref. [14] with permission from Springer Nature)

hot isopropanol (80–85°C) was added to the cell pellet, which was homogenized for 2 min at the highest speed to lyse cells. 3 ml of hexane and 2.5 ml of 12% sodium sulfate solution were added, then vortexed.

2. The organic and the aqueous layers in the tube phase separate. The top organic layer is collected in a tube using a pipette.
3. To the first tube containing the aqueous layer, 2 ml of hexane:isopropanol (7:2, v/v) was added, vortexed, and the top organic layer was pooled into the previous tube (*see* **Note 22**).
4. The pooled organic layer was dried under a stream of nitrogen using a nitrogen concentrator to obtain lipid (*see* **Note 23**).
5. Fatty acid methyl esters (FAMEs) of the lipid are prepared. The lipid is taken in a glass tube, and 1 ml of boron trifluoride methanol (*see* **Note 24**) is added.
6. The mixture is heated in a 75°C water bath for 2 h, after which it is cooled in ice. On cooling, 1 ml of hexane, 1 ml of water, and 50 μl of acetic acid are added, then vortexed. The contents of the tube separate into the top organic layer and the bottom aqueous layer.
7. Collect the top organic layer containing FAMEs into a fresh tube. The FAMEs are analyzed using a GC-FID system (Fig. 3).

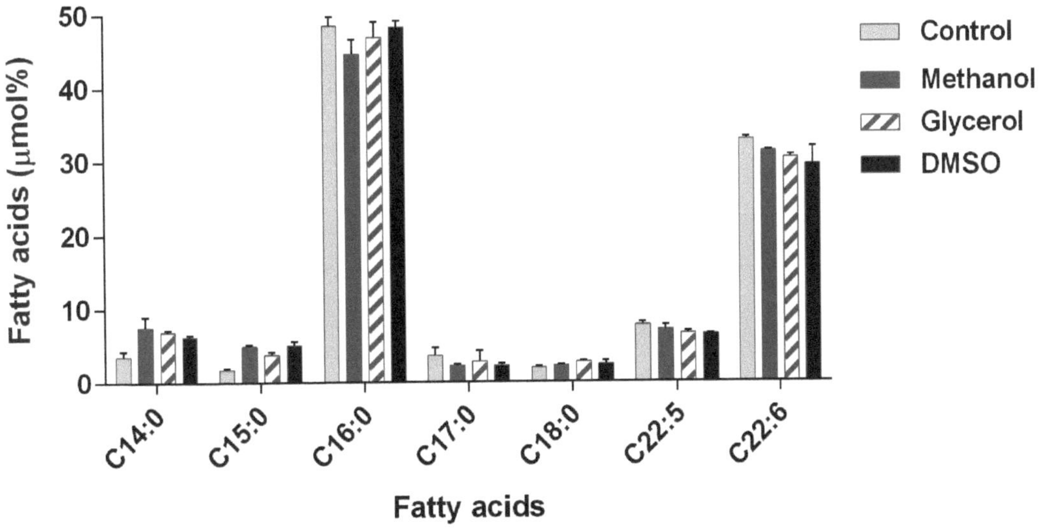

Fig. 3 Fatty acid profile of cells on revival. The cells preserved in liquid nitrogen were revived after 12 months of storage and inoculated in fresh medium. Upon sufficient growth, lipid was extracted from the cells, and fatty acid methyl esters were prepared, which were further analyzed by GC-FID. Nonadecanoic acid was used as an internal standard, and fatty acid percentages were calculated. Values are mean ± standard deviation ($n = 3$). No significant difference in the fatty acid profile was seen among the treatments and control ($p < 0.05$) before preservation, control, methanol, glycerol, and DMSO. (Reproduced from Ref. [14] with permission from Springer Nature)

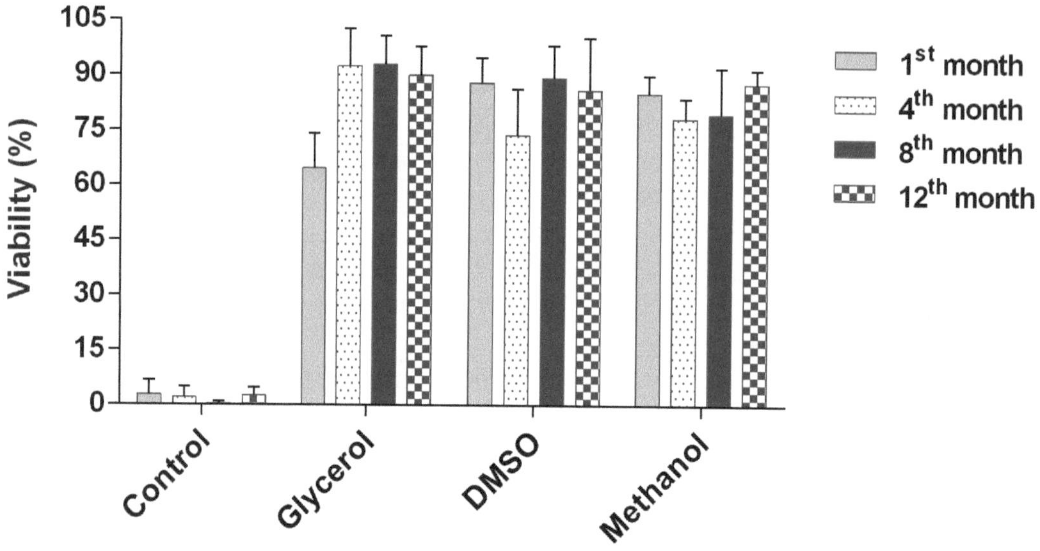

Fig. 4 Viability percentages of cells stored in liquid nitrogen. The cells preserved in glycerol, DMSO, ethanol, and control were revived at 1, 4, 8, and 12 months of storage. The cells were stained with the cell-exclusion dye Evans blue, and cell viability was estimated using a hemocytometer. Values are mean ± standard deviation ($n = 3$). "*" indicates values that are significantly different ($p < 0.05$) first month, fourth month, eighth month, and 12th month

3.6 Viability

1. To 100 μl of thawed cells, add 10 μl of Evans blue (0.1% in phosphate-buffered saline, w/v) and mix gently using a pipette. After 10 min of incubation, load 10 μl of the mixture into a hemocytometer and count live and dead cells.
2. Dead cells take up the stain and appear blue, while live cells remain unstained. Viability percentage is calculated and reported (Fig. 4).

4 Notes

1. 1 N NaOH can be prepared by dissolving 10 g of NaOH in 100 ml of water, then making up the volume in a measuring cylinder. NaOH is highly corrosive, and dissolving NaOH in water is an exothermic reaction. Protective gloves should be worn while handling NaOH.
2. The autoclaved ASW medium is cooled to room temperature before using or storing. Autoclaved ASW medium can be stored at 4°C for up to a month. When stored at 4°C, the medium was brought to room temperature before using.
3. DMSO and methanol are polar organic solvents and are volatile; therefore, they cannot be autoclaved. It vaporizes on heating and becomes highly flammable.

4. PTFE (polytetrafluoroethylene) is nonreactive and suitable for filter sterilizing most solvents, including dimethyl sulfoxide and methanol.
5. Cryotubes are specialized tubes for storing samples at ultra-low temperatures, such as −80°C and −196°C.
6. A cryogenic freezing container is a passive container used for controlled-rate freezing of biological samples. It is made of polycarbonate material and has a compartment to hold isopropanol and cryo vials. It maintains precise cooling of 1°C/min when used with isopropanol in a mechanical −80°C freezer.
7. 12% (w/v) sodium sulfate is prepared by dissolving 12 g of anhydrous sodium sulfate (Na_2SO_4) in 70 ml of distilled water, then making up the volume to 100 ml.
8. Dissolve 50 mg of Evans Blue powder in 25 ml of PBS. Make up to 50 ml with PBS to obtain 0.1% Evans Blue.
9. All steps for culture and cryoprotectant handling for preservation and revival are carried out in a laminar flow chamber under aseptic conditions, and aseptic conditions were maintained. All sterile plasticware is to be used, and all glassware is autoclaved and dried.
10. ASW media is handled aseptically whenever it is used for growing or preserving cells.
11. Equilibration is the process of incubating cells with a cryoprotectant so that it is loaded into the cells before freezing, reducing osmotic stress and ice damage during freezing.
12. 1 degree Celsius cooler is filled with isopropanol till the indicated line, after keeping the samples in the container. The cooler is kept in a mechanical −80°C freezer, and the samples are cooled at 1°C per minute.
13. Using a 1-degree Celsius cooler helps with controlled, slow cooling, thereby preventing the formation of large crystals and cell damage.
14. Isopropanol in a 1°C cooler needs to be replaced after every five cooling–warming cycles. Repeated cycles of cooling and warming cause moisture to accumulate through vapor condensation, diluting the isopropanol and impairing its thermal properties, thereby disrupting the critical −1°C/min cooling rate.
15. Cryocans are cans designed to hold biological samples for long-term storage in liquid nitrogen (LN2). They have multilayer vacuum insulation that prevents boil-off of LN_2 and helps retain LN_2 for longer.
16. Make sure the cryocan never runs out of LN_2 and that it is refilled periodically in the can. It is recommended to store the cryocan in a cool room to minimize evaporation.

17. Cryocane is an aluminium cane that holds cryovials for storage in LN_2. Alternatively, if the cryocan is large and there are more samples, cardboard or PC cryoboxes can be used to store samples in an LN_2 canister.
18. For thawing, the tubes should be placed in the water bath as soon as possible to minimize cell damage.
19. It usually takes 1–2 min for the visible ice crystals to melt.
20. Fatty acid analysis, morphological observation under a microscope, and viability are tools to evaluate the efficiency of preservation and revival and are not part of the preservation or revival process. These need not be carried out each time preservation or revival is done.
21. For pelleting cells, the culture is centrifuged at $5000 \times g$ for 6 min.
22. Hexane has a density lower than that of water; therefore, it floats on water and forms the top layer upon phase separation. Lipids, or the FAMEs, being soluble in hexane, remain in the organic layer.
23. The tip of the nitrogen concentrator should be kept slightly above the sample's liquid surface, and the nitrogen stream should be at a speed that does not splash the sample.
24. Boron trifluoride methanol should be handled in a fume hood, and protective gloves should be used.

Conflict of Interest

The authors have no conflicts of interest to declare that are relevant to the content of this chapter.

References

1. Rosa SM, Soria MA, Vélez CG, Galvagno MA (2010) Improvement of a two-stage fermentation process for docosahexaenoic acid production by *Aurantiochytrium limacinum SR21* applying statistical experimental designs and data analysis. Bioresour Technol 101:2367–2374. https://doi.org/10.1016/j.biortech.2009.11.056
2. Huang TY, Lu WC, Chu IM (2012) A fermentation strategy for producing docosahexaenoic acid in *Aurantiochytrium limacinum* SR21 and increasing C22:6 proportions in total fatty acid. Bioresour Technol 123:8–14. https://doi.org/10.1016/j.biortech.2012.07.068
3. Li J, Liu R, Chang G, Li X et al (2015) A strategy for the highly efficient production of docosahexaenoic acid by *Aurantiochytrium limacinum* SR21 using glucose and glycerol as the mixed carbon sources. Bioresour Technol 177:51–57. https://doi.org/10.1016/j.biortech.2014.11.046
4. Trevathan-Tackett SM, Treby S, Gleason FH, Macreadie PI (2018) Cryopreservation methods are effective for long-term storage of Labyrinthula cultures. Dis Aquat Organ 130:65–70. https://doi.org/10.3354/dao03266
5. Abreu L, Borges L, Marangoni J, Abreu PC (2012) Cryopreservation of some useful microalgae species for biotechnological exploitation. J Appl Phycol 24:1579–1588. https://doi.org/10.1007/s10811-012-9818-0

6. Stock W, Pinseel E, De Decker S et al (2018) Expanding the toolbox for cryopreservation of marine and freshwater diatoms. Sci Rep 8:1–9. https://doi.org/10.1038/s41598-018-22460-0
7. Kapoore RV, Huete-Ortega M, Day JG et al (2019) Effects of cryopreservation on viability and functional stability of an industrially relevant alga. Sci Rep 9:1–12. https://doi.org/10.1038/s41598-019-38588-6
8. Nakanishi K, Deuchi K, Kuwano K (2012) Cryopreservation of four valuable strains of microalgae, including viability and characteristics during 15 years of cryostorage. J Appl Phycol 24:1381–1385. https://doi.org/10.1007/s10811-012-9790-8
9. Brand JJ, Diller KR (2004) Application and theory of algal cryopreservation. Nov Hedwigia 79:175–189. https://doi.org/10.1127/0029-5035/2004/0079-0175
10. Bui TVL, Ross IL, Jakob G, Hankamer B (2013) Impact of procedural steps and cryopreservation agents in the cryopreservation of chlorophyte microalgae. PLoS One 8:1–9. https://doi.org/10.1371/journal.pone.0078668
11. Taylor R, Fletcher RL (1999) Cryopreservation of eukaryotic algae—a review of methodologies. J Appl Phycol 10:481–501. https://doi.org/10.1023/A:1008094622412
12. Gwo JC, Chiu JY, Chou CC, Cheng HY (2005) Cryopreservation of a marine microalga, *Nannochloropsis oculata* (Eustigmatophyceae). Cryobiology 50:338–343. https://doi.org/10.1016/j.cryobiol.2005.02.001
13. Yang D, Li W (2016) Methanol-promoted lipid remodelling during cooling sustains cryopreservation survival of *Chlamydomonas reinhardtii*. PLoS One 11:1–17. https://doi.org/10.1371/journal.pone.0146255
14. Dalmia A, Tumaney AW (2021) Optimisation and evaluation of cryopreservation method for *Aurantiochytrium limacinum*. J Appl Phycol 33:869–878. https://doi.org/10.1007/s10811-020-02346-3

Part IV

Microbial Characterization and Identification

Chapter 20

Metabolomics in Microbial Genetic Resource Conservation: From Strain Identity to Functional Preservation

Jorge David Cadena-Zamudio, Daniel Alejandro Cadena-Zamudio, Jorge Cadena-Iñiguez, Marco A. Ramírez-Mosqueda, Ramón Ignacio Arteaga-Garibay, María Isabel Iñiguez-Luna and José Martín Ruvalcaba-Gómez

Abstract

The conservation of microbial genetic resources (MGR) is increasingly recognized as a strategic pillar for food security, biotechnology, ecosystem resilience, and climate change adaptation. Traditionally, *ex situ* conservation efforts have prioritized strain identity, viability, and genetic stability, relying mainly on cryopreservation and lyophilization as long-term storage strategies. However, growing evidence indicates that cellular survival and genomic integrity do not necessarily guarantee the preservation of microbial functionality, as preservation procedures may induce subtle yet biologically relevant metabolic and physiological alterations. In this context, metabolomics emerges as a critical tool for advancing microbial conservation beyond identity-based paradigms toward a framework of functional conservation. By directly capturing the biochemical phenotype, metabolomics provides an integrated readout of microbial physiological state, enabling the assessment of post-preservation metabolic stability, the identification of functionally relevant biomarkers, and the evaluation of preservation-induced metabolic reprogramming. This chapter critically reviews the role of metabolomics in the conservation of microbial genetic resources, highlighting its contributions to functional characterization, post-preservation quality assessment, protocol optimization under quality-by-design principles, and strain prioritization based on functional diversity and resilience. Methodological and conceptual challenges, including standardization, reproducibility, and metabolite identification, are discussed in the context of integrating metabolomics into biobanking and quality assurance frameworks. Finally, future perspectives are outlined, emphasizing the integration of metabolomics with other omics layers, regulatory standards, and predictive approaches to support functional, resilient, and application-oriented microbial conservation under global change.

Keywords Microbial genetic resources, Metabolomics, Functional conservation, Microbial biobanks, Cryopreservation, Functional stability

Marco A. Ramírez-Mosqueda et al. (eds.), *Methods for The Conservation of Microorganisms*, Methods in Molecular Biology, vol. 3090, https://doi.org/10.1007/978-1-0716-5623-5_20,

1 Introduction

The conservation of microbial genetic resources (MGR) has become a strategic component for food security, public health, biotechnology, and ecosystem resilience [1]. Unlike other biological resources, microorganisms constitute a substantial fraction of the planet's functional biodiversity, engaging in critical processes such as nutrient cycling, biological suppression of pathogens, plant–microorganism symbiosis, pollutant degradation, and the production of high-value metabolites, including antibiotics, siderophores, enzymes, volatile compounds, and signaling molecules [2, 3]. Nevertheless, microbial conservation faces a fundamental structural challenge: preserving not only taxonomic identity or the genome but also functionality understood as the set of biochemical and physiological capacities that underpin their ecological and biotechnological utility. In practice, *ex situ* conservation programs such as collections, banks, and biological resource centers have historically prioritized the genetic stability and overall viability of strains, primarily relying on methods such as cryopreservation and lyophilization [4]. Although these approaches are considered robust and widely accepted, it is now recognized that cellular viability does not necessarily equate to the preservation of functional phenotypes [5]. Recent evidence from biobanks and microbial collections indicates that factors such as precultivation conditions, the type and concentration of cryoprotectants, freezing and thawing rates, dehydration processes, and long-term storage can induce cellular stress, compromise membrane integrity, and profoundly reconfigure metabolism, directly impacting key traits such as growth, tolerance to abiotic stress, secondary metabolite production, and symbiotic performance [6, 7]. Consequently, the most recent international guidelines and regulatory frameworks emphasize the need for quality systems, traceability, and functional verification of conserved biological material, beyond mere survival [8]. Within this context, metabolomics emerges as an integrative platform for the study and conservation of MGR, as it enables the direct measurement of the biochemical phenotype, that is, the set of metabolites that reflect the physiological state and functional response of a microorganism under defined conditions. This approach positions metabolomics as both a conceptual and methodological bridge between genomic potential, transcriptional and proteomic regulation, and the effective cellular function expressed in the metabolome. Furthermore, it offers robust tools to characterize functional diversity, assess post-preservation metabolic stability, optimize conservation protocols, and prioritize strains of high ecological or biotechnological value, in line with modern trait-based approaches in microbial ecology [2,3]. The use of functional approaches is particularly relevant in the current scenario, where climate change imposes increasing conditions of water, thermal, and chemical stress that reshape the structure, metabolism, and function of microbial communities impacting both natural ecosystems and production systems [9, 10]. Within this framework, MGR conservation cannot be limited to passive strain preservation; instead, it must be oriented toward functional conservation, understood as the ability of microorganisms to maintain or recover their metabolic and ecological properties under changing environmental conditions.

Accordingly, this work critically examines the role of metabolomics as a key tool for the conservation of microbial genetic resources, highlighting its contributions to functional characterization, post-preservation stability assessment, and strategic decision-making in biobanks and conservation programs. Additionally, it discusses the main methodological and conceptual challenges, as well as future perspectives for integrating metabolomics into functional conservation frameworks aimed at microbial resilience and the sustainable use of biodiversity in the context of global change.

2 Microbial Genetic Resources and Conservation

2.1 From the "Strain-Identity" Paradigm to the "Strain-Function" Paradigm

Microbial genetic resources (MGR) encompass cultivable microorganisms such as bacteria, filamentous fungi, yeasts, microalgae, and actinobacteria as well as intraspecific variants and, increasingly, complex microbial communities preserved as consortia or microbiomes [11, 12]. This diversity represents not only a wealth of taxonomic variation but also a broad array of metabolic and ecological capabilities that underpin essential processes in both natural and productive systems. Consequently, the conservation of MGR has evolved toward approaches that acknowledge the functional performance of a strain or consortium as the primary determinant of its utility, rather than its isolated taxonomic identity [13, 14]. Within this framework, biological resource centers (BRCs) have transitioned from being passive strain repositories to becoming integrated platforms that combine custodianship, authentication, documentation, biosafety, distribution, and quality assurance. Recent studies underscore that the strategic value of BRCs lies in their capacity to ensure that conserved microbial resources retain properties aligned with their intended use, particularly in agricultural, environmental, and biomedical applications [15, 16]. This perspective reinforces the necessity of incorporating functional criteria into microbial collection management. Complementarily, it has been increasingly recognized that regulatory frameworks and biobanking best practices are expanding their scope to include the functional adequacy of biological material [16, 17]. Recent research on microbial biobanks highlights that the quality of a resource cannot be assessed solely in terms of viability or purity, but must also account for its physiological and metabolic stability over time especially when destined for applied research or technology transfer [18]. In the realm of culture collections, it has been noted that while taxonomic identity verification through molecular tools is essential, it is insufficient to guarantee the conservation of key functional traits [19]. Genetically identical microorganisms may display divergent metabolic profiles as a result of cultivation history, preservation conditions, or sublethal adaptation processes, potentially leading to significant variations in their functional performance [20]. This phenomenon highlights the limitations inherent in the paradigm focused solely on "strain–identity." In response to this scenario, the "strain–function" paradigm has emerged as a robust alternative, proposing that microorganisms be evaluated and conserved based on measurable functional traits such as the production of bioactive metabolites, tolerance to abiotic stress, metabolic efficiency, or symbiotic interaction capacity [3, 21]. From an ecological standpoint, this approach aligns with contemporary frameworks based on microbial functional traits, which enable the linkage of microbial diversity with ecosystem processes and environmental resilience [22, 23]. In this context, metabolomics assumes a central role as a tool for operationalizing the "strain–function" paradigm [24, 25]. By providing direct characterization of the metabolic state and functional stability of conserved microorganisms, metabolomics allows conservation strategies to move beyond binary viability assessments toward functionally grounded evidence, capable of capturing subtle yet relevant changes in microbial physiology [25]. In this way, it becomes a strategic component for modernizing the management of MGR and for reinforcing their conservation and sustainable use in the face of increasingly dynamic environmental and productive scenarios.

3 *Ex situ* Methods for Microbial Conservation

Ex situ conservation strategies constitute the operational core of most microbial genetic resource preservation programs. Their primary objective is to maintain the genetic and phenotypic stability of strains over extended periods, while minimizing the need for successive subculturing, which may favor adaptation processes, genetic drift, and the loss of functional traits [26]. However, recent evidence suggests that, even under controlled conditions, *ex situ* preservation methods can induce physiological and metabolic alterations with significant implications for the functionality of conserved material [27, 28]. From this general framework, it is necessary to conduct a differentiated analysis of the main *ex situ* conservation strategies currently in use, considering not only their effectiveness in terms of viability and genetic stability, but also their physiological and metabolic impacts.

3.1 Cryopreservation

Cryopreservation is widely regarded as the method of choice for long-term microbial storage, due to its ability to reduce metabolic activity and limit genetic changes associated with repeated propagation. Several recent studies highlight that, when properly implemented, cryopreservation enables the preservation of viability and genomic stability across a wide range of microbial taxa, including bacteria, fungi, and actinobacteria [29, 30]. From a metabolic standpoint, cryopreservation is not a physiologically neutral process, even when cellular viability and genomic stability are retained after thawing. Various studies have demonstrated that freeze–thaw cycles can induce sublethal alterations in the metabolome, reflected in changes to energy balance, redox status, and membrane integrity effects that do not always result in immediate losses in growth or colony-forming capacity [31–33]. These metabolic disturbances may persist during the initial recovery phases post-cryopreservation and condition the subsequent functional performance of preserved strains. In particular, metabolomic analyses have revealed that cryopreservation can differentially affect central metabolic pathways such as glycolysis, the tricarboxylic acid (TCA) cycle, and the pentose phosphate pathway, as well as metabolites associated with redox homeostasis, including glutathione, NAD(H)/NADP(H), and antioxidant derivatives [34, 35]. These modifications reflect adaptive responses to oxidative and osmotic stress induced during freezing, and their extent depends on both the type of cryoprotectant used and the cell's preexisting metabolic state.

Moreover, the presence of cryoprotectants such as glycerol or DMSO can directly influence the basal metabolic profile after thawing, either through their transient incorporation into metabolic pathways, their effects on membrane fluidity, or their modulation of intracellular osmotic gradients [36–38]. In microorganisms capable of endogenously accumulating compatible sugars or polyols, cryopreservation may induce compensatory metabolic reprogramming, characterized by the differential accumulation of osmoprotectants and stabilizing metabolites. While this response supports survival, it may also temporarily alter the functional phenotype [39, 40]. These findings suggest that assessing cryopreservation solely based on viability recovery is insufficient to ensure the preservation of the expected biochemical phenotype [40]. From this perspective, the integration of metabolomic analyses enables the identification of functional biomarkers of post-cryopreservation stability, the discrimination between protocols that preserve only survival versus those that maintain metabolic integrity, and the rational optimization of freezing and thawing conditions to minimize functional deviations [41]. Consequently, cryopreservation should be understood not merely as a strategy of metabolic arrest, but as a process that induces specific metabolic responses whose characterization is essential for truly functional microbial conservation.

3.2 Lyophilization (freeze-drying)

Lyophilization is widely employed for bacteria and certain fungi and yeasts, primarily due to its logistical advantages, such as ease of transport and storage at room temperature. However, this method exposes cells to severe dehydration and oxidative stress, which can compromise the integrity of membranes, proteins, and enzymatic systems [33]. Experimental studies have shown that resistance to lyophilization is closely linked to the prior physiological and metabolic state of the cells, with the growth phase emerging as a critical determinant of survival and functional stability after drying [42]. In particular, cells in the stationary phase tend to exhibit greater dehydration tolerance than those in the exponential phase, due to a prior metabolic reprogramming characterized by the accumulation of protective compounds, such as compatible sugars (especially trehalose), osmoprotective amino acids, and endogenous antioxidant systems [39, 42, 43]. From a biochemical perspective, lyophilization induces structural and functional damage to membranes, proteins, and enzymatic systems as a result of the loss of structural water, increased production of reactive oxygen species, and alterations in the cellular microenvironment [43]. It has been documented that the presence or absence of protective metabolites directly influences the stability of phospholipid membranes, the preservation of protein conformation, and the ability of cells to resume metabolism after rehydration [44, 45]. These changes do not always manifest as immediate viability loss, but they can lead to reduced metabolic activity, altered metabolite production, and functional variability among genetically identical strains. In this context, it has been noted that lyophilization is not a metabolically neutral process; rather, it selects for specific physiological states and may amplify preexisting metabolic differences between cultures subjected to slightly different pre-drying conditions [46, 47]. These findings underscore the need to incorporate metabolic assessments whether through targeted analysis of key metabolites or comparative metabolomic profiling as part of protocol optimization and validation, particularly when the conservation objective is to preserve functional traits and not merely cellular viability. Taken together, the *ex situ* strategies currently employed have proven effective in preserving viability and, in many cases, genetic stability of microorganisms. However, recent evidence indicates that these methods can induce subtle yet functionally relevant physiological and metabolic alterations that remain undetected by conventional evaluations. This limitation highlights the need to integrate analytical tools capable of assessing the functional stability of conserved material, thereby establishing a more robust framework for the validation and optimization of microbial preservation protocols.

3.3 Contributions of Metabolomics to the Study and Conservation of Microorganisms: Analytical Approaches and Platforms

Metabolomics is defined as the systematic and comprehensive analysis of low-molecular-weight metabolites present within a cell, organism, or biological community under defined conditions, providing a direct readout of the system's functional physiological state [48,49]. In microbiology, its relevance is particularly high due to the substantial metabolic plasticity of microorganisms, their rapid environmental adaptability, and their central role in ecological and biotechnological processes such as competition, chemical communication, niche colonization, and the production of secondary metabolites with applied value [50, 51]. Unlike genomic or transcriptomic approaches, which describe functional potential or gene regulation, metabolomics captures the integrated outcome of metabolic activity, offering an immediate reflection of responses to environmental, nutritional, or physical changes [52]. This characteristic is especially valuable in the context of microbial preservation, where storage procedures (e.g., cryopreservation, lyophilization, cold storage, or prolonged subculturing) can induce osmotic, oxidative, or energetic stress that may not result in overt viability loss but can trigger early metabolic alterations [53, 54]. Within this framework, metabolomics should not be viewed as an isolated exploratory technique, but rather as a strategic tool for assessing post-preservation physiological stability, identifying compromised metabolic pathways, and proposing functional quality biomarkers that complement classical viability- or growth-based criteria [50, 55].

From a methodological perspective, metabolomics applied to microorganisms is primarily organized into two complementary approaches: untargeted and targeted. The untargeted approach captures a broad range of metabolic signals without prior bias, making it especially useful for discovering metabolites associated with preservation-induced stress, metabolic reprogramming, or physiological instability [49, 56]. In microbial conservation, this approach is critical when evaluating poorly characterized strains, microbial consortia, or germplasm collections, as it allows for the identification of global metabolic signatures that distinguish pre- and post-preservation states without requiring restrictive prior hypotheses [57, 58]. Conversely, targeted metabolomics focuses on the precise quantification of specific, physiologically relevant metabolites such as osmoprotectants, antioxidants, or energy intermediates [59]. In the context of microbial preservation, this approach is particularly valuable for validating hypotheses generated by untargeted studies and for establishing reproducible quality control (QA/QC) indicators, such as trehalose, proline, reduced glutathione, or organic acids associated with stress tolerance [55, 60]. The sequential combination of untargeted and targeted approaches is widely recommended as the optimal strategy for translating exploratory insights into practical tools for microbial banks and long-term conservation schemes [61, 62]. However, the effective implementation of these approaches requires careful attention to reproducibility and standardization an especially critical factor when metabolomics is to be integrated into preservation protocols. In this regard, the guidelines set forth by the Metabolomics Standards Initiative (MSI), and their ongoing updates through FAIR initiatives, remain the benchmark for ensuring transparency in sample preparation, data acquisition, metabolite identification, and statistical analysis [63, 64]. These standards are particularly relevant in microbiology, where minor variations in culturing or sampling can result in significant metabolic differences [65].

The analytical platforms supporting microbial metabolomics mainly liquid chromatography–mass spectrometry (LC–MS), gas chromatography–mass spectrometry (GC–MS), and nuclear magnetic resonance (NMR) should not be considered in isolation but as interconnected components within an integrated analytical strategy [63–67]. LC–MS offers high sensitivity and broad coverage for polar and semipolar metabolites, making it particularly suitable for untargeted studies and targeted validation of stress-related metabolites; GC–MS provides high reproducibility and excellent structural identification capabilities for volatile compounds and central metabolites of primary metabolism; whereas NMR delivers absolute quantification, robust inter-laboratory reproducibility, and exceptional value as a reference tool for analytical quality control [68, 69]. In the context of microbial preservation, the strategic selection and combination of metabolomic platforms is a critical step for translating analytical data into functional insights. Proper integration of these technologies not only enables the detection of subtle metabolic changes induced by preservation procedures but also facilitates the linking of these changes to specific metabolic pathways thereby supporting the identification of robust functional biomarkers and the implementation of QA/QC frameworks grounded in reproducible metabolomic evidence [65, 69]. Nonetheless, the ability of these platforms to generate biologically interpretable information directly depends on rigorous methodological coordination throughout the entire experimental workflow. In microorganisms, factors such as standardized cultivation (including medium composition, growth phase, oxygenation, and cell density), rapid metabolic quenching to prevent post-sampling artifacts, efficient and consistent extraction methods, and systematic use of analytical quality controls (e.g., pooled QC samples, internal standards, and instrumental drift monitoring) are essential to ensure data comparability and reproducibility [58, 70]. These methodological practices not only enhance analytical robustness but also define the real capacity of metabolomics to capture subtle yet biologically significant metabolic changes that would otherwise go unnoticed through traditional viability or growth criteria [71]. Taken together, metabolomics applied to the study and conservation of microorganisms emerges as an integrative tool capable of linking molecular, physiological, and functional processes transforming microbial conservation assessment from a merely descriptive to a predictive, quantitative, and functionally informative approach, aligned with the current needs of conservation banks and quality management systems.

4 Key Contributions of Metabolomics to Microbial Conservation

Microbial conservation has historically been grounded in taxonomic and classical phenotypic criteria and, more recently, in genomic approaches aimed at authenticating the identity and genetic stability of preserved strains. These tools have been essential for ensuring traceability and accurate classification of microbial resources; however, they prove insufficient when the goal of conservation extends beyond mere taxonomic identity toward the preservation of the functional potential of microorganisms [4, 11]. In this context, metabolomics emerges as a key discipline to complement traditional approaches by providing direct information on the physiological status and effective functional expression of conserved genotypes.

4.1 Functional Characterization for Describing and Differentiating Microbial Genetic Resources

Traditional microbial conservation relies primarily on taxonomy and genomics to authenticate strain identity and assess long-term stability. Nevertheless, there is ample evidence that genetically similar microorganisms can exhibit significant functional differences, stemming from regulatory mechanisms, cultivation history, environmental adaptation, microbial epigenetics, or variations in physiological state at the time of preservation [49, 72]. These differences may not be reflected in conserved genetic markers but are clearly observable at the metabolic level. In this scenario, metabolomics provides a complementary functional characterization, capable of capturing the integrated expression of active metabolic pathways and generating comparative profiles or "metabolic fingerprints" that allow for the description and differentiation of microbial genetic resources beyond their genomic identity [73]. These metabolic fingerprints constitute reproducible chemical signatures that reflect the operational physiology of a strain under defined conditions and have proven particularly useful in distinguishing strains with high genetic similarity but divergent functional behavior [74, 75]. The incorporation of metabolomics into microbial conservation becomes especially relevant when analyzed from the perspective of regulatory frameworks and quality assurance/control systems (QA/QC) [70]. Current guidelines for microbial collections and Biological Resource Centres such as those established by the OECD focus primarily on ensuring taxonomic identity, purity, viability, and genetic stability of strains, leaving a substantial gap in the systematic assessment of the functional and metabolic status of conserved resources [76, 77]. In this regard, metabolomics offers a unique opportunity to complement these frameworks by incorporating quantitative functional criteria capable of reflecting the true physiological stability of strains over time.

The Metabolomics Standards Initiative (MSI) provides a structured framework of minimum information for reporting metabolomic studies, covering sample preparation, instrumental acquisition, data processing, and metabolite identification elements essential for ensuring the reproducibility, comparability, and interoperability of metabolic profiles across laboratories and collections [78–80]. Additionally, the adoption of FAIR principles (Findable, Accessible, Interoperable, and Reusable) strengthens the value of metabolomics as a quality tool by ensuring that metabolic data associated with conserved strains are traceable, reusable, and comparable over the long term facilitating integrated functional information management within microbial banks [65]. From a legal and genetic resource protection standpoint, the Budapest Treaty on the International Recognition of the Deposit of Microorganisms for the Purposes of Patent Procedure ensures the preservation and availability of deposited microorganisms, but it does not consider criteria for evaluating the conservation of their metabolic functionality [81–83]. In this respect, metabolomics can serve as a technical-scientific complement that documents and supports the functional status of deposited strains, thereby enhancing the description of biological material associated with intellectual property rights and technology transfer processes [84]. The systematic integration of standardized metabolomic profiles would thus enable the development of functional quality systems, in which microbial conservation ensures not only the existence of the genetic resource but also the preservation of its metabolic, ecological, and applied potential aligning with current trends in microbial biodiversity management and modern biotechnology.

4.2 Post-Preservation Assessment: Stability of the Biochemical Phenotype

One of the most direct and applicable contributions of metabolomics to microbial conservation lies in its ability to objectively assess the physiological changes induced by preservation processes, particularly following cryopreservation and lyophilization [85, 86]. Traditionally, post-preservation evaluation has relied on growth recovery and cell viability; however, there is growing evidence that these criteria are insufficient to ensure the stability of the biochemical and functional phenotype of preserved microorganisms [87, 88]. Metabolomics overcomes this limitation by offering an integrated readout of the metabolic state, revealing sublethal alterations that may compromise long-term functional performance [89]. In the case of cryopreservation, cellular stability critically depends on the use of cryoprotectants, whose classical role includes the mitigation of damage from ice crystal formation, osmotic dehydration, and oxidative stress [90]. Foundational studies and recent reviews have documented the function of compounds such as glycerol, sugars, and polyols in protecting membranes and macromolecules during freezing and thawing [91–93]. From a metabolomic perspective, these mechanisms can be quantitatively assessed through the monitoring of oxidative stress indicators and membrane damage, such as alterations in redox metabolites; changes in core energy pathways including glycolysis, the tricarboxylic acid (TCA) cycle, and the pentose phosphate pathway (PPP); as well as the accumulation or depletion of osmoregulatory and cellular stabilizing compounds, including trehalose, compatible amino acids, and polyols [54, 75, 94]. This approach not only enables a detailed description of the metabolic impact of cryopreservation but also facilitates the rational optimization of preservation protocols by comparing different cryoprotectants, concentrations, and cooling conditions based on quantifiable metabolic responses rather than solely survival rates [29, 95]. Furthermore, the identification of consistently affected metabolites or pathways may lead to the definition of functional biomarkers for post-cryopreservation stability, useful for quality control in microbial biobanks [96, 97].

In the context of lyophilization, metabolomics has proven particularly valuable for understanding how the prior physiological state of microorganisms determines their tolerance to the process. Studies in probiotic microorganisms have shown that cells in the stationary phase exhibit greater resistance to desiccation, associated with metabolic reprogramming that promotes the accumulation of protective compounds such as trehalose and specific compatible amino acids [39, 98]. This provides a functional basis for selecting optimal cultivation conditions prior to lyophilization [95].

The implication for microbial conservation is both clear and conceptually significant: it is not sufficient to recover viable colonies after preservation. It is essential to verify that the overall metabolic profile or a set of key biomarkers is consistent with the expected functional phenotype of the strain. Accordingly, metabolomics enables a shift from binary survival assessments toward functional post-preservation evaluations that reflect the actual quality of the preserved microbial resource and support strategic decision-making in the management, prioritization, and long-term use of microbial collections.

4.3 Metabolomics for Protocol Optimization: A "Quality by Design" Approach

The application of metabolomics to microbial conservation enables the optimization of preservation protocols from a systematic and quantitative perspective, aligned with the "Quality by Design" (QbD) paradigm. Rather than evaluating preservation methods solely based on final outcomes such as viability recovery, metabolomics allows preservation conditions to be treated as a multivariable experimental problem, in which factors such as the type and concentration of cryoprotectants, freezing or warming rates, drying conditions during lyophilization, protective formulations (sugars, proteins, polymers), and reactivation protocols are compared in a controlled manner [99–102]. In microbial collections, cryopreservation is widely regarded as the method of choice for long-term storage due to its capacity to minimize repeated subculturing and reduce genetic drift [103]. However, its effectiveness critically depends on the appropriate selection and validation of preservation conditions, as well as the implementation of best practices that ensure the physiological and functional stability of conserved strains [104, 105]. Within this context, metabolomics adds a layer of quantitative assessment that enables the detection of metabolic deviations induced by suboptimal conditions, even when apparent cell survival is high. From a QbD perspective, metabolomics facilitates the identification of critical functional quality attributes, defined as metabolites or metabolic pathways whose stability is directly associated with the preservation of the expected phenotype [99, 106]. For example, monitoring metabolites involved in redox balance, energy metabolism, or osmoregulation makes it possible to distinguish between protocols that merely preserve viability and those that maintain the microorganism's metabolic integrity [107, 108]. In this way, the selection of cryoprotectants or protective formulations can be guided by their capacity to minimize specific metabolic alterations, rather than by empirical or historical criteria [106]. Moreover, metabolomics allows for an integrated evaluation of post-preservation reactivation conditions, a stage often underestimated yet critical for functional recovery. Differences in reactivation media, temperature, or nutrient availability can trigger metabolic reprogramming that affects the functional expression of strains, phenomena that can be detected and quantified through comparative metabolomic profiling [109–111]. This information is essential to close the QbD loop, ensuring that selected conditions not only preserve the cell during storage but also support consistent and reproducible metabolic recovery [111]. Taken together, integrating metabolomics into preservation protocol optimization enables a transition from a reactive approach focused on correcting failures to a predictive and preventive model, in which preservation conditions are deliberately designed to minimize functional loss. This quality by-design paradigm strengthens the robustness of microbial conservation systems and lays the foundation for quality assurance frameworks that incorporate identity, viability, and functional metabolic stability as complementary and quantifiable criteria.

4.4 Strain Prioritization and Conservation Based on Functional Diversity

Within the operational context of microbial collections and conservation banks, limitations in space, budget, and personnel inevitably necessitate decision-making regarding which strains to preserve, prioritize, or discard [112]. Traditionally, such prioritization processes have relied on taxonomic criteria, geographic origin, frequency of use, or historical relevance. However, these criteria are insufficient when the goal is to maximize the functional value of a collection under real-world constraints [113]. In this scenario, metabolomics emerges as a strategic tool to underpin microbial conservation with quantifiable functional criteria, in alignment with recent advances in functional microbial ecology [114]. The aim is to estimate the metabolic breadth of a strain by characterizing the number, diversity, and distribution of metabolites and active metabolic pathways, thereby providing a direct proxy for its functional versatility [115, 116]. Strains exhibiting broad metabolomic profiles tend to demonstrate greater adaptability to environmental changes and enhanced potential to fulfill multiple ecological or biotechnological roles, making them priority candidates for long-term conservation [25, 117]. This approach moves beyond the simple enumeration of genes or functional annotations, instead reflecting effective metabolic expression under defined conditions [118]. Additionally, metabolomic analysis facilitates the detection of rare or high-value metabolites, including compounds with industrial, pharmaceutical, agricultural, or ecological potential that may not be evident through routine genomic analyses [118]. Identifying such metabolites enables the recognition of strains with unique functional value, even within taxonomically redundant groups, thus providing an objective criterion for their prioritization within microbial collections [119, 120]. From this perspective, metabolomics supports the preservation not only of taxonomic diversity but also of the chemical and functional uniqueness of microorganisms. A third key contribution of metabolomics to strain prioritization lies in the identification of resilient microorganisms capable of maintaining essential metabolic functions under environmental stress or following preservation processes [119, 120]. The stability of metabolomic profiles in response to perturbations serves as a functional indicator of resilience, particularly relevant in contexts of environmental change, field applications, or long-term storage [121]. This criterion distinguishes strains that not only survive but also sustain their functional performance an essential attribute for use-oriented conservation.

The incorporation of these metabolomic criteria directly aligns with conceptual advances in functional microbial ecology, which has evolved from purely taxonomic approaches to frameworks based on functional traits, redundancy, and resilience [122, 123]. These frameworks emphasize the need for a unified approach to studying microbial functions, recognizing that multiple taxa may perform similar roles (functional redundancy), yet not all contribute equally to system stability and resilience [124, 125]. These concepts are directly transferable to microbial conservation, where metabolomics enables the operationalization of functional diversity and redundancy in concrete management and prioritization decisions. Taken together, metabolomics provides a quantitative foundation for transitioning toward a microbial conservation paradigm based on functional diversity, in which strain prioritization is guided by metabolic breadth, chemical uniqueness, and functional resilience. This approach not only optimizes the use of limited resources but also enhances the capacity of microbial collections to preserve key functions, anticipate future needs, and respond to emerging challenges in biotechnology, agriculture, and environmental sustainability.

5 Current Challenges and Limitations in the Application of Metabolomics to Microbial Conservation

Despite its high potential, the application of metabolomics to microbial conservation faces significant challenges, one of the most critical being the extreme dependence of the metabolome on the experimental context [126]. Unlike other omics layers, the metabolome responds immediately to environmental and cultivation variations, meaning that so-called biological noise associated with cultivation can overshadow the biological signal of interest unless rigorous standardization is implemented [127, 128]. In studies comparing metabolic profiles between strains or between pre- and post-preservation states, it is essential to strictly control variables such as culture medium composition, pH, carbon and nitrogen sources, trace elements, growth phase, cell density, and oxygenation and agitation conditions [129–131]. Without such standardization, technical or environmental variations may be misinterpreted as true biological differences, weakening the validity of conclusions and limiting the usefulness of metabolomics as a quality assurance and control (QA/QC) tool [132, 133]. This biological challenge is compounded in the absence of robust analytical standardization particularly relevant when metabolomics is to be integrated into formal QA/QC frameworks within microbial banks. The reproducibility of metabolomic analyses especially in untargeted approaches critically depends on the systematic use of internal standards, pooled quality control (QC) samples, instrumental drift monitoring, and transparent data acquisition and processing strategies [70, 132–134].

In this context, the guidelines proposed by the Metabolomics Standards Initiative (MSI) remain a fundamental reference by establishing minimum reporting requirements for metabolomic studies, including key aspects of sample preparation, instrumental acquisition, metabolite identification, and data processing. Although not specifically designed for microbial conservation applications, these guidelines provide an indispensable foundation for ensuring data comparability, reproducibility, and reusability essential conditions for the reliable integration of metabolomics into quality management systems and decision-making processes [58, 77, 80]. However, the adoption of these standards also highlights one of the main methodological tensions in applying metabolomics to microorganisms: the inherent difficulty of achieving high and uniform levels of structural identification, particularly for secondary metabolites [135–137]. One of the most persistent and limiting challenges in this field is the reliable identification of metabolites, as a significant portion of the microbial metabolome consists of lineage-specific compounds often cultivation-dependent or stress-induced that are poorly represented in commercial libraries or general-use spectral databases [58, 77, 135, 136]. This limitation represents a critical bottleneck, as it hinders unequivocal structural assignment and restricts the functional interpretation of metabolic changes observed after preservation processes [138]. Unlike primary metabolites, which are broadly conserved across taxa, microbial secondary metabolites exhibit extraordinary structural diversity modulated by specific regulation, cultivation history, and environmental cues [139, 140]. This complexity increases the likelihood of incomplete or ambiguous annotations in untargeted studies, even when rigorous reporting standards are followed [77, 135, 136]. As a result, a significant proportion of signals detected by LC–MS or GC–MS remain without definitive structural identification, limiting their direct integration into formal QA/QC frameworks for conservation purposes [57]. In response to this scenario, the most pragmatic and widely adopted strategy involves a hybrid identification approach combining tentative annotations based on exact mass, isotopic patterns, and fragmentation spectra with the targeted validation of key metabolites directly relevant to microbial conservation [141, 142]. This approach acknowledges that, from an operational standpoint, it is not necessary to identify the entire metabolome exhaustively in order to obtain valuable functional information. Instead, it is more efficient to prioritize the structural confirmation of metabolites associated with oxidative stress, energy metabolism, membrane integrity, and osmoregulation metabolites that can act as functional biomarkers of post-preservation stability [57, 77, 136]. Furthermore, metabolite identification in microorganisms is reinforced through the use of specialized databases, hierarchical annotation strategies, and the integration of complementary information from genomics and transcriptomics, which helps to narrow the plausible chemical space and increase annotation confidence [141–143]. Nonetheless, even with these approaches, full structural identification remains a challenge underscoring the need for clearly defined confidence levels and identification criteria, as recommended by MSI standards.

All of the above makes it clear that, within the framework of microbial conservation, this limitation does not invalidate the use of metabolomics; rather, it redefines its objectives. Instead of seeking exhaustive metabolome identification, what is needed is functionally informative, reproducible, and comparable identification across preservation cycles. The combination of robust metabolic fingerprints with targeted validation of a limited set of critical metabolites allows us to overcome the identification bottleneck and facilitates the integration of metabolomics into quality control and decision-making frameworks, without compromising scientific rigor or operational applicability.

6 Future Perspectives: Toward Functional, Predictive, and Resilient Microbial Conservation

Future directions in microbial conservation clearly point toward a paradigm shift: moving from identity- and viability-centered schemes to a model of "functional conservation" that preserves not only the existence of genetic resources but also their metabolic fitness and performance under changing conditions [144, 145]. In this scenario, the integration of metabolomics with other omics layers, particularly genomics and transcriptomics will enable the construction of an operational framework that combines identity (what the strain is), potential (what it could do), and function (what it is actually doing) [146, 147]. For microbial collections, this integration represents a strategic opportunity, as it allows not only for strain authentication but also for the certification of functional suitability for specific applications such as biofertilizers, biocontrol agents, pollutant biodegradation, or high-value metabolite production thus narrowing the gap between conservation and applied use [84]. In parallel, the strengthening of regulatory frameworks and international standards in biobanks and biological collections such as ISO 20387:2018 (Biotechnology-Biobanking) and the guidelines promoted by the OECD and the World Federation for Culture Collections (WFCC) creates the possibility for metabolomics to become established as a tool for functional certification within quality assurance and control (QA/QC) frameworks [18, 148, 149]. In this context, it is reasonable to anticipate that, for certain high-value or regulated microbial resources, minimum metabolomic panels may be incorporated as part of quality criteria [150]. This does not imply exhaustive metabolome profiling in every evaluation, but rather the strategic definition of key functional biomarkers such as osmoprotectants (e.g., trehalose and compatible amino acids), redox indicators, or membrane lipid components and comparative stability profiles that enable reproducible assessment of the conservation of the expected biochemical phenotype [27, 151]. The rapid advancement of artificial intelligence (AI) and computational approaches applied to omics data further reinforces the future role of metabolomics in microbial conservation [152]. Machine learning models have demonstrated the ability to infer metabolomic profiles from genomic, transcriptomic, or community-level data, as well as to predict unobserved metabolites and functional relationships within complex microbiomes [153, 154]. These approaches enable a shift from descriptive metabolomic profiling toward mechanistic prediction, facilitating functional evaluation even in scenarios where exhaustive sampling is unfeasible such as complex environmental systems or large-scale conservation studies [155].

Finally, the role of metabolomics gains strategic relevance in light of the challenges posed by climate change, which is generating increasing scenarios of water, thermal, and chemical stress capable of reshaping microbial communities and profoundly altering their ecosystem functions [156]. The capacity of metabolomics to directly measure physiological state and altered metabolic pathways under stress positions it as a key tool for both in situ conservation for example, in agricultural soils or degraded ecosystems and for the *ex situ* selection and preservation of resilient microbial resources [157, 158]. Notably, community-level metabolomic profiles have proven useful for evaluating microbial functional status and resilience, opening direct pathways for functional monitoring and restoration based on metabolomic evidence [158, 159]. Taken together, these trends outline a future in which metabolomics becomes a cross-cutting axis of microbial conservation, integrated with other omics, quality standards, computational tools, and climate adaptation strategies. This approach will enable the development of more informed, predictive, and functionally oriented conservation systems capable of responding to the scientific, regulatory, and environmental demands of the twenty-first century.

Conflict of Interest

The authors have no conflicts of interest to declare that are relevant to the content of this chapter.

References

1. Onen OI, Aboh AA, Mfam AN, Akor MO, Nweke CN, Osuagwu AN (2020) Microbial diversity: values and roles in ecosystems. Asian J Biol 9(1):10–22
2. Trivedi P, Leach JE, Tringe SG, Sa T, Singh BK (2020) Plant–microbiome interactions: from community assembly to plant health. Nat Rev Microbiol 18(11):607–621
3. Malik AA, Martiny JBH, Brodie EL, Martiny AC, Treseder KK, Allison SD (2020) Defining trait-based microbial strategies with consequences for soil carbon cycling under climate change. ISME J 14:1–9
4. Geweely NS (2023) New frontiers review of some recent conservation techniques of organic and inorganic archaeological artefacts against microbial deterioration. Front Microbiol 14:1146582
5. Gilbert JA, Peixoto RS, Scholz AH, Dominguez Bello MG, Korsten L, Berg G, et al. (2025) Launching the IUCN Microbial Conservation Specialist Group as a global safeguard for microbial biodiversity. Nat Microbiol 10(10):2359–2360.
6. Alkhatib R, Gaede KI (2024) Data management in biobanking: strategies, challenges, and future directions. BioTech 13(3):34
7. Smith D, Clayton S, Ryan MJ, Day JG, Green P (2001) Preservation methodology. In: Smith D, Ryan MJ, Day JG (Eds.) The UK National Culture Collection (UKNCC) biological resource: properties, maintenance and management. UK National Culture Collection, 72–119
8. Solecki R, Schumacher DM, Pfeil R, Bhula R, MacLachlan DJ (2017) OECD guidance documents and test guidelines. In: Ambrus A, Hamilton D (Eds.) Food Safety Assessment of Pesticide Residues. World Scientific Publishing, Singapore, 13–36
9. Cavicchioli R, Ripple WJ, Timmis KN, Azam F, Bakken LR, Baylis M et al (2019) Scientists' warning to humanity: microorganisms and climate change. Nat Rev Microbiol 17(9):569–586
10. Delgado-Baquerizo M, Reich PB, Trivedi C, Eldridge DJ, Abades S, Alfaro FD et al (2020) Multiple elements of soil biodiversity drive ecosystem functions across biomes. Nat Ecol Evol 4:210–220
11. Sharma SK, Singh SK, Ramesh A, Sharma PK, Varma A, Ahmad E et al (2018) Microbial genetic resources: status, conservation, and access and benefit-sharing regulations. In: Microbial resource conservation: conventional to modern approaches. Springer International Publishing, Cham, 1–33
12. Moreno-Espíndola IP, Gutiérrez-Navarro A, Franco-Vásquez DC, Vega-Martínez D (2024) Reflections on microbial genetic resources in agricultural systems. Curr Res Microb Sci 8, 100337.
13. Hugenholtz P, Tyson GW (2008) Microbiology: metagenomics. Nature 455:481–483
14. Garg D, Patel N, Rawat A, Rosado AS (2024) Cutting edge tools in the field of soil microbiology. Curr Res Microb Sci 6:100226
15. Smith D, Ryan M (2012) Implementing best practices and validation of cryopreservation techniques for microorganisms. Sci World J 2012(1):805659
16. Lima N (2025) Preserving and utilizing microbial diversity for innovation and sustainability. Microbiology 171(3):001544
17. Cadena-Zamudio JD, Ramírez-Mosqueda MA, Cadena-Zamudio DA (2024) Agrobiodiversity and food security: challenges and sustainable solutions. Agro Prod 17(10):101–113
18. Martins A, daSilva JFM, Sampaio P, Lima N (2022) Quality and competence management

in microbial biobanks. Importance of microbiology teaching and microbial resource management for sustainable futures. Academic Press, 157–192
19. Lajoie G, Kembel SW (2019) Making the most of trait-based approaches for microbial ecology. Trends Microbiol 27(10):814–823
20. Overmann J, Abt B, Sikorski J (2017) Present and future of culturing bacteria. Annu Rev Microbiol 71(1):711–730
21. Wan J, Crowther TW (2022) Uniting the scales of microbial biogeochemistry with trait-based modelling. Funct Ecol 36(6):1457–1472
22. Allison SD, Martiny JB (2008) Resistance, resilience, and redundancy in microbial communities. Proc Natl Acad Sci USA 105(Suppl 1):11512–11519
23. Feng J, Zeng XM, Zhang Q, Zhou XQ, Liu YR, Huang Q (2021) Soil microbial trait-based strategies drive metabolic efficiency along an altitude gradient. ISME Commun 1(1):71
24. Yakushev AV (2015) Integral structural-functional method for characterizing microbial populations. Eurasian Soil Sci 48(4):378–394
25. Zhou Z, Tran PQ, Breister AM, Liu Y, Kieft K, Cowley ES et al (2022) METABOLIC: high-throughput profiling of microbial genomes for functional traits, metabolism, biogeochemistry, and community-scale functional networks. Microbiome 10(1):33
26. Gams W (2002) *Ex situ* conservation of microbial diversity. Microorganisms in plant conservation and biodiversity. Springer, Netherlands, Dordrecht, 269–283
27. Agrawal R (2024) Preservation of microorganisms. Textbook of industrial microbiology. Springer Nature Singapore, Singapore, 65–76
28. Izah SC, Odubo TC, Ogwu MC, Hait M (2024) Aspects of microorganisms in the food industry. Food safety and quality in the global south. Springer Nature Singapore, Singapore, 399–425
29. Singh SK, Baghela A (2017) Cryopreservation of microorganisms. Modern tools and techniques to understand microbes. Springer International Publishing, Cham, 321–333
30. Redkina VV, Temraleeva AD (2025) Effectiveness of microalgae and cyanobacteria cryopreservation (for the strains from the All-Russian Collection of Microorganisms). Microbiology 94(4):505–514
31. Palermo A (2023) Metabolomics- and systems-biology-guided discovery of metabolite lead compounds and druggable targets. Drug Discov Today 28(2):103460
32. Baust JG, Gao D, Baust JM (2009) Cryopreservation: an emerging paradigm change. Organogenesis 5(3):90–96
33. Krivitsky V, Zverzhinetsky M, Krivitsky A, Hsiung LC, Naddaka V, Gabriel I et al (2019) Cellular metabolomics by a universal redox-reactive nanosensors array: from the cell level to tumor-on-a-chip analysis. Nano Lett 19(4):2478–2488
34. Whaley D, Damyar K, Witek RP, Mendoza A, Alexander M, Lakey JR (2021) Cryopreservation: an overview of principles and cell-specific considerations. Cell Transplant 30:0963689721999617
35. Taylor MJ, Weegman BP, Baicu SC, Giwa SE (2019) New approaches to cryopreservation of cells, tissues, and organs. Transfus Med Hemother 46(3):197–215
36. Pegg DE (2010) The relevance of ice crystal formation for the cryopreservation of tissues and organs. Cryobiology 60(3):S36–S44.
37. Warner RM, Brown KS, Benson JD, Eroglu A, Higgins AZ (2022) Multiple cryoprotectant toxicity model for vitrification solution optimization. Cryobiology 108:1–9
38. Zalomova LV, Fesenko JrEE (2024) FBS-based cryoprotective compositions for effective cryopreservation of gut microbiota and key intestinal microorganisms. BMC Res Notes 17(1):168
39. Carvalho AS, Silva J, Ho P, Teixeira P, Malcata FX, Gibbs P (2004) Relevant factors for the preparation of freeze-dried lactic acid bacteria. Int Dairy J 14(10):835–847
40. Moos M, Korbelová J, Štětina T, Opekar S, Šimek P, Grgac R, Koštál V (2022) Cryoprotective metabolites are sourced from both external diet and internal macromolecular reserves during metabolic reprogramming for freeze tolerance in drosophilid fly, *Chymomyza costata*. Metabolites 12(2):163
41. Murray KA, Gibson MI (2022) Chemical approaches to cryopreservation. Nat Rev Chem 6(8):579–593
42. Rocha IV, Bezerra MF, Sobreira M, Leal NC, de Almeida AMP (2024) Lyophilization for bacteria preservation: a promising approach for *Yersinia pestis* strains from an unique collection in Brazil (Fiocruz-CYP). Antonie Van Leeuwenhoek 117(1):61
43. Broeckx G, Vandenheuvel D, Claes IJ, Lebeer S, Kiekens F (2016) Drying techniques of probiotic bacteria as an important step towards the development of novel pharmabiotics. Int J Pharm 505(1–2):303–318
44. Gaidhani KA, Harwalkar M, Bhambere D, Nirgude PS (2015) Lyophilization/freeze drying–a review. World J Pharm Res 4(8):516–543
45. Prado NO, Lalli LA, Blanes L, Zanette DL, Aoki MN (2023) Lyophilization of molecular biology reactions: a review. Mini Rev Med Chem 23(4):480–496

46. Lee E, Lee JE (2025) Alterations in intracellular metabolites of Saccharomyces cerevisiae during freeze-drying and strategies for enhancing viability. LWT 234, 118567.
47. Musi V, Aiello E, Arena MP, De Vero L, Pulvirenti A, Gullo M (2026) Freeze-drying effects on viability and cellular stability in a subset of sourdough lactic acid bacteria strains. Curr Microbiol 83(1):76
48. Wishart DS (2011) Advances in metabolite identification. Bioanalysis 3(15):1769–1782
49. Johnson CH, Ivanisevic J, Siuzdak G (2016) Metabolomics: beyond biomarkers and towards mechanisms. Nat Rev Mol Cell Biol 17(7):451–459
50. Pinu FR, Goldansaz SA, Jaine J (2019) Translational metabolomics: current challenges and future opportunities. Metabolites 9(6):108
51. Saunders KD, Lewis HM, Beste DJ, Cexus O, Bailey MJ (2023) Spatial single cell metabolomics: current challenges and future developments. Curr Opin Chem Biol 75:102327
52. Baharum SN, Azizan KA (2018) Metabolomics in systems biology. Omics applications for systems biology. Springer International Publishing, Cham, 51–68
53. Berendsen RL, Pieterse CM, Bakker PA (2012) The rhizosphere microbiome and plant health. Trends Plant Sci 17(8):478–486
54. Thukral M, Allen AE, Petras D (2023) Progress and challenges in exploring aquatic microbial communities using non-targeted metabolomics. ISME J 17(12):2147–2159
55. Shyam K, Kumar N, Chandel H, Dogra AS, Sharma G, Saxena G (2023) Omics technologies in environmental microbiology and microbial ecology: insightful applications in bioremediation research. In: Genomics approach to bioremediation: principles, tools, and emerging technologies, pp 433–454
56. Liu P, Gao R, Gao L, Bi J, Jiang Y, Zhang X, Wang Y (2022) Distinct quality changes of asparagus during growth by widely targeted metabolomics analysis. J Agric Food Chem 70(50):15999–16009
57. Evans AM, O'Donovan C, Playdon M, Beecher C, Beger RD, Bowden JA, Metabolomics Quality Assurance and Quality Control Consortium (mQACC) et al (2020) Dissemination and analysis of the quality assurance (QA) and quality control (QC) practices of LC–MS based untargeted metabolomics practitioners. Metabolomics 16(10):113
58. Lippa KA, Aristizabal-Henao JJ, Beger RD, Bowden JA, Broeckling C, Beecher C et al (2022) Reference materials for MS-based untargeted metabolomics and lipidomics: a review by the metabolomics quality assurance and quality control consortium (mQACC). Metabolomics 18(4):24
59. Theodoridis G, Gika H, Raftery D, Goodacre R, Plumb RS, Wilson ID (2023) Ensuring fact-based metabolite identification in liquid chromatography–mass spectrometry-based metabolomics. Anal Chem 95(8):3909–3916
60. Singh D, Geat N, Mehriya M, Rajawat MVS, Prasanna R, Kumar A et al (2021) Omics (genomics, proteomics, metabolomics, etc.) tools to study the environmental microbiome and bioremediation. Waste to energy: prospects and applications. Springer, Singapore, 235–260
61. Chen J, Li T, Qin X, Du G, Zhou Y (2022) Integration of non-targeted metabolomics and targeted quantitative analysis to elucidate the synergistic antidepressant effect of Bupleurum chinense DC–*Paeonia lactiflora* Pall herb pair by regulating purine metabolism. Front Pharmacol 13:900459
62. Yang X, Yang X, Li B, Zhang J, Yan Z (2024) Combined non-targeted and targeted metabolomics reveals the mechanism of delaying aging of ginseng fibrous root. Front Pharmacol 15:1368776
63. Du X, Dastmalchi F, Ye H, Garrett TJ, Diller MA, Liu M et al (2023) Evaluating LC–HRMS metabolomics data processing software using FAIR principles for research software. Metabolomics 19(2):11
64. Powell CD, Moseley HN (2023) The metabolomics workbench file status website: a metadata repository promoting FAIR principles of metabolomics data. BMC Bioinformatics 24(1):299
65. Zulfiqar M, Crusoe MR, König-Ries B, Steinbeck C, Peters K, Gadelha L (2024) Implementation of FAIR practices in computational metabolomics workflows—a case study. Metabolites 14(2):118
66. Plumb RS, Gethings LA, Rainville PD, Isaac G, Trengove R, King AM, Wilson ID (2023) Advances in high throughput LC/MS based metabolomics: a review. TrAC Trends Anal Chem 160:116954
67. Bajo-Fernández M, Souza-Silva ÉA, Barbas C, Rey-Stolle MF, García A (2024) GC–MS-based metabolomics of volatile organic compounds in exhaled breath: applications in health and disease. A review. Front Mol Biosci 10:1295955
68. Nagana Gowda GA, Raftery D (2021) NMR-based metabolomics. Cancer metabolomics: methods and applications. Springer International Publishing, Cham, 19–37
69. Homobono Brito de Moura P, Leleu G, Da Costa G, Marti G, Pétriacq P, Valls Fonayet J, Richard T (2025) Integrating NMR and MS

for improved metabolomic analysis: from methodologies to applications. Molecules 30(12):2624

70. Mosley JD, Schock TB, Beecher CW, Dunn WB, Kuligowski J, Lewis MR et al (2024) Establishing a framework for best practices for quality assurance and quality control in untargeted metabolomics. Metabolomics 20(2):20
71. Dunn WB, Kuligowski J, Lewis M, Mosley JD, Schock T, Ulmer Holland C, Metabolomics Quality Assurance and Quality Control Consortium (mQACC) et al (2023) Metabolomics 2022 workshop report: state of QA/QC best practices in LC–MS-based untargeted metabolomics, informed through mQACC community engagement initiatives. Metabolomics 19(11)::93
72. Bhosle A, Wang Y, Franzosa EA, Huttenhower C (2022) Progress and opportunities in microbial community metabolomics. Curr Opin Microbiol 70:102195
73. Feussner K, Abreu IN, Klein M, Feussner I (2023) Metabolite fingerprinting: a powerful metabolomics approach for marker identification and functional gene annotation. In: Methods in enzymology, vol 680, Academic Press, 325–350
74. Withers E, Hill PW, Chadwick DR, Jones DL (2020) Use of untargeted metabolomics for assessing soil quality and microbial function. Soil Biol Biochem 143, 107758.
75. Chandran H, Meena M, Sharma K (2020) Microbial biodiversity and bioremediation assessment through omics approaches. Front Environ Chem 1:570326
76. Harrill JA, Viant MR, Yauk CL, Sachana M, Gant TW, Auerbach SS et al (2021) Progress towards an OECD reporting framework for transcriptomics and metabolomics in regulatory toxicology. Regul Toxicol Pharmacol 125:105020
77. Malinowska JM, Whelan M (2025) The important role of standards for the uptake of transcriptomics and metabolomics based *in vitro* methods in regulatory toxicology. Arch Toxicol 99(10):3865–3875
78. Fiehn O, Robertson D, Griffin J, van der Werf M, Nikolau B, Morrison N et al (2007) The metabolomics standards initiative (MSI). Metabolomics 3(3):175–178
79. Sumner LW, Amberg A, Barrett D, Beale MH, Beger R, Daykin CA et al (2007) Proposed minimum reporting standards for chemical analysis: chemical analysis working group (CAWG) metabolomics standards initiative (MSI). Metabolomics 3(3):211–221
80. Spicer RA, Salek R, Steinbeck C (2017) A decade after the metabolomics standards initiative it's time for a revision. Sci Data 4(1):1–3
81. Bussas V, Sharma A, Shouche Y (2017) IP and the Budapest Treaty—depositing biological material for patent purposes. Microbial resources. Academic Press, 275–292
82. Parashar A (2017) International depository authority and its role in microorganism's deposition. J Clin Diagn Res 11(8):DE01
83. Nwakoby IP, Iheukwumere IH, Iheukwumere CM, Nwakoby NE, Idigo MA, Ike VE (2025) Intellectual property law and microbiology: patenting microorganisms and genetic material. IPS J Educ 2(1):43–47
84. Sharma S, Devi U, Mishra G, Dwivedi N (2024) Intellectual property rights (IPR) in biotechnology. Collab Insight 2:182–207.
85. Merlo ME, Jankevics A, Takano E, Breitling R (2011) Exploring the metabolic state of microorganisms using metabolomics. Bioanalysis 3(21):2443–2458
86. Marvasi M (2020) Integrative applications of microbial biotechnology to cultural heritages conservation and promotion. In: IOP Conf Ser: Mater Sci Eng 949(1):012095
87. Hoefman S, Van Hoorde K, Boon N, Vandamme P, De Vos P, Heylen K (2012) Survival or revival: long-term preservation induces a reversible viable but non-culturable state in methane-oxidizing bacteria. PLoS One 7(4):e34196
88. Sharma SK, Kumar R, Vaishnav A, Sharma PK, Singh UB, Sharma AK (2017) Microbial cultures: maintenance, preservation and registration. Modern tools and techniques to understand microbes. Springer International Publishing, Cham, 335–367
89. Pinu FR, Villas-Boas SG (2017) Extracellular microbial metabolomics: the state of the art. Metabolites 7(3):43
90. Jahed KR, Saini AK, Sherif SM (2023) Coping with the cold: unveiling cryoprotectants, molecular signaling pathways, and strategies for cold stress resilience. Front Plant Sci 14:1246093
91. Len JS, Koh WSD, Tan SX (2019) The roles of reactive oxygen species and antioxidants in cryopreservation. Biosci Rep 39(8):BSR20191601
92. Neri L, Faieta M, Di Mattia C, Sacchetti G, Mastrocola D, Pittia P (2020) Antioxidant activity in frozen plant foods: effect of cryoprotectants, freezing process and frozen storage. Foods 9(12):1886
93. Kushkevych I, KovářováA, Dordevic D, Gaine J, Kollar P, Vítězová M, Rittmann SKM (2021) Distribution of sulfate-reducing bacteria in the environment: cryopreservation techniques and their potential storage application. Processes 9(10):1843

94. Zhao Y, Yao Y, Xu H, Xie Z, Guo J, Qi Z, Jiang H (2022) Soil metabolomics and bacterial functional traits revealed the responses of rhizosphere soil bacterial community to long-term continuous cropping of Tibetan barley. PeerJ 10:e13254
95. Noufeu T, Li Y, Toure NF, Yao H, Zeng X, Du Q, Pan D (2025) Overview of glycometabolism of lactic acid bacteria during freeze-drying: changes, influencing factors, and application strategies. Foods 14(5):743
96. Araújo R, Bento LF, Fonseca TA, Von Rekowski CP, da Cunha BR, Calado CR (2022) Infection biomarkers based on metabolomics. Metabolites 12(2):96
97. García CJ, Alacid V, Tomás-Barberán FA, García C, Palazón P (2022) Untargeted metabolomics to explore the bacteria exo-metabolome related to plant biostimulants. Agronomy 12(8):1926
98. Zhen N, Zeng X, Wang H, Yu J, Pan D, Wu Z, Guo Y (2020) Effects of heat shock treatment on the survival rate of *Lactobacillus acidophilus* after freeze-drying. Food Res Int 136:109507
99. Jacyna J, Kordalewska M, Markuszewski MJ (2019) Design of experiments in metabolomics-related studies: an overview. J Pharm Biomed Anal 164:598–606
100. Mohd Kamal K, Mahamad Maifiah MH, Abdul Rahim N, Hashim YZHY, Abdullah Sani MS, Azizan KA (2022) Bacterial metabolomics: sample preparation methods. Biochem Res Int 2022(1):9186536
101. Sørensen MB, Møller JK, Strube ML, Gotfredsen CH (2024) Designing optimal experiments in metabolomics. Metabolomics 20(4):69
102. Chen Y, Cao X, Xie L, Tang J, Liu L, Wang D et al (2025) Comparative transcriptomics and metabolomics provide insight into degeneration-related physiological mechanisms of *Morchella importuna* after long-term preservation. Microb Biotechnol 18(1):e70045
103. Bolhuis H, Grego M (2024) Cryopreservation and recovery of a complex hypersaline microbial mat community. Cryobiology 114:104859
104. Patejko M, Jacyna J, Markuszewski MJ (2017) Sample preparation procedures utilized in microbial metabolomics: an overview. J Chromatogr B 1043:150–157
105. Senaratne WMTN, Jayaweera JAAS (2024) Comparison of microbial preservation methods: a narrative review. Germs 14(4):375
106. Adhatrao M (2024) Contemporary trends in analytical spectroscopy and integrating QBD principles for enhanced method development. J Pharma Insights Res 2(5):155–172
107. Oyedeji AB, Green E, Adebiyi JA, Ogundele OM, Gbashi S, Adefisoye MA et al (2021) Metabolomic approaches for the determination of metabolites from pathogenic microorganisms: a review. Food Res Int 140:110042
108. AgorsorPI , EzeMO (2025) Early detection of infectious diseases: a review of recent advances in pathogen identification, molecular tools, and metabolomics-driven biomarker discovery. J Proteome Res
109. Eleutherio E, Panek A, De Mesquita JF, Trevisol E, Magalhães R (2015) Revisiting yeast trehalose metabolism. Curr Genet 61(3):263–274
110. Rizvi A, Shankar A, Chatterjee A, More TH, Bose T, Dutta A et al (2019) Rewiring of metabolic network in *Mycobacterium tuberculosis* during adaptation to different stresses. Front Microbiol 10:2417
111. Yuan Z, Ge Z, Fu Q, Wang F, Wang Q, Shi X, Wang B (2024) Investigation of cold-resistance mechanisms in cryophylactic yeast *Metschnikowia pulcherrima* based on comparative transcriptome analysis. Front Microbiol 15:1476087
112. Vitale GA, Geibel C, Minda V, Wang M, Aron AT, Petras D (2024) Connecting metabolome and phenotype: recent advances in functional metabolomics tools for the identification of bioactive natural products. Nat Prod Rep 41(6):885–904
113. Fiorini F, Bajerski F, Jeske O, Lepleux C, Overmann J, Broenstrup M (2022) A metabolomics-based toolbox to assess and compare the metabolic potential of unexplored, difficult-to-grow bacteria. Mar Drugs 20(11):713
114. Le Loarer A, Dufossé L, Bignon J, Frédérich M, Ledoux A, Fouillaud M, Gauvin-Bialecki A (2023) OSMAC method to assess impact of culture parameters on metabolomic diversity and biological activity of marine-derived Actinobacteria. Mar Drugs 22(1):23
115. Lu Y, Pang Z, Xia J (2023) Comprehensive investigation of pathway enrichment methods for functional interpretation of LC–MS global metabolomics data. Brief Bioinform 24(1):bbac553
116. Nguyen QH, Nguyen H, Oh EC, Nguyen T (2024) Current approaches and outstanding challenges of functional annotation of metabolites: a comprehensive review. Brief Bioinform 25(6):bbae498
117. van der Hooft JJJ, Mohimani H, Bauermeister A, Dorrestein PC, Duncan KR, Medema MH (2020) Linking genomics and metabolomics to chart specialized metabolic diversity. Chem Soc Rev 49(11):3297–3314

118. Han JR, Li S, Li WJ, Dong L (2024) Mining microbial and metabolic dark matter in extreme environments: a roadmap for harnessing the power of multi-omics data. Adv Biotechnol 2(3):26
119. Zammit G, Zammit MG, Buttigieg KG (2023) Emerging technologies for the discovery of novel diversity in cyanobacteria and algae and the elucidation of their valuable metabolites. Diversity 15(11):1142
120. Wang Z, Yu J, Wang C, Hua Y, Wang H, Chen J (2025) The deep mining era: genomic, metabolomic, and integrative approaches to microbial natural products from 2018 to 2024. Mar Drugs 23(7):261
121. Tiwari P, Dufossé L (2023) Focus and insights into the synthetic biology-mediated chassis of economically important fungi for the production of high-value metabolites. Microorganisms 11(5):1141
122. Louca S, Polz MF, Mazel F, Albright MB, Huber JA, O'Connor MI et al (2018) Function and functional redundancy in microbial systems. Nat Ecol Evol 2(6):936–943
123. Yang Y (2021) Emerging patterns of microbial functional traits. Trends Microbiol 29(10):874–882
124. Wood JL, Franks AE (2018) Understanding microbiomes through trait-based ecology. Microbiol Aust 39(1):53–56
125. Bittleston LS, Freedman ZB, Bernardin JR, Grothjan JJ, Young EB, Record S et al (2021) Exploring microbiome functional dynamics through space and time with trait-based theory. mSystems 6(4):10–1128
126. Ivashko M, Burmei S, Yusko L, Chaikovska T, Boyko N (2023) Microbiological diagnostics: from traditional to molecular genetic methods: a literature review. Bull Med Biol Res 5(4):34–41
127. Prince EK, Pohnert G (2010) Searching for signals in the noise: metabolomics in chemical ecology. Anal Bioanal Chem 396(1):193–197
128. Aretz I, Meierhofer D (2016) Advantages and pitfalls of mass spectrometry based metabolome profiling in systems biology. Int J Mol Sci 17(5):632
129. Lin Y, Li W, Sun L, Lin Z, Jiang Y, Ling Y, Lin X (2019) Comparative metabolomics shows the metabolic profiles fluctuate in multi-drug resistant *Escherichia coli* strains. J Proteomics 207:103468
130. Duan W, Guan Q, Zhang HL, Wang FZ, Lu R, Li DM et al (2023) Improving flavor, bioactivity, and changing metabolic profiles of goji juice by selected lactic acid bacteria fermentation. Food Chem 408:135155
131. Shen H, Rösch P, Thieme L, Pletz MW, Popp J (2023) Comparison of bacteria in different metabolic states by micro-Raman spectroscopy. J Mol Struct 1277:134831
132. Kirwan JA, Gika H, Beger RD, Bearden D, Dunn WB, Goodacre R et al (2022) Quality assurance and quality control reporting in untargeted metabolic phenotyping: mQACC recommendations for analytical quality management. Metabolomics 18(9):70
133. Kirwan JA, Bruning U, Mosley JD (2025) Quality assurance in metabolomics and metabolic profiling. Metabolic profiling: methods and protocols. Springer US, New York, 15–51
134. Maroli AS, Powers R (2023) Closing the gap between *in vivo* and *in vitro* omics: using QA/QC to strengthen *ex vivo* NMR metabolomics. NMR Biomed 36(4):e4594
135. Bhatla SC, Lal MA (2023) Secondary metabolites. Plant physiology, development and metabolism. Springer Nature Singapore, Singapore, 765–808
136. DiBello M, Healy AR, Nikolayevskiy H, Xu Z, Herzon SB (2023) Structure elucidation of secondary metabolites: current frontiers and lingering pitfalls. Acc Chem Res 56(12):1656–1668
137. Assouguem A, Annemer S, Mohammed K, Lazraq A (2025) Innovative approaches in the extraction, identification, and application of secondary metabolites from plants. Phyton 94(6):1631
138. Dinglasan JLN, Otani H, Doering DT, Udwary D, Mouncey NJ (2025) Microbial secondary metabolites: advancements to accelerate discovery towards application. Nat Rev Microbiol 23(6):338–354
139. McNeal KS, Herbert BE (2009) Volatile organic metabolites as indicators of soil microbial activity and community composition shifts. Soil Sci Soc Am J 73(2):579–589
140. Vrzalová A, Pečinková P, Illés P, Gurská S, Džubák P, Szotkowski M et al (2022) Mixture effects of tryptophan intestinal microbial metabolites on aryl hydrocarbon receptor activity. Int J Mol Sci 23(18):10825
141. Ribbenstedt A, Ziarrusta H, Benskin JP (2018) Development, characterization and comparisons of targeted and non-targeted metabolomics methods. PLoS One 13(11):e0207082
142. Chaleckis R, Meister I, Zhang P, Wheelock CE (2019) Challenges, progress and promises of metabolite annotation for LC–MS-based metabolomics. Curr Opin Biotechnol 55:44–50
143. de la Fuente AG, Traldi F, Siroka J, Kretowski A, Ciborowski M, Otero A et al (2018) Characterization and annotation of oxidized glycerophosphocholines for non-targeted metabolomics with LC–QTOF–MS data. Anal Chim Acta 1037:358–368

144. Kalwani M, Chakdar H, Srivastava A, Pabbi S, Shukla P (2022) Effects of nanofertilizers on soil and plant-associated microbial communities: emerging trends and perspectives. Chemosphere 287:132107
145. Tan C, Kalhoro MT, Faqir Y, Ma J, Osei MD, Khaliq G (2022) Climate-resilient microbial biotechnology: a perspective on sustainable agriculture. Sustainability 14(9):5574
146. Ferrocino I, Rantsiou K, McClure R, Kostic T, De Souza RSC, Lange L, Microbiome Support Consortium et al (2023) The need for an integrated multi-OMICs approach in microbiome science in the food system. Compr Rev Food Sci Food Saf 22(2):1082–1103
147. Satya S, Sharma S, Choudhary G, Kaushik G (2024) Advances in environmental microbiology: a multi-omic perspective. In: Microbial omics in environment and health. Springer Nature Singapore, Singapore, 175–204
148. Dagher G (2022) Quality matters: international standards for biobanking. Cell Prolif 55(8):e13282
149. Messerschmidt K (2023) The transition from sample storage to a biobank—a risk assessment and gap analysis in preparation for the ISO 20387 certification
150. Ingsriswang S, Eurwilaichitr L, Chunhametha S, Harnpicharnchai P, Likhitrattanapisal S, Kitikhun S (2022) Microbial preservation for biotechnological development. In: Importance of microbiology teaching and microbial resource management for sustainable futures. Academic Press, 193–227
151. Sionek B, Szydłowska A, Kołożyn-Krajewska D (2024) The role of microorganisms and their antibacterial compounds in food biopreservation. Appl Sci 14(13):5557
152. Chaturvedi N, Yadav MK, Sharma M (2024) Applications of artificial intelligence and machine learning in microbial diagnostics and identification. Methods in microbiology, vol 55. Academic Press, 213–230
153. Qureshi MDA, Ramzan MF, Amjad F, Haider N (2024) Artificial intelligence in metabolomics for disease profiling: a machine learning approach to biomarker discovery. Indus J Biosci Res 2(02):87–96
154. Hasan ME, Osman MA, Hemeida AA, ElHefnawi M (2025) Bioinformatics and artificial intelligence approaches in metabolic pathways. In: Metabolic dynamics in host-microbe interaction. Springer Nature Singapore, Singapore, 17–32
155. Surekha S, Kumar D, Reddy AD, Gupta S (2024) Functional profiling of microbial communities through AI approaches. Genomic Intelligence. CRC Press, 104–118
156. Srivastava P, Sachan K, Baskar P, Saikanth DRK, Lytand W, Kumar RKM, Singh BV (2023) Soil microbes expertly balancing nutrient demands and environmental preservation and ensuring the delicate stability of our ecosystems—a review. Int J Plant Soil Sci 35(18):989–1000
157. Dawson HM, Connors E, Erazo NG, Sacks JS, Mierzejewski V, Rundell SM et al (2023) Microbial metabolomic responses to changes in temperature and salinity along the western Antarctic Peninsula. ISME J 17(11):2035–2046
158. Olanrewaju OS, Glick BR, Babalola OO (2024) Metabolomics-guided utilization of beneficial microbes for climate-resilient crops. Curr Opin Chem Biol 79:102427
159. Buenrostro-Muñoz J, Jarmusch SA, Souza V, Martínez-Cárdenas A, Fajardo-Hernández CA, Yeverino IR et al (2024) Metabolomic diversity in microbial mats under different environmental conditions: a tool to test microbial ecosystem chemical change. Chem Biodivers 21(6):e202300829

Chapter 21

A Simple and Efficient Method for DNA Extraction from Leachate Samples and Its Validation by High-Throughput Microbial Sequencing

Ana M. Escalante-Mañé, Luis A. Sáenz-Carbonell, Rubén Ríos-Medina, Geny I. López-Arellano, Rafael Rivera-Bustamante, Julio Vega-Arreguin, Iván Oros-Ortega, Julio Aguilar-Caamal and Carlos A. Puch-Hau

Abstract

In this study, we report the successful extraction of DNA from landfill leachates using the cetyltrimethylammonium bromide (CTAB) technique and its validation for high-throughput sequencing of bacteria and fungi. This protocol provides a simple and efficient method to obtain high-quality DNA from complex samples such as leachates and represents a valuable tool for molecular studies and diversity profiling of microorganisms in sanitary landfills.

Keywords DNA extraction, High-throughput sequencing, Microorganisms, Next-generation sequencing, Wastewater

1 Introduction

Landfill leachates are a highly toxic mixture of pollutants consisting of dissolved organic matter, heavy metals, inorganic macro-components, xenobiotic organic compounds [1], and microplastics [2], among others. Therefore, the development of efficient and sustainable treatments to reduce the contaminant load of landfill leachates is an environmental priority. Biological treatments based on the use of autochthonous microorganisms (adapted to these extreme environments) could represent a promising alternative. However, because it is estimated that less than 2% of bacteria can be cultured in the laboratory [3], the use of new technologies for microorganisms' diversity profiling through high-throughput DNA sequencing is required. A fundamental step in this endeavor is the implementation of a protocol that allows the successful extraction of high-quality DNA from complex samples such as leachates.

Marco A. Ramírez-Mosqueda et al. (eds.), *Methods for The Conservation of Microorganisms*, Methods in Molecular Biology, vol. 3090, https://doi.org/10.1007/978-1-0716-5623-5_21,

There are some reports on the massive sequencing of microorganisms in leachates; however, the methodologies used for DNA extraction have been based on commercial kits, such as the Power Soil® DNA kit (Mobio Laboratories, Inc., Carlsbad, CA, USA) [4], Ultraclean Soil DNA Kit (Mobio Laboratories Inc) [5], FastDNA™ Spin Kit for Soil (MP Biomedicals, Santa Ana, CA, USA) [6,7], and DNeasy® PowerSoil® Kit (Qiagen, Valencia, CA, USA) [8]. In many countries, these commercial kits are expensive and may require time-consuming customs processing and clearance. Thus, having a DNA isolation method for leachate samples that is low-cost and accessible would be a useful tool for obtaining good-quality DNA for molecular studies. The cetyltrimethylammonium bromide (CTAB) protocol is currently the most used procedure for DNA extraction [9–11], because it is less expensive than commercial kits and it breaks down the cells walls, allowing for effective deproteination [12]. This protocol has been successfully used to extract DNA from diverse samples such as marine sediments [12], different plant species [13], animal manure [14], insects [15], among others.

In this study, we aimed to evaluate the effectiveness of the CTAB protocol for extracting high-quality DNA from both fresh and mature landfill leachates and to validate its applicability for high-throughput sequencing of bacterial and fungal communities. Our findings highlight the relevance of implementing robust, low-cost DNA extraction methods as indispensable tools to advance microbial diversity profiling and support the development of biotechnological strategies for landfill leachate treatment.

2 Materials

2.1 Collection of Samples

Samples are collected from the leachate storage basins of a sanitary landfill in Merida, Yucatan, Mexico (20°58'04"N, 89°42'22"W) (Fig. 1). In total, 10 landfill leachate samples (1 L per sample) are collected, corresponding to mature leachate (>10 years old, $n = 6$) and fresh leachate (leachate from the beginning of the liquid-waste stream emanating from the waste source before any storage or treatment process, $n = 4$). After collection, the samples are transported to the laboratory facilities in high-density polyethylene containers and stored at 4°C (*see* **Note 1**).

2.2 DNA Extraction

The reagents and chemicals used for DNA extraction are the following:

1. Extraction buffer: 2% CTAB (cetyl trimethyl ammonium bromide), 1% PVP (polyvinylpyrrolidone), 1.4 M NaCl, 100 mM Tris–HCl (tris-hydrochloride), 20 mM EDTA (disodium salt), 1% SDS (sodium dodecyl sulfate), and 0.1 M thioglycerol.
2. Phenol:chloroform:isoamyl alcohol (25:24:1 v/v/v).

Fig. 1 Leachate storage basins of the sanitary landfill from Merida, Yucatan, Mexico

3. 3 M sodium acetate (pH 5.2).
4. Isopropanol.
5. Absolute and 70% ethanol.
6. RNase (Ribonucleasa A Sigma-Aldrich, USA).
7. Ultrapure water.

The whole process of DNA extraction is carried out at room temperature. The material used for the process of DNA extraction is previously sterilized.

3 Methods

3.1 Preparation of Solutions

Composition of 2% CTAB Extraction Buffer for 100 mL (Double-Distilled Water)

1. 2% m/v CTAB (cetyl trimethyl ammonium bromide)—2 g.
2. 1.4 M NaCl)— 8.17 g.
3. 100 mM Tris–HCl —1.575 g.
4. 20 mM EDTA (ethylenediaminetetraacetic acid)—0.75 g.
5. 1% m/v PVP-40 (polyvinylpyrrolidone)—1.0 g. pH = 8 (adjusted by 1 M NaOH).
6. 0.1 M Thioglycerol — 0.86 mL.

Phenol:Chloroform:Isoamyl Alcohol (25:24:1 v/v/v) (50 mL)

1. 25 ml Phenol.
2. 24 ml chloroform.
3. 1 ml isoamyl alcohol.

3 M Sodium Acetate (pH 5.2)

1. Add 800 mL of double-distilled water in a glass beaker.
2. Weight 246.1 g of sodium acetate and add to the solution.
3. Adjust the pH to 5.2 with glacial acetic acid. Allow the solution to cool overnight.
4. Adjust the pH once more to 5.2 with glacial acetic acid.
5. Add double-distilled water to 1 L.

3.2 DNA Isolation

DNA extraction is performed using a CTAB extraction buffer. The detailed protocol for DNA extraction from leachate samples included the following steps:

1. A 120 mL aliquot of landfill leachate is centrifuged (in three portions of 40 mL each) at 4000 x g for 15 min at 4°C. The pellets from each portion are combined, yielding a pellet that is approximately 0.2 g (*see* **Note 2**).
2. The pellet is resuspended with 500 μL of hot (65°C) CTAB buffer (*see* **Note 3**).
3. The samples are incubated at 65°C for 30 min and subsequently cooled to room temperature.
4. The samples are emulsified (in 2-mL tubes) with an equal volume of phenol:chloroform:isoamyl alcohol (25:24:1 v/v/v) and centrifuged at 14,000 x g for 5 min at room temperature.
5. The aqueous phase from each sample is recovered and transferred to a new 1.5-mL tube.
6. Immediately, the samples are mixed with 0.1 volumes of 3 M sodium acetate (pH 5.2) and 0.6 volumes of cold isopropanol. Subsequently, the samples are incubated at −20°C for 1 h to precipitate the DNA.
7. The samples are centrifuged at 14,000 rpm for 10 min at 4°C to pellet the DNA.
8. The supernatant is discarded, and the DNA pellet is washed with 500 μL of cold 70% ethanol. The samples are centrifuged at 14,000 x g for 10 min at room temperature.
9. Each DNA pellet is resuspended in 30 μL of ultrapure deionized sterile water.
10. Next, 2 μL of RNase A is added to each sample and is incubated for 1 h at 37°C.
11. The concentration and integrity of total DNA are is determined using a NanoDrop™ 1000 (Thermo Scientific, Waltham, MA, USA) and 2% agarose gel electrophoresis.

As mentioned by Foysal et al. [16], the quantity and quality of DNA isolated from environmental samples are crucial for obtaining robust high-throughput sequencing data commonly used for microbial community analysis.

The results showed that high-quality DNA is successfully extracted from both fresh and mature landfill leachate samples using the CTAB buffer and protocol described here (Fig. 2). DNA concentration ranged from 248 to 1733 ng/μL in mature leachate and from 234 to 1097 ng/μL in fresh leachate. Additionally, the absorbance ratios (A260/280 and A260/230) indicated that the extracted DNA was of sufficient purity for sequencing and molecular studies (see Table 1).

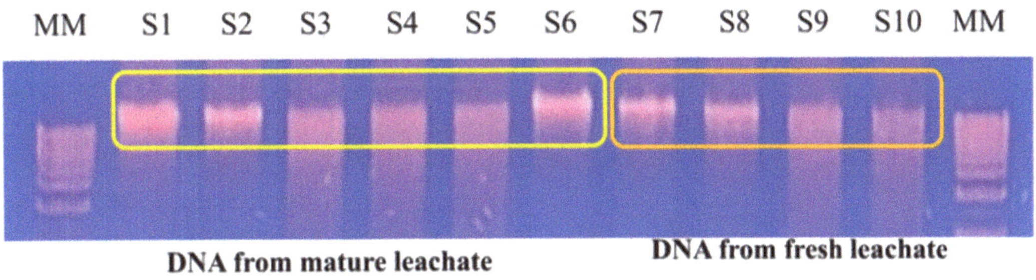

Fig. 2 DNA extracted from landfill leachate samples using the cetyltrimethylammonium bromide (CTAB) protocol. DNA was visualized on a 2% agarose gel. S1–S6: mature leachate samples; S7–S10: fresh leachate samples. MM: molecular mass marker (1 kb ladder)

Table 1
DNA concentrations in landfill leachate samples

Sample type	Sample number	Concentration (ng/μL)	Absorbance ratio (A260/280)	Absorbance ratio (A260/230)
ML	S1	1733.10	2.00	2.06
	S2	248.50	1.62	1.34
	S3	1042.20	1.90	1.83
	S4	990.55	1.70	1.90
	S5	645.10	1.85	1.72
	S6	1387.65	1.90	1.85
FL	S7	881.25	1.80	1.60
	S8	665.50	1.70	1.92
	S9	1097.70	1.90	1.84
	S10	234.10	1.71	1.44

ML mature leachate, *FL* fresh leachate

3.3 Massive Sequencing

To assess whether DNA quality is suitable for bacterial and fungal analyses, samples are subjected to high-throughput sequencing by Zymo Research (Irvine, CA, USA; https://zymoresearch.eu/pages/16s-its-amplicon-sequencing).

Prior to shipment for sequencing, samples are homogenized to a concentration of 50 ng/μL in a final volume of 30 μL. Subsequently, they are sealed with a plastic film and placed in a polystyrene cooler for shipment to Zymo Research company. The general sequencing protocol performed by Zymo Research is described below:

1. For the construction of the bacterial 16S rRNA sequencing library, the Quick-16S™ NGS Library Prep Kit (Zymo Research) is used.
2. The primers used amplified the V3–V4 region of the 16S rRNA gene (Zymo Research).
3. For fungal samples, the Quick-16S™ NGS Library Prep Kit is adapted using custom ITS primers (Zymo Research).
4. For both bacterial and fungal samples, the library is prepared using real-time polymerase chain reaction (PCR) to control the number of cycles and minimize PCR chimera formation.
5. The final PCR products are quantified using fluorescence measurements and pooled at equimolar concentrations.
6. The final pooled library is purified using the Select-a-Size DNA Clean & Concentrator™ (Zymo Research) and subsequently is quantified with a TapeStation® (Agilent Technologies, Santa Clara, CA) and a Qubit® fluorometer (Thermo Fisher Scientific, Waltham, MA, USA).
7. Sequencing is performed on an Illumina® NextSeq™ instrument using a P1 reagent kit (600 cycles; Illumina, San Diego, CA, USA).
8. Amplicon sequence variants (ASVs) are inferred from raw reads using the DADA2 pipeline [17].

As a result of the bacterial sequencing process, a total of 5,650,276 raw reads are obtained from the ten leachate samples, of which 1,892,613 passed quality filtering, yielding an overall retention rate of 33.5% (see Table 2). The number of unique reads (amplicon sequence variants, ASVs) is consistently higher in FL (mean = 1141) than in ML (mean = 662). For fungal samples, similar results are observed, although the difference in ASVs between ML and FL is less pronounced than for bacteria (ML mean = 103 and FL mean = 116) (see Table 2).

Table 2
Raw and processed sequencing reads obtained from landfill leachate samples

	Leachate	Samples	Raw sequences reads	Processed sequence reads	Proportion (processed/ raw)	ASV for sample	Mean ASV
Bacteria	ML	S1	596,832	202,368	0.34	690	662
		S2	540,602	177,177	0.33	661	
		S3	594,784	206,285	0.35	624	
		S4	639,536	225,381	0.35	652	
		S5	605,926	214,164	0.35	703	
	FL	S6	542,798	186,586	0.34	641	1141
		S7	524,118	155,900	0.30	1152	
		S8	580,516	177,520	0.31	1222	
		S9	523,462	169,712	0.32	1106	
		S10	501,702	142,486	0.28	1082	
Fungi	ML	S1	502,604	220,857	0.44	108	103
		S2	606,662	267,533	0.44	152	
		S3	521,262	226,801	0.44	107	
		S4	670,652	292,683	0.44	83	
		S5	539,094	236,402	0.44	84	
	FL	S6	581,982	257,916	0.44	86	116
		S7	547,904	236,991	0.43	122	
		S8	526,320	224,022	0.43	111	
		S9	436,092	188,240	0.43	100	
		S10	606,230	254,271	0.42	134	

ML mature leachate, FL fresh leachate, *ASV* amplicon sequence variants

4 Notes

1. Because municipal solid waste leachates are toxic and may contain pathogenic microorganisms, DNA/RNA Shield (Cat # R1100-50-E, Zymo Research, USA) is added at a 1:1 ratio to inactivate pathogens and protect DNA for degradation. After the reagent is added, the samples were mixed by inversion approximately ten times prior to further laboratory processing.
2. The 120 mL sample is centrifuged in three 40 mL aliquots in a single 50 mL Falcon tube. After each centrifugation at

4000 rpm for 10 min at 4°C, the supernatant is carefully removed to avoid disturbing the pellet.

3. The pellet is diluted with 500 µL CTAB buffer preheated to 65°C in water bath CTAB buffer using a 1000 µL micropipette. The tip is trimmed to facilitate sample aspiration and enhance efficient dissolution of the pellet.

Acknowledgments

This study was supported by the TecNM with project number 18310.23-PD awarded to Carlos A. Puch-Hau: "Massive DNA sequencing of microorganisms in leachates of municipal solid waste in Yucatan: in search of biological alternatives for detoxification". We thank Ivan Cordova Lara, Avel González Sánchez, Alejandra Cañizares Martínez, Mercedes Quintanilla Mena and Marcela Del Río García for their technical assistance.

Conflict of Interest Statement The authors have no conflicts of interest to declare that are relevant to the content of this chapter.

References

1. Luo H, Zeng Y, Cheng Y, He D, Pan X (2020) Recent advances in municipal landfill leachate: a review focusing on its characteristics, treatment, and toxicity assessment. Sci Total Environ 703:135468. https://doi.org/10.1016/j.scitotenv.2019.135468
2. Wang L, Wang H, Huang Q, Yang C, Wang L, Lou Z, Zhou Q, Wang T, Ning C (2023) Microplastics in landfill leachate: a comprehensive review on characteristics, detection, and their fates during advanced oxidation processes. Water 15:2. https://doi.org/10.3390/w15020252
3. Wade W (2002) Unculturable bacteria—the uncharacterized organisms that cause oral infections. J R Soc Med 95:2. https://doi.org/10.1177/014107680209500207
4. Rajasekar A, Sekar R, Medina-Roldan E, Bridge J, Moy CK, Wilkinson S (2018) Next-generation sequencing showing potential leachate influence on bacterial communities around a landfill in China. Can J Microbiol 64:537–549. https://doi.org/10.1139/cjm-2017-0543
5. Yang S, Song L (2019) Succession of bacterial community structure and metabolic function during solid waste decomposition. Bioresour Technol 291:121865. https://doi.org/10.1016/j.biortech.2019.121865
6. Xu S, Lu W, Liu Y, Ming Z, Liu Y, Meng R, Wang H (2017) Structure and diversity of bacterial communities in two large sanitary landfills in China as revealed by high-throughput sequencing (MiSeq). Waste Manage 63:41–48. https://doi.org/10.1016/j.wasman.2016.07.047
7. Zhao R, Liu J, Feng J, Li X, Li B (2021) Microbial community composition and metabolic functions in landfill leachate from different landfills of China. Sci Total Environ 767:144861. https://doi.org/10.1016/j.scitotenv.2020.144861
8. Passarini MRZ, Moreira JVF, Gomez JAM, Bonugli-Santos RC (2021) DNA metabarcoding of the leachate microbiota from sanitary landfill: potential for bioremediation process. Arch Microbiol 203:4847–4858. https://doi.org/10.1007/s00203-021-02471-8
9. Doyle JJ, Doyle JL (1987) A rapid DNA isolation procedure for small quantities of fresh leaf tissue. Phytochem Bull 19:11–15
10. Doyle JJ, Doyle JL (2019) CATB isolation: The true story. Plant Sci Bull 65:15–18
11. Schenk JJ, Becklund LE, Carey SJ, Fabre PP (2023) What is the "modified" CTAB protocol? Characterizing modifications to the CTAB DNA extraction protocol. Appl Plant Sci 11(3):e11517. https://doi.org/10.1002/aps3.11517

12. Kachiprath B, Jayanath G, Solomon S, Sarasan M (2018) CTAB influenced differential elution of metagenomic DNA from saltpan and marine sediments. 3 Biotech 8(44):1–15. https://doi.org/10.1007/s13205-017-1078-x
13. Aboul-Maaty NAF, Oraby HAS (2019) Extraction of high-quality genomic DNA from different plant orders applying a modified CTAB-based method. Bull Natl Res Cent 43(1):1–10. https://doi.org/10.1186/s42269-019-0066-1
14. Kouakou JL, Gonedelé-Bi S, Assamoi JB, Assanvo N'Guetta SPA (2022) Optimization of the Cetyltrimethylammonium bromide (CTAB) DNA extraction protocol using forest elephant dung samples. MethodsX 9:101867. https://doi.org/10.1016/j.mex.2022.101867
15. Suganthi M, Abirami G, Jayanthi M, Kumar KA, Karuppanan K, Palanisamy S (2023) A method for DNA extraction and molecular identification of Aphids. MethodsX 10:102100. https://doi.org/10.1016/j.mex.2023.102100
16. Foysal MJ, Salgar-Chaparro SJ (2024) Improving the efficiency of DNA extraction from iron incrustations and oilfield-produced water. Sci Rep 14(1):2954. https://doi.org/10.1038/s41598-024-53134-9
17. Callahan BJ, McMurdie PJ, Rosen MJ, Han AW, Johnson AJ, Holmes SP (2016) DADA2: high resolution sample inference from Illumina amplicon data. Nat Methods 13(7):581–583. https://doi.org/10.1038/nmeth.3869

Chapter 22

Optimized DNA Extraction from Coffee Roots Colonized by Arbuscular Mycorrhizal Fungi Using a Commercial Kit

Lorena Jacqueline Gómez-Godínez, Alejandra Ortiz-Gasca, José Luis Aguirre-Noyola, Magali Ruiz-Rivas, Sergio De los Santos-Villalobos and Dora Trejo-Aguilar

Abstract

The molecular identification of arbuscular mycorrhizal fungi (AMF) from root tissues can often be limited by the co-extraction of polysaccharides, phenolic compounds, and other PCR inhibitors, particularly in woody or highly lignified roots, where cell lysis is also often difficult. Although commercial extraction kits offer standardized workflows and reduce the handling of harmful reagents, the manufacturer's protocols are insufficient for root samples with high levels of secondary metabolites. This protocol describes the optimization of DNA extraction from coffee (*Coffea arabica* L.) roots colonized by AMF, based on the use of the commercial ZymoBIOMICS™ Quick-DNA fecal/soil Microbe Kit, designed for the rapid and simple isolation of microbial DNA and host cells from fecal and soil samples, incorporating a pre-maceration and pre-digestion step with proteinase K and β-mercaptoethanol. This modification significantly improves yield, DNA integrity, and PCR amplification of Glomeromycota-specific ribosomal markers (SSU/18S).

Keywords Glomeromycota, *Coffea arabica*, Optimization of commercial kits, Root-associated fungi, SSU ribosomal DNA

1 Introduction

Arbuscular mycorrhizal fungi (AMF) are obligate biotrophic fungi capable of establishing symbiosis with approximately 80% of vascular plants [1]. They play a fundamental role in nutrient and water acquisition for the plant, stress tolerance, and improvements in soil structure in agroecosystems [2,3]. Their colonization is intracortical and is characterized by the formation of structures such as vesicles, hyphae, and arbuscules [4].

Molecular characterization is essential for understanding diversity, community dynamics, and potential responses to management practices [5]. In community analyses based on PCR or sequencing,

Marco A. Ramírez-Mosqueda et al. (eds.), *Methods for The Conservation of Microorganisms*, Methods in Molecular Biology, vol. 3090, https://doi.org/10.1007/978-1-0716-5623-5_22,

the ratio of fungal DNA to plant DNA does not reflect the total extraction yield, as it depends on colonization, extraction efficiency, and the presence of inhibitors [6]. Extracting high-quality DNA from colonized roots remains a significant technical challenge, as fungal DNA is typically present at low levels relative to plant DNA and is often co-extracted with compounds that inhibit subsequent enzymatic reactions, particularly PCR [7,8].

When extracting DNA from root tissues, the spatial distribution of Fungal Amino Acids (FAAs) is heterogeneous, which can introduce variability between subsamples even under similar colonization levels [9]. These biological characteristics increase the sensitivity of molecular analysis to the quality of the extracted DNA, underscoring the need for high-quality, reproducible extracts.

Perennial crops, particularly coffee, contain high concentrations of lignin, polyphenols, polysaccharides, and metal ions that affect both DNA yield and purity [10,11]. The presence of polysaccharides is associated with viscous extracts and has been documented to inhibit Taq polymerase and restriction enzyme activity, reducing amplification efficiency and reliability [12]. In turn, oxidized polyphenols can covalently bind to DNA, generating extracts with brownish hues that decrease the stability or functionality of the DNA for molecular analysis [13].

Extracts of suboptimal quality can lead to biases in species detection, underestimation of mycorrhizal diversity, and low reproducibility between replicates, especially when ribosomal markers are used for community analyses [14]. The residual presence of these inhibitors can differentially affect amplification efficiency among taxa, favoring the detection of those with a higher number of ribosomal copies or greater affinity for the primers used, thus distorting the relative representation of communities [15]. In the case of Glomeromycota, where intra-organism variation in ribosomal regions has been described, this methodological sensitivity can amplify intragenomic variants unevenly, increasing the apparent richness or modifying the observed community structure [16]. Therefore, obtaining high-purity DNA is not only a technical requirement but also a critical factor for the correct ecological and functional interpretation of AMF communities in comparative studies [17–19].

The performance of commercial protocols or those strictly adhering to the manufacturer's instructions can be inconsistent between replicates and treatments when the rootlet contains high levels of inhibitory compounds or is partially desiccated [7,20], prompting adjustments prior to lysis/purification to improve fungal DNA recovery.

Based on these considerations, this protocol describes the optimized extraction of DNA from *Coffea arabica* roots colonized by AMF using a commercial kit for complex samples, incorporating a predigestion step to improve root tissue lysis and reduce inhibitory load. The DNA obtained was validated using spectrophotometric metrics.

2 Materials

2.1 Biological Material

1. Fine roots of *Coffea arabica* L. var. Garnica, collected under field conditions, 250 mg of root tissue per extraction.

2.2 Reagents

1. ZymoBIOMICS™ DNA Miniprep Kit (Zymo Research)
2. TE 50:20 Buffer (10 mM Tris–HCl, 1 mM EDTA, pH 8.0)
3. Proteinase K (20 mg/mL)
4. β-Mercaptoethanol
5. Nuclease-free water

2.3 Equipment

1. Mortar and pestle
2. Liquid nitrogen
3. Thermoblock
4. Microcentrifuge
5. 1.5 mL microcentrifuge tubes
6. Vortex mixer
7. Micropipettes (2 to 1000 μL)
8. Agarose gel electrophoresis system
9. NanoDrop™ spectrophotometer or equivalent

3 Methods

3.1 Sample Preservation

1. During collection, carefully remove soil and other materials adhering to the roots with a dry brush as quickly as possible, avoiding washing with water (*See* **Note 1**).
2. Place in sterile Falcon tubes with silica gel.
3. Store samples at 4°C until processing.

3.2 Predigestion Treatment

1. Macerate 250 mg of root tissue in a sterile mortar, gradually adding liquid nitrogen until a fine powder is obtained (*See* **Note 2**).
2. Hydrate the powdered material with 2.3 mL of TE solution (50:20) (*See* **Note 3**).
3. Continue maceration until a paste is obtained.
4. Transfer the powdered material to a sterile 1.5 mL microcentrifuge tube using a micropipette, cutting 2 to 3 mm from the sterile tip to facilitate suction.
5. Add 6 μL of proteinase K (final concentration 20 mg/mL) and 30 μL of β-mercaptoethanol (*See* **Note 4**).

6. Vortex for 10 s.
7. Incubate at 56°C for 60 minutes with gentle shaking (300 rpm) in a thermoblock to promote protein digestion.
8. Subsequently, incubate at 65°C for 30 minutes to enhance the reducing action of β-mercaptoethanol on phenolic compounds and resistant cellular structures (see Fig. 1).
9. Centrifuge at 10,000 × *g* for 1 min.
10. Recover the supernatant for processing with the kit.

3.3 DNA Extraction Using the ZymoBIOMICS™ Kit

1. Substitute the manufacturer's steps for the BashingBead™ buffer with the extracted sample (TE 50:20). This substitution was made due to the previously pulverized nature of

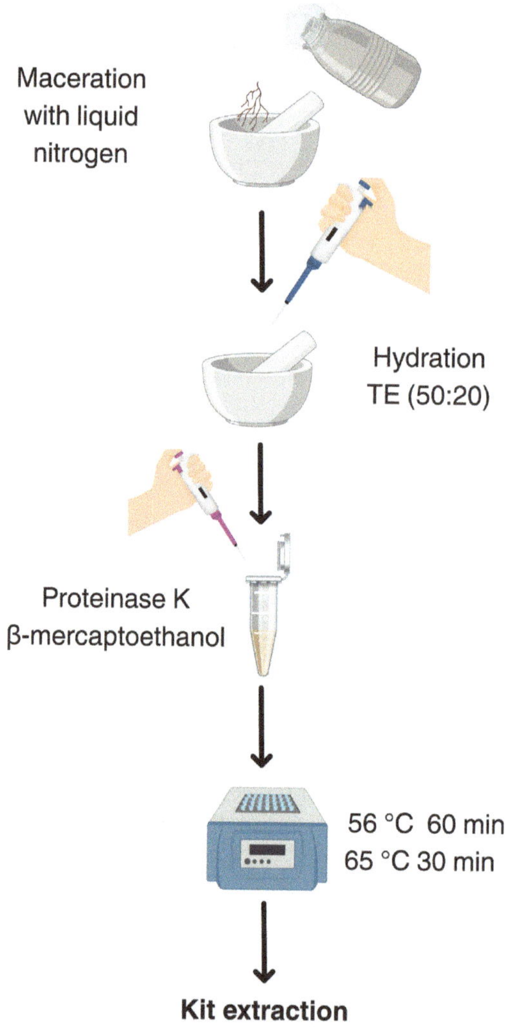

Fig. 1 Implemented predigestion treatment

the tissue in liquid nitrogen, ensuring equivalent mechanical disruption without the need for additional bead beating (*See* **Notes 5 y 6**).

2. Transfer up to 400 µL of the supernatant to a Zymo-Spin™ III-F filter with a collection tube.
3. Centrifuge at 8000 × *g* for 1 min.
4. To the filtrate, add 800 µL of Genomic Lysis Buffer and 400 µL of 95% ethanol to the collection tube from Step 12, mix well, and let stand for 10 min.
5. Transfer 800 µL of the mixture to a Zymo-Spin™ IICR column in its collection tube.
6. Centrifuge at ≥10,000 × *g* for 1 min.
7. Discard the flow and repeat Step 15 until the entire volume has been processed.
8. Add 200 µL of DNA Pre-Wash Buffer to the Zymo-Spin™ IICR column in a new collection tube.
9. Centrifuge at ≥10,000 × *g* for 1 min.
10. Add 500 µL of g-DNA Wash Buffer to the Zymo-Spin™ IICR column.
11. Centrifuge at ≥10,000 × *g* for 1 min.
12. Transfer the column to a clean 1.5 mL tube and add 100 µL (50 µL minimum) of DNA Elution Buffer, preheated to 65°C for 15 min, directly to the membrane. Let it stand for 5 min.
13. Centrifuge at ≥10,000 × *g* for 30 s to elute the DNA.
14. Place a Zymo-Spin™ III-HRC filter in a collection tube and add 600 µL of Prep Solution. Centrifuge at 8000 × *g* for 3 min.
15. Transfer the elution from Step 23 onto the prepared filter and centrifuge at exactly 16,000 × *g* for 3 min.
16. The DNA filtrate is suitable for PCR.
17. Quantify the quantity and quality using NanoDrop™, with Elution Buffer as a blank (*See* **Note 7**).

The results obtained after applying the optimized protocol showed that incorporating a predigestion step prior to lysis and purification substantially improves the quality of the extracted DNA. A260/A280 absorbance values greater than 1.8 indicate efficient protein removal. At the same time, optimal A260/A230 ratios (> 1.8) indicate a significant reduction in polysaccharides, phenolic compounds, and residual salts, which are frequently responsible for partial or complete PCR inhibition.

In roots colonized by AMF, extraction often shows a trade-off between yield, purity, and PCR performance. This balance depends on both the method and the sample condition. In direct comparisons of mycorrhizal roots, CTAB-based protocols recovered greater amounts of DNA than simpler approaches such as TE buffers [21] or commercial kits; however, higher recovery does not necessarily translate into better performance, since co-extraction of inhibitors can reduce amplification reliability [22].

The addition of agents such as PVP/PVPP or activated carbon has been associated with improvements in the purity and performance of nested PCRs targeting the SSU region [23]. Additionally, factors such as tissue preservation and condition (including desiccation and storage method) can affect the total DNA recovered from AMF-colonized roots, potentially leading to replicate-to-replicate variation if these pre-analytical steps are not controlled [24]. Commercial kits have shown greater consistency in amplification [25], although they are sensitive to the chemical nature of the sample when the manufacturer's instructions are strictly followed [26].

It has been documented that, in molecular studies of AMF, the total DNA yield alone is not a sufficient indicator of quality for subsequent analyses, as the presence of co-extracted compounds can affect amplification efficiency and consistency, even when concentrations appear adequate [27]. In this context, DNA quality and the effective removal of inhibitors are crucial for reproducibility between replicates and for the stability of profiles obtained with ribosomal markers.

Similarly, variations in DNA quality and in the methodological steps associated with extraction and amplification have been observed, as these can introduce biases in the detection of mycorrhizal taxa, leading to differences in observed community composition and even underestimating the actual diversity in studies based on PCR and massive sequencing [28].

The specific amplification of the SSU region corroborated the functional suitability of the obtained extracts (18S rDNA) from Glomeromycota, confirming that the DNA quality was sufficient for molecular applications. This protocol offers an intermediate alternative between classical high-recovery methods and commercial high-reproducibility kits for obtaining high-quality mycorrhizal DNA from complex roots, thereby improving the reliability of molecular identification studies and the analysis of arbuscular mycorrhizal fungal communities in perennial agricultural systems (*See* **Note 8**).

4 Notes

1. Root samples should not be washed with water before drying when molecular community analysis is planned, as washing can remove loosely attached fungal structures. Dry brushing is recommended to minimize contamination and preserve fungal material .
2. The efficiency of this protocol depends largely on the complete mechanical disruption of the root tissue in liquid nitrogen.

Incomplete spraying can lead to heterogeneous lysis and reduced DNA recovery, especially in lignified perennial roots.

3. Dried roots tend to absorb the extraction buffer rapidly. If excessive absorption is observed, the volume of TE buffer can be adjusted slightly (±10%–15%) to ensure sufficient recovery of the supernatant after centrifugation.
4. The predigestion step with proteinase K and β-mercaptoethanol is essential when working with roots containing high levels of phenolic compounds or structural polymers. Shortening incubation times may negatively affect the A260/A230 ratios.
5. Replacing the manufacturer's BashingBead™ step is only suitable when prior cryogenic spraying has achieved fine homogenization. For fresh or fibrous tissues that are not completely disturbed, mechanical bead beating may be necessary.
6. Due to the heterogeneous spatial distribution of intraradical fungal structures (hyphae, arbuscules, vesicles), composite sampling and thorough homogenization are recommended to reduce variability between extractions.
7. The expected DNA yield ranges from 70 to 300 ng/μL, depending on the intensity of colonization, the condition of the root and the degree of lignification.
8. The protocol was optimized for *Coffea arabica* roots under field conditions. Application to other perennial species may require minor adjustments to the predigestion time or buffer volumes, depending on the tissue chemistry.

Conflict of Interest

The authors have no conflicts of interest to declare that are relevant to the content of this chapter.

References

1. Wipf D, Krajinski F, van Tuinen D, Recorbet G, Courty PE (2019) Trading on the arbuscular mycorrhiza market: from arbuscules to common mycorrhizal networks. New Phytol 223(3):1127–1142
2. Bao-Dong Chen, Wei Fu, Song-Lin Wu, Yong-Guan Zhu (2024) Involvements of mycorrhizal fungi in terrestrial ecosystem carbon cycling. Chin J Plant Ecol 48(1):1–20
3. Chen L, Wang M, Shi Y, Ma P, Xiao Y, Yu H, Ding J (2023) Soil phosphorus form affects the advantages that arbuscular mycorrhizal fungi confer on the invasive plant species, Solidago canadensis, over its congener. Front Microbiol 14:1160631
4. Cargill RI, Shimizu TS, Kiers ET, Kokkoris V (2025) Cellular anatomy of arbuscular mycorrhizal fungi. Curr Biol 35(11):R545–R562

5. Rastogi G, Sani RK (2011) Molecular Techniques to assess microbial community structure, function, and dynamics in the environment. In Ahmad I, Ahmad F, Pichtel J. (eds) Microbes and Microbial Technology. Springer, New York, NY pp 29–57. https://doi.org/10.1007/978-1-4419-7931-5_2
6. Corona Ramírez A, Symanczik S, Gallusser T, Bodenhausen N (2023) Quantification of arbuscular mycorrhizal fungi root colonization in wheat, tomato, and leek using absolute qPCR. Mycorrhiza 33(5):387–397
7. Buljević N, Preiner D, Šikuten I, Tomaz I (2025) Comparison of different DNA isolation methods from grapevine (Vitis vinifera) leaves. Separations 12(11):316. https://doi.org/10.3390/separations12110316
8. de los Monteros-Silva NE, Montesdeoca ÁO, Zambrano AA, Carrera JC, Fukushima EO, Llerena S, Mora MC (2025) A streamlined plant DNA extraction method with liquid nitrogen-free approach. J Trop Biodiversity Biotechnol 10(2):jtbb12766. https://doi.org/10.22146/jtbb.12766
9. Rasmussen PU, Hugerth LW, Blanchet FG, Andersson AF, Lindahl BD, Tack AJ (2018) Multiscale patterns and drivers of arbuscular mycorrhizal fungal communities in the roots and root-associated soil of a wild perennial herb. New Phytol 220(4):1248–1261
10. De Andrade FHA, Ferreira AMO, Azevedo LM, de Oliveira Santos M, Carvalho GR, De Resende MLV, Silva VA (2023) IBA and melatonin increase trigonelline and caffeine during the induction and initiation of adventitious roots in *Coffea arabica* L. cuttings. Sci Rep 13(1):15151
11. Zagoskina NV, Zubova MY, Nechaeva TL, Kazantseva VV, Goncharuk EA, Katanskaya VM, Aksenova MA (2023) Polyphenols in plants: structure, biosynthesis, abiotic stress regulation, and practical applications. Int J Mol Sci 24(18):13874
12. Salehi Z, Amirahmadi A, Rezaei A, Farrokh P, Ghasemian J (2023) Comparison of five DNA extraction methods in three medicinal plants: Peganum harmala L., Tamarix ramosissima Ledeb. and Potentilla reptans L. Mol Biol Res Commun. 12(1):1–16. https://doi.org/10.22099/mbrc.2023.45131.1798
13. Yurkov AP, Kryukov AA, Gorbunova AO, Kojemyakov AP, Stepanova GV, Machs EM, Shishova MF (2018) Molecular genetic identification of arbuscular mycorrhizal fungi. Ecol Genet 16(2):11–23
14. Lopez LMS, Gonzales HL, Garcia WEV, de Jesus Velez Chicoma RL, Greta MC (2025) Rapid and efficient DNA extraction protocol from Peruvian native cotton (Gossypium barbadense L.) Lambayeque, Peru. Methods Protoc 8(3):50
15. Tedersoo L, Bahram M, Zinger L, Nilsson RH, Kennedy PG, Yang T, Mikryukov V (2022) Best practices in metabarcoding of fungi: from experimental design to results. Mol Ecol 31(10):2769–2795
16. Stefani F, Laterrière M, Abdellatif L, Banchini C, Mader Stevens G, Séguin S, Dalpé Y (2025) The pitfalls of r DNA-based AMF identification: a comparative analysis of rDNA and protein-coding genes. New Phytol 248(3):1501–1515
17. Allen TR, Millar T, Berch SM, Berbee ML (2003) Culturing and direct DNA extraction find different fungi from the same ericoid mycorrhizal roots. The new Phytologist 160(1):255–272. http://www.jstor.org/stable/1514222
18. Ye X, Lei B (2023) The current status and trends of DNA extraction. BioEssays 45(8):2200242
19. Demkina A, Slonova D, Mamontov V, Konovalova O, Yurikova D, Rogozhin V, Isaev A (2023) Bencmarking DNA isolation methods for marine metagenomics. Sci Rep 13(1):22138
20. Carli M, Pedrelli A, Panattoni A, Pellegrini E, Nali C, Cotrozzi L, Rizzo D (2025) An optimized DNA extraction protocol for reliable PCR-based detection and characterization of grapevine flavescence dorée phytoplasma. Plant Methods 21(1):131
21. Diédhiou AG, Borges WL, Sadio O, de Faria SM (2014) Assessment of DNA extraction methods for detection of arbuscular mycorrhizal fungi in plant roots by nested-PCR. Acta Sci Biol Sci 36(4):433–441
22. Anderson IC, Cairney JW (2004) Diversity and ecology of soil fungal communities: increased understanding through the application of molecular techniques. Environ Microbiol 6(8):769–779
23. Bainard LD, Klironomos JN, Hart MM (2010) Differential effect of sample preservation methods on plant and arbuscular mycorrhizal fungal DNA. J Microbiol Meth 82(2):124–130
24. Kohout P, Sudová R, Janoušková M, Čtvrtlíková M, Hejda M, Pánková H et al. (2014) Comparison of commonly used primer sets for evaluating arbuscular mycorrhizal fungal communities: is there a universal solution?. Soil Biol Biochem 68:482–493
25. Krüger M, Stockinger H, Krüger C, Schüßler A (2009) DNA-based species level detection of Glomeromycota: one PCR primer set for all arbuscular mycorrhizal fungi. New Phytol 183(1):212–223
26. Thonar C, Erb A, Jansa J (2012) Real-time PCR to quantify composition of arbuscular

mycorrhizal fungal communities—marker design, verification, calibration and field validation. Mol Ecol Resour 12(2):219–232

27. Tedersoo L, Anslan S, Bahram M, Põlme S, Riit T, Liiv I, Abarenkov K (2015) Shotgun metagenomes and multiple primer pair-barcode combinations of amplicons reveal biases in metabarcoding analyses of fungi. MycoKeys 10:1–43

28. Chagnon PL, Bainard LD (2015) Using molecular biology to study mycorrhizal fungal community ecology: Limits and perspectives. Plant Signaling & Behavior 10(7):1–3. https://doi.org/10.1080/15592324.2015.1046668

Part V

Conclusions and Perspectives

Check for updates

Chapter 23

Conclusions and Perspectives on the Conservation of Microbial Genetic Resources

Marco A. Ramírez-Mosqueda and Lizandro Ramírez-Trejo

Abstract

Microbial genetic resources (MGRs) play a significant role in global ecological stability and in the development of modern biotechnology. In this context, in situ conservation strategies are essential to preserve the ecological interactions of the vast majority of non-culturable microorganisms, while ex situ conservation strategies ensure their availability in biobanks under controlled standards. This book compiles and systematizes the main strategies, methodological criteria, and experimental protocols aimed at the conservation of MGRs. In particular, effective methodologies for the conservation of isolated strains and complex microbiomes are presented, together with strategies for their functional characterization. Likewise, the reviews included current trends and the conceptual frameworks that support the conservation of microbial diversity. The integration of omics technologies, systems biology, and the digitization of biological information is emerging as a necessary tool for the development of new biotechnological solutions. Nevertheless, the consolidation of international collaboration networks, together with investment in specialized infrastructure, will be decisive in safeguarding microbial biocapital as a renewable resource of high scientific, ecological, and economic value.

Keywords Microbial genetic resources, Conservation strategies, Biobanks, Biodiversity, Environmental sustainability

1 Introduction

Microorganisms constitute the most diverse group on the planet and have developed a wide range of biogeochemical functions that support complex interaction networks across the different levels of organization of living matter. This functional diversity represents the foundation that maintains the habitability of the biosphere and sustains the evolution and persistence of multicellular organisms [1]. Despite their ecosystem relevance, their conservation has been biased toward microorganisms with direct utilitarian value for humanity, disregarding the so-called "microbial dark matter," of which only about 1% is currently known [2]. Furthermore, microbial reservoirs and collections (in situ and ex situ) have been progressively eroded:

Marco A. Ramírez-Mosqueda et al. (eds.), *Methods for The Conservation of Microorganisms*, Methods in Molecular Biology, vol. 3090, https://doi.org/10.1007/978-1-0716-5623-5_23,

on the one hand, due to increasing anthropogenic pressures [3], and due to funding limitations [4]. In this context, the protection of microbial biocapital emerges as a global strategic priority to maintain ecological sustainability and future biotechnological development.

Microbial Genetic Resources (MGRs) are defined as any material of microbial origin (including bacteria, archaea, fungi, viruses, and protists) that contains functional units of heredity and possesses real or potential value for humanity. At present, their study is strongly driven by bioprospecting, contributing significantly to the development of modern biotechnology [5]. However, to enable their utilization, it is necessary to transition from their characterization to their conservation. In this regard, the conservation of MGR requires an integrative approach that articulates in situ strategies, aimed at preserving evolutionary processes in their natural habitats, with ex situ strategies developed under international standards of quality and biosafety. In turn, the functional characterization of MGR demands the implementation of omics technologies, together with the establishment of international collaboration networks, in order to document and guide their biotechnological valuation.

2 Conclusions

This book provides a comprehensive overview of conservation strategies for microbial genetic resources (MGRs), spanning from the foundations that underpin current practices to highly specialized protocols. In its first part, the theoretical bases and methodological criteria governing MGR conservation are established, highlighting the importance of integrating in situ and ex situ strategies, as well as adopting international standards that ensure the quality, traceability, and reproducibility of the conserved resources. The chapters devoted to experimental protocols demonstrate that cryopreservation and lyophilization remain the most robust methodologies for long-term conservation, proving effective across a wide diversity of microorganisms. Complementarily, the work incorporates methodological advances in microbial isolation, cultivation, and characterization, including strategies for handling recalcitrant taxa and the development of optimized DNA extraction protocols and high-throughput analyses. Finally, the integration of omics technologies emerges as a key component for moving toward conservation centered on biological functionality, beyond taxonomic identity. This work provides the foundation for the strategic utilization of MGR in contexts aligned with current demands in biotechnology, public health, and environmental sustainability.

3 Perspectives

Despite the advances presented, significant challenges persist, including the limited cultivability of most microorganisms and the need to preserve complex ecological interactions. In this context, future conservation strategies should be oriented toward the study and maintenance of complete microbiomes. The development of approaches such as culturomics, systems biology, and synthetic biology will expand the spectrum of conserved microorganisms and optimize their utilization. Likewise, the incorporation of artificial intelligence tools will facilitate the integration and analysis of large omics datasets, driving the identification of new functions and biotechnological applications. Furthermore, strengthening international regulatory frameworks and collaboration networks will be essential to ensure equitable access, strategic conservation, and the sustainable use of MGR. Finally, the consolidation of global initiatives will position microbial genetic resources as a central component in responding to current challenges such as climate change, food security, and public health.

Conflict of Interest

The authors have no conflicts of interest to declare that are relevant to the content of this chapter.

References

1. Sessitsch A, Wakelin S, Schloter M, Maguin E, Cernava T, Champomier-Verges MC, Charles TC, Cotter PD, Ferrocino I, Kriaa A, Lebre P, Cowan D, Lange L, Kiran S, Markiewicz L, Meisner A, Olivares M, Sarand I, Schelkle B, Selvin J, Smidt H, Overbeek LV, Berg G, Cocolin L, Sanz Y, Fernandes JrWL, Liu SJ, Ryan M, Singh B, Kostic T (2023) Microbiome interconnectedness throughout environments with major consequences for healthy people and a healthy planet. Microbiol Mol Biol Rev 87(3):e00212-22. https://doi.org/10.1128/mmbr.00212-22
2. Redford KH (2023) Extending conservation to include Earth's microbiome. Conserv Biol 37(3):e14088. https://doi.org/10.1111/cobi.14088
3. Gilbert JA, Scholz AH, Dominguez Bello MG, Korsten L, Berg G, Singh BK, Boetius A, Wang F, Greening C, Wrighton K, Bordenstein SR, Jansson J, Lennon JT, Souza V, Allard SM, Thomas T, Cowan D, Crowther TW, Nguyen N, Harper L, Haraoui LP, Ishaq SL, McFall-Ngai M, Redford KH, Peixoto R (2025) Safeguarding microbial biodiversity: microbial conservation specialist group within the species survival commission of the International Union for Conservation of Nature. FEMS Microbiol Ecol 101(12):fiaf107. https://doi.org/10.1093/femsec/fiaf107
4. McCluskey K, Alvarez A, Bennett R, Bokati D, Boundy-Mills K, Brown D, Bull CT, Coffey M, Dreaden T, Duke C, Dye G, Ehmke E, Eversole K, Fenstermacher K, Geiser D, Glaeser JA, Greene S, Gribble L, Griffith MP, Hanser K, Humber R, Johnson BW, Kermode A, Krichevsky M, Laudon M, Leach J, Leslie J, May M, Melcher U, Nobles D, Risso Fonseca N, Robinson S, Ryan M, Scott J, Silflow C, Vidaver A, Webb KM, Wertz JE, Yentsch S, Zehr S (2016) The US culture collection network lays the foundation for progress in preservation of valuable microbial resources. Phytopathology

106(6):532–540. https://doi.org/10.1094/PHYTO-02-16-0074-RVW
5. Vuong P, Chong S, Kaur P (2022) The little things that matter: How bioprospecting microbial biodiversity can build towards the realization of United Nations Sustainable Development Goals. npj Biodiversity 1:4. https://doi.org/10.1038/s44185-022-00006-y

Kolibri 360 GmbH
Ferdinand-Jühlke-Straße 7
99095 Erfurt, Deutschland
produktsicherheit@kolibri360.de